I0888142

Digital Technologies, Ethics, and Decentralization in the Digital Era

Balraj Verma
Chitkara Business School, Chitkara University, Punjab, India

Babita Singla
Chitkara Business School, Chitkara University, Punjab, India

Amit Mittal
Chitkara Business School, Chitkara University, Punjab, India

IGI Global
PUBLISHER of TIMELY KNOWLEDGE

A volume in the Advances in Web Technologies and Engineering (AWTE) Book Series

Published in the United States of America by
IGI Global
Engineering Science Reference (an imprint of IGI Global)
701 E. Chocolate Avenue
Hershey PA, USA 17033
Tel: 717-533-8845
Fax: 717-533-8661
E-mail: cust@igi-global.com
Web site: http://www.igi-global.com

Library of Congress Cataloging-in-Publication Data

Names: Verma, Balraj, 1985- editor. | Singla, Babita, 1988- editor. | Mittal, Amit, 1974- editor.
Title: Digital technologies, ethics, and decentralization in the digital era / edited by Balraj Verma, Babita Singla, Amit Mittal.
Description: Hershey, PA : Engineering Science Reference, [2024] | Includes bibliographical references and index. | Summary: "The book explores how the digital revolution is reshaping society, economy, and international relations. It aims to understand and illuminate the role of digitalization in disrupting traditional systems, empower through decentralized technologies, analyze economic shifts, redefine privacy norms, address ethics, and offer insights into reshaped global cooperation"-- Provided by publisher.
Identifiers: LCCN 2023051958 (print) | LCCN 2023051959 (ebook) | ISBN 9798369317624 (hardcover) | ISBN 9798369317631 (ebook)
Subjects: LCSH: Information technology--Economic aspects. | Information technology--Social aspects.
Classification: LCC HC79.I55 D5739 2024 (print) | LCC HC79.I55 (ebook) | DDC 303.48/33--dc23/eng/20240306
LC record available at https://lccn.loc.gov/2023051958
LC ebook record available at https://lccn.loc.gov/2023051959

This book is published in the IGI Global book series Advances in Web Technologies and Engineering (AWTE) (ISSN: 2328-2762; eISSN: 2328-2754)

British Cataloguing in Publication Data
A Cataloguing in Publication record for this book is available from the British Library.

For electronic access to this publication, please contact: eresources@igi-global.com.

Advances in Web Technologies and Engineering (AWTE) Book Series

Ghazi I. Alkhatib
The Hashemite University, Jordan
David C. Rine
George Mason University, USA

ISSN:2328-2762
EISSN:2328-2754

Mission

The **Advances in Web Technologies and Engineering (AWTE) Book Series** aims to provide a platform for research in the area of Information Technology (IT) concepts, tools, methodologies, and ethnography, in the contexts of global communication systems and Web engineered applications. Organizations are continuously overwhelmed by a variety of new information technologies, many are Web based. These new technologies are capitalizing on the widespread use of network and communication technologies for seamless integration of various issues in information and knowledge sharing within and among organizations. This emphasis on integrated approaches is unique to this book series and dictates cross platform and multidisciplinary strategy to research and practice.

The **Advances in Web Technologies and Engineering (AWTE) Book Series** seeks to create a stage where comprehensive publications are distributed for the objective of bettering and expanding the field of web systems, knowledge capture, and communication technologies. The series will provide researchers and practitioners with solutions for improving how technology is utilized for the purpose of a growing awareness of the importance of web applications and engineering.

Coverage

- Quality of service and service level agreement issues among integrated systems
- Case studies validating Web-based IT solutions
- IT readiness and technology transfer studies
- Virtual teams and virtual enterprises: communication, policies, operation, creativity, and innovation
- Web systems performance engineering studies
- Software agent-based applications
- Security, integrity, privacy, and policy issues
- Mobile, location-aware, and ubiquitous computing
- Integrated Heterogeneous and Homogeneous Workflows and Databases within and Across Organizations and With Suppliers and Customers
- Competitive/intelligent information systems

IGI Global is currently accepting manuscripts for publication within this series. To submit a proposal for a volume in this series, please contact our Acquisition Editors at Acquisitions@igi-global.com or visit: http://www.igi-global.com/publish/.

The Advances in Web Technologies and Engineering (AWTE) Book Series (ISSN 2328-2762) is published by IGI Global, 701 E. Chocolate Avenue, Hershey, PA 17033-1240, USA, www.igi-global.com. This series is composed of titles available for purchase individually; each title is edited to be contextually exclusive from any other title within the series. For pricing and ordering information please visit http://www.igi-global.com/book-series/advances-web-technologies-engineering/37158. Postmaster: Send all address changes to above address.

Titles in this Series

For a list of additional titles in this series, please visit:
www.igi-global.com/book-series/advances-web-technologies-engineering/37158

Internet of Behaviors Implementation in Organizational Contexts
Luísa Cagica Carvalho (Instituto Politécnico de Setúbal, Portugal) Clara Silveira (Polytechnic Institute of Guarda, Portugal) Leonilde Reis (Instituto Politecnico de Setubal, Portugal) and Nelson Russo (Universidade Aberta, Portugal)
Engineering Science Reference • copyright 2023 • 471pp • H/C (ISBN: 9781668490396) • US $270.00 (our price)

Supporting Technologies and the Impact of Blockchain on Organizations and Society
Luís Ferreira (Polytechnic Institute of Cávado and Ave, Portugal) Miguel Rosado Cruz (Polytechnic Institute of Viana do Castelo, Portugal) Estrela Ferreira Cruz (Polytechnic Institute of Viana do Castelo, Portugal) Hélder Quintela (Polytechnic Institute of Cavado and Ave, Portugal) and Manuela Cruz Cunha (Polytechnic Institute of Cavado and Ave, Portugal)
Engineering Science Reference • copyright 2023 • 337pp • H/C (ISBN: 9781668457474) • US $270.00 (our price)

Concepts, Technologies, Challenges, and the Future of Web 3
Pooja Lekhi (University Canada West, Canada) and Guneet Kaur (University of Stirling, UK & Cointelegraph, USA)
Engineering Science Reference • copyright 2023 • 602pp • H/C (ISBN: 9781668499191) • US $360.00 (our price)

Perspectives on Social Welfare Applications' Optimization and Enhanced Computer Applications
Ponnusamy Sivaram (G.H. Raisoni College of Engineering, Nagpur, India) S. Senthilkumar (University College of Engineering, BIT Campus, Anna University, Tiruchirappalli, India) Lipika Gupta (Department of Electronics and Communication Engineering, Chitkara University Institute of Engineering and Technology, Chitkara University, India) and Nelligere S. Lokesh (Department of CSE-AIML, AMC Engineering College, Bengaluru, India)
Engineering Science Reference • copyright 2023 • 336pp • H/C (ISBN: 9781668483060) • US $270.00 (our price)

Advancements in the New World of Web 3 A Look Toward the Decentralized Future
Jane Thomason (UCL London Blockchain Centre, UK) and Elizabeth Ivwurie (British Blockchain and Frontier Technology Association, UK)
Engineering Science Reference • copyright 2023 • 323pp • H/C (ISBN: 9781668466582) • US $240.00 (our price)

Architectural Framework for Web Development and Micro Distributed Applications
Guillermo Rodriguez (QuantiLogic, USA)
Engineering Science Reference • copyright 2023 • 268pp • H/C (ISBN: 9781668448496) • US $250.00 (our price)

IGI Global
PUBLISHER of TIMELY KNOWLEDGE

701 East Chocolate Avenue, Hershey, PA 17033, USA
Tel: 717-533-8845 x100 • Fax: 717-533-8661
E-Mail: cust@igi-global.com • www.igi-global.com

Table of Contents

Detailed Table of Contents

The growth aspirations of developing countries supported by industrial prowess of developed economies, while leading to phases of rapid globalisation across the world, also resulted in skewed power and trade dynamics in favour of developed nations. Later, the world economy got hit hard by deglobalization. Today, nations are realigning their policies and national priorities through engaging in collaborative partnership with other economies to overcome the current phase of deglobalization. This chapter will track the phases of global growth and its reverse internationalization from the lens of various black swan events and will also analyse the changing global economic landscape resulting in the emergence of new pillars of global trade and cooperation in the form of economic trade blocs and regional trade agreements leading to a multipolar world and new reality of global trade. From a business perspective, the chapter will dwell on how corporates are realigning their international business practices, in this era of deglobalisation, to maintain their growth trajectory.

This chapter explores how digital technologies are reshaping decentralization in our fast-evolving technological era. It defines decentralization and examines its diverse aspects in politics, economics, and technology. The chapter emphasizes blockchain technology, highlighting its immutability, decentralized consensus, and smart contracts as pivotal elements driving decentralized applications and disrupting traditional paradigms. To provide context, it conducts a thorough literature review, tracing decentralization's historical evolution, exploring its impact on various sectors like decentralized finance, governance, financial inclusion, and system resilience. The chapter also addresses challenges such as regulatory uncertainties, scalability issues, and security concerns, drawing from authoritative sources and academic research. In conclusion, it sets the stage for a deeper exploration of decentralization's applications and impact across sectors, acknowledging both its potential and complexity as a transformative paradigm shift.

In today's rapidly evolving business landscape, the emergence of disruptive technologies and innovative business models have revolutionized traditional market structures. One such transformation is the shift towards decentralized marketplaces, driven by blockchain technology, smart contracts, and decentralized finance (DeFi) solutions. This chapter aims to explore the concept of disruptive business models and their profound impact on market dynamics, focusing on the rise of decentralized marketplace. This chapter addresses a pressing need for comprehensive, up-to-date, and insightful information on a topic that is rapidly changing the business landscape. It provides a valuable resource for a diverse audience, ranging from industry professionals to academics and policymakers. It can be highly beneficial for several reasons, given the ongoing evolution and significance of decentralized marketplaces.

The aim of this chapter is to do a comprehensive examination of the existing scholarly literature pertaining to artificial intelligence (AI) along with the service quality of AI tools within the framework of digital transformation, managing the effects of deglobalization, and fostering the use of technology. An in-depth study was carried out on a dataset consisting of 156 articles that were taken from the Scopus database. For the purpose of conducting this inquiry, bibliometric methods were utilized, and more specifically, VOSviewer software was utilized, in order to assess the performance analysis and science mapping of the literature that was under investigation. The findings of the study imply that there is a taxonomical representation of current scientific research on the incorporation of artificial intelligence with technological adoption across a variety of fields, with a particular emphasis on online shopping and the level of service quality provided by AI tools.

In the business world, the digital transformation has ushered in an era of unprecedented change. The explosive growth of digital technology over the past three decades has profoundly changed how businesses operate, compete, and engage with their customers. Businesses across industries have been forced to navigate this constantly changing environment due to the adoption of cloud computing, data analytics, the growth of e-commerce, and the rise of artificial intelligence. Organizations are forced to reconsider their strategies, operations, and consumer engagement models as a result of this significant instability. This chapter discusses the role of artificial intelligence in aiding customised and personalized marketing strategies, acknowledging the diverse preferences and behaviour of consumers. It shares insights on branding and customer management in an AI-driven environment. It also emphasises on online and technology-mediated customer engagement.

usage and preferences to accurately target, increasing the efficiency and effectiveness of advertising campaigns mobile, Web 2.0. Implicit communication, mobile-related advertising, and mobile-specific content have become important parts of social media advertising. Researchers from various fields have drawn on rapidly evolving social media platforms, each with its own unique perspective.

The issue of digital divide in India continues to be one of the critical issues in the 21st century, impacting the social, economic, and the various educational opportunities available for the citizens. This study seeks to examine the current state of digital divide in India by focusing on disparity in internet access, mobile ownership, and digital literacy. This study also aims to identify the numerous factors responsible for digital divide and to assess the key initiatives undertaken by the Government of India to mitigate the disparities. The study employs secondary data gathered from National Family Health Survey, GSMA, TRAI, and NSO. The findings of this study reveal the disparity in internet access, digital literacy, and mobile ownership rooted due to socio-economic factors, education, and geographic locations. This study underscores the need of addressing the digital gap in India and provides valuable insights for the policymakers, stakeholders, and the practitioners to take initiatives towards digital inclusion and equity in the country.

This chapter explores the intersection of digital identity and global data sovereignty, focusing on blockchain's impact. The authors analyze its transformative potential, security, and challenges concerning data regulations. This chapter advocates for a user-centric, ethical, and secure approach to digital identity, addressing societal implications and emphasizing user sovereignty. The chapter calls for interdisciplinary collaboration and outlines future directions for a secure digital identity framework.

In the midst of the 21st-century digital revolution, we find ourselves navigating a complex landscape where our analog instincts clash with our digital dependencies. Our smartphones, compact marvels of technology, house the entirety of our daily existence, from groceries to fashion and medicine. Yet, amidst this convenience, fundamental questions about data safety, security controls, and data ownership linger ominously in our minds. This chapter delves into the vital trifecta of privacy, security, and data ownership in our increasingly digitized world. In an era of ever-evolving technology, the ordinary user constantly frets about the sanctity of their data. While technology has undoubtedly bestowed countless benefits upon humanity, it also bears a dark side, epitomized by the surge in digital frauds and scams. Such concerns propel individuals to ponder deeply: Is my data truly private? and Who ultimately lays claim to, or safeguards, my data? This exploration ventures into the realms of artificial intelligence, virtual reality, the metaverse, online payments, and virtual classrooms.

In the rapidly evolving technological landscape, the pervasive integration of artificial intelligence (AI) has brought to light the pressing need to address the ethical dimensions associated with its widespread adoption. This study comprehensively explores AI ethics for ethical decision-making in the digital era. It offers a structured guide for aligning AI with ethical principles, emphasizing transparency, bias mitigation, and interdisciplinary collaboration in AI deployment. Additionally, it delves into the evolving AI landscape, highlighting potential societal impacts. It calls upon policymakers and stakeholders to engage in persistent dialogue and to remain adaptable in the face of a continuously transforming technological environment, advocating for the continuous refinement and adaptation of regulatory frameworks. This framework acts as a compass for ethically sound AI decisions, fostering a responsible, human-centric approach. It aims to forge a symbiotic relationship where AI uplifts society while upholding ethical values, making it a tool for societal betterment.

This chapter delves into the intersection of ethics and big data, with a primary focus on the ethical concerns arising from AI. The primary objective of this chapter is to highlight a novel approach that researchers might employ throughout the process of conducting a systematic literature review (SLR) to enhance efficiency and reduce costs associated with data synthesis and abstraction. Further, the conclusion emphasizes the need to navigate the intersection of ethics and big data, particularly concerning AI, presents a complex landscape of ethical concerns.

Fairness is threatened by algorithm bias, systematic and unfair disparities in machine learning results. Amazon's AI-driven hiring tool favoured men. AI promised data-driven, impartial decision-making, but it has revealed sector-wide prejudice, perpetuating systematic imbalances. The algorithm's bias is data and design. Biassed historical data and feature selection and pre-processing can bias algorithms. Development is harmed by human biases. Algorithm prejudice impacts money, education, employment, and crime. Diverse and representative data collection, understanding complicated "black box" algorithms, and legal and ethical considerations are needed to address this bias. Despite these issues, algorithm bias elimination techniques are emerging. This chapter uses secondary data to study algorithm bias. Algorithm bias is defined, its origins, its prevalence in data, examples, and issues are discussed. The chapter also tackles bias reduction and elimination to make AI a more reliable and impartial decision-maker.

Preface

Digital Technologies, Ethics, and Decentralization in the Digital Era, edited by Balraj Verma, Babita Singla, and Amit Mittal, emerges as a comprehensive exploration into the transformative forces of the digital revolution upon society, economics, and international relations. In this edited volume, our collective aim has been to dissect the intricate tapestry of digitalization, illuminating its multifaceted impacts, disruptions, and the profound reshaping of global paradigms.

This book serves as a guiding beacon for scholars, policymakers, industry professionals, and curious minds eager to navigate the labyrinth of the digital age. It embarks upon an odyssey, delving into the disruptive nature of digital technologies, probing how decentralization and de-globalization become catalyzed by these very innovations.

Within these pages, the thematic journey takes us through the landscape of digital transformation, spotlighting the disruptive potentials of decentralized technologies like blockchain and the far-reaching implications of decentralized finance (DeFi). It scrutinizes the redefined contours of economic systems, as digital disintermediation reconfigures traditional marketplaces and global economic relationships.

Moreover, our exploration extends to the nuanced realms of digital identity, data sovereignty, and the ethical dimensions of this digital epoch. Here, the book probes the ethical conundrums embedded in privacy, algorithmic bias, and the societal impacts of decentralization, emphasizing the imperatives of inclusivity and equitable technological access.

The final leg of this intellectual expedition propels us into the horizon of global cooperation, reimagined in the digitized world. Here, digital diplomacy and the evolving nature of international collaborations converge, beckoning us to contemplate multilateral approaches to address global challenges amid decentralization and de-globalization.

Each chapter, meticulously crafted, unfolds a tapestry of insights, offering a panoramic view of the intricate interplay between digital technologies and the reconfiguration of our societal, economic, and diplomatic architectures. Our hope is that this collective endeavor will not only inform but also inspire discourse and action, propelling us towards a more nuanced understanding of the digital revolution's profound impacts and the possibilities it unfurls for our collective future.

ORGANIZATION OF THE BOOK

Chapter 1: The Evolving Nature of International Collaboration and Partnership

Authored by Amanpreet Chopra, Sridhar Manohar, Artur Zawadski, Mayank Jain, this chapter tracks the phases of global growth and its reverse internationalization, analyzing the changing economic landscape and the emergence of new pillars of global trade and cooperation amid deglobalization. It delves into how nations are realigning policies through collaborative partnerships and examines how corporations are adapting their international business practices in this era to sustain growth.

Chapter 2: Decentralization in the Digital Age

In this chapter by Swaty Sharma, the focus is on digital technologies' impact on decentralization. It defines and explores decentralization in politics, economics, and technology, particularly emphasizing blockchain's role in driving decentralized applications. The chapter reviews historical evolution, challenges, and potential applications of decentralization across sectors.

Chapter 3: Disruptive Business Models and the Shift Towards Decentralized Marketplaces

Authored by Vinoth S and Nidhi Srivastava, this chapter examines how disruptive technologies and innovative business models are reshaping traditional markets, emphasizing the rise of decentralized marketplaces driven by blockchain and DeFi solutions. It explores the impact on market dynamics and addresses the need for updated insights in this rapidly evolving landscape.

Chapter 4: Unpacking the Role of Service Quality of AI Tools in Catalyzing Digital Transformation

Authors Namita Sharma and Urvashi Tandon conducted a bibliometric analysis examining the service quality of AI tools within the framework of digital transformation. They explore the representation of scholarly research on AI's integration across various fields and its impact on technology adoption, especially in areas like online shopping and service quality.

Chapter 5: The Digital Transformation: Crafting Customer Engagement Strategies for Success

Authored by Gajalakshmi N S Yadav and Seranmadevi R, this chapter discusses how the digital transformation has affected customer engagement strategies. It delves into the role of AI in customized marketing strategies, branding, and online customer engagement.

Chapter 6: DeFi's Transformative Influence on the Global Financial Landscape

Saravanakrishnan V examines the multifaceted impact of DeFi on the global financial industry. It explores its disruption of traditional banking models, financial inclusion, tokenization, advantages, and security challenges, presenting a comprehensive view of DeFi's implications.

Chapter 7: The Digital Disruption of Distribution – An Automobile Industry Perspective

Kirtikumar Tolani, Asif Saraiya, and Sridhar Manohar focus on the digitalization of automobile distribution, discussing its impact on customer acquisition, retail experience, and customer lifecycle management. The chapter explores how digitalization is reshaping the automobile industry's distribution phase.

Chapter 8: Social Media Advertising – A Dimensional Change Creator in Consumer Purchase Intention

Authored by Bharathi S. and Dr. Deep Gurung, this chapter highlights the role of social media advertising in influencing Generation Z's consumer purchase intentions. It covers methods of business expansion through social media, analyzing characteristics affecting consumer behavior.

Chapter 9: Bridging the Digital Divide – Navigating the Landscape of Digital Equity

Authors Priya Gupta and Anjali Verma examine India's digital divide, addressing disparities in internet access, mobile ownership, and digital literacy. They highlight socio-economic factors contributing to this gap and discuss initiatives for digital inclusion and equity.

Chapter 10: Digital Identity and Data Sovereignty – Redefining Global Information Flows: Data Sovereignty in a Digital World

Kaushikkumar Patel explores the intersection of digital identity, data sovereignty, and blockchain's impact. The chapter advocates for a secure and user-centric approach to digital identity, addressing societal implications and outlining future directions for a robust digital identity framework.

Chapter 11: Navigating the Digital Era – Exploring Privacy, Security, and Ownership of Personal Data

Authored by Atul Grover, this chapter delves into the crucial aspects of privacy, security, and data ownership in the digitized world. It discusses the challenges posed by technology while exploring various digital domains and their implications on user data.

Chapter 12: Ethics and Artificial Intelligence – A Theoretical Framework for Ethical Decision Making in the Digital Era

Yashpal Azad and Amit Kumar offer a comprehensive exploration of AI ethics and ethical decision-making. The chapter emphasizes transparency, bias mitigation, societal impacts, and regulatory adaptability, presenting a structured framework for ethical AI deployment.

Chapter 13: The Intersection of Ethics and Big Data – Addressing Ethical Concerns in the Digital Age of Artificial Intelligence

Divya Goswami and Balraj Verma delve into the ethical concerns stemming from the intersection of Big Data and AI. They present a systematic literature review approach, enhancing data synthesis efficiency, and highlight the complex landscape of ethical concerns.

Chapter 14: Comprehending Algorithmic Bias and Strategies for Fostering Trust in Artificial Intelligence

Authored by Sidhi Menon U, Theresa Siby, Natchimuthu Natchimuthu, this chapter examines algorithmic bias and its impacts. It addresses issues arising from biased algorithms, explores reduction strategies, and advocates for a more reliable and impartial AI decision-making process.

Chapter 15: A Critical Data Ethics Analysis of Algorithmic Bias and the Mining/Scraping of Heirs' Property Records

Robin Throne explores the data ethics surrounding AI, machine learning, and data mining/scraping in the context of heirs' property rights. It examines the risks, privacy implications, and ongoing land dispossession issues, calling for further research and policy scrutiny.

IN SUMMARY

The compilation you hold in your hands, *Digital Technologies, Ethics, and Decentralization in the Digital Era*, is a testament to the vibrant mosaic that is the digital revolution. Across its pages, this book breathes life into the multifaceted tapestry of technological advancements reshaping our world.

Our authors, experts in their respective domains, have meticulously navigated the labyrinth of digital transformation. From the evolution of international collaborations to the disruptive force of decentralized technologies and the ethical nuances of artificial intelligence, each chapter illuminates a distinct facet of our digital age.

This book, a labor of collective wisdom and scholarly rigor, transcends boundaries. It beckons policymakers, industry professionals, academics, and the curious minds of the general public into a shared dialogue on the metamorphosis unfolding in our midst.

As editors, our aspiration was to curate a comprehensive compendium that not only dissects the intricacies of digitalization but also sparks conversations, inspires innovation, and shapes future narratives. We aimed not only to analyze the present but to provide guideposts for the road ahead.

In these pages, we've traversed the disruptive waves of change—examining decentralized finance's transformative prowess, unraveling the complexities of digital identity and sovereignty, and dissecting the ethical quandaries inherent in our AI-driven world.

Our hope is that this anthology will serve as a beacon, illuminating pathways for informed decision-making, policy formulation, and scholarly exploration. It stands as a testament to the resilience, adaptability, and transformative potential of human ingenuity in an era of rapid digital metamorphosis.

In conclusion, *Digital Technologies, Ethics, and Decentralization in the Digital Era* isn't merely a book; it's a testament to the collective journey we undertake as a global society, navigating the uncharted waters of a digital epoch. May it be a compass guiding us toward a future where technology serves humanity's greater good.

Balraj Verma
Chitkara Business School, Chitkara University, Punjab, India

Babita Singla
Chitkara Business School, Chitkara University, Punjab, India

Amit Mittal
Chitkara Business School, Chitkara University, Punjab, India

Chapter 1
The Evolving Nature of International Collaboration and Partnership

Amanpreet Singh Chopra
https://orcid.org/0009-0009-7181-8794
Chitkara Business School, Chitkara University, Punjab, India

Sridhar Manohar
https://orcid.org/0000-0003-0173-3479
Chitkara Business School, Chitkara University, Punjab, India

Artur Zawadski
Sunrise CSP Pty., Australia

Mayank Jain
University of Exeter, UK

ABSTRACT

The growth aspirations of developing countries supported by industrial prowess of developed economies, while leading to phases of rapid globalisation across the world, also resulted in skewed power and trade dynamics in favour of developed nations. Later, the world economy got hit hard by deglobalization. Today, nations are realigning their policies and national priorities through engaging in collaborative partnership with other economies to overcome the current phase of deglobalization. This chapter will track the phases of global growth and its reverse internationalization from the lens of various black swan events and will also analyse the changing global economic landscape resulting in the emergence of new pillars of global trade and cooperation in the form of economic trade blocs and regional trade agreements leading to a multipolar world and new reality of global trade. From a business perspective, the chapter will dwell on how corporates are realigning their international business practices, in this era of deglobalisation, to maintain their growth trajectory.

DOI: 10.4018/979-8-3693-1762-4.ch001

1. INTRODUCTION

Looking back into the history of wars, one of the most epic ones was the 'Battle of Thermopylae' between Persians and Sparta, famously projected in the 2007 movie "300". The battle was not about the existing dominance of Athens but more about their fear of the rise of Sparta as the future dominant force, which could endanger the current status quo of power dynamics, tilting in favour of Sparta. More than 2000 years ago, historians from Athens and Greece dubbed this the "Thucydides trap." A trap where decisions are taken in the time horizon leading to decision dynamics of "now or later", as Do we implement strategic decisions now or wait for the competitor's move Today, as United States of America is well into the seventh year of a trade war with China which is challenging its global dominance, a few questions need to be explored - Are we witnessing the new 'Thucydides trap'? Are we well past the golden era of globalisation? Is deglobalisation a real and apparent danger for the world economy? Is the great reset of Covid 19 leading us to a new era of Cold War? Which countries or economic blocs would rise as new poles of global trade dominance in this multipolar world?

In the contemporary literature on international business, the concept of globalisation has been defined through the lens of trade, tariffs, transactions, collaborations, treaties, and, at times, through their integration as Eichengreen (2023) defined globalisation as an "*Integrated global economy in which cross-border transactions and GDP rise in tandem*". However, it is well recognised in international business practices that globalisation is more than trade; it is also about openness, interconnectedness of people, exchange of culture, knowledge, and intellect (Grunstein, 2022).

The world economy, which hit its peak Purchasing Power Parity (PPP) of $163.51 trillion in 2022 (IMF, 2023), has witnessed its share of phases of globalisation starting from the first phase of integration of the global economy a century before World War I, during which global trade increased annually by 4% (García-Herrero and Tan, 2020), to the second phase after World War II, resulting in the establishment of General Agreement on Tariff and Trade (GATT) in 1948 and the World Trade Organization (WTO) in 1999. Increased global capital flow, industrialisation driven by technological exchange, and the rise of the middle class in the US and Europe resulted in the peak of globalisation in 2007 before it converged to a gradual decline after the great financial crisis of 2008 (García-Herrero and Tan, 2020). On the other hand, the ascend of the global economy was noted to be maximum during 30 year period from 1970-2000, which is also termed a phase of hyper-globalization (Miśkiewicz and Ausloos, 2010).

In the 2023 World Trade Report, the World Trade Organization defined the end of the 1st stage of globalisation with the onset of World War I, resulting in three decades of deglobalisation, during which world trade shrieked from $ 29.50 billion in 1919 to $ 21.5 billion in 1938 (World Trade Organization, 2023) and subsequently, the next phase of globalisation sets in reaching a peak in global trade between 2007-2009 and a peak of inward Foreign Direct Investment (FDI) between 2007-2011 (Witt, 2019).

On the other hand, Kim et al., (2020) suggested that the pendulum of globalisation and deglobalisation swinging much before that, with the first globalisation phase dated from 1840-1929 followed by ten years of deglobalisation between 1929-1939 and the second phase of globalisation from 1979-2008 before the recent deglobalisation set in after the great financial crisis and US-China trade war (Kim et al., 2020).

2. LIMIT TO GLOBALIZATION

Pathway: Globalization to De-Globalization

There are natural limits to globalisation that vary with countries, their historical growth and the timespan. In contemporary literature, the concept of deglobalization is defined as *"Reduced -number of exchanges in trade, investment, and movement of people*" (García-Herrero and Tan, 2020) or "*the process of weakening interdependence among nations"* (Witt, 2019). On the other hand, the current trade war between the US and China led to the coining of a new phrase, 'Decoupling', which is defined in line with deglobalisation as *"the process of weakening interdependence between two nations or bloc of nations (Witt et al., 2023)"*. While noting the introduction of the Euro in 1999 as the first trigger toward deglobalisation, it was established that 20 countries reached their natural limit of global growth by the beginning of the 21st century (Miśkiewicz and Ausloos, 2010).

The global geopolitical scenarios are dynamically driven by 'black swan' events, i.e., phenomena that are characterized as rare and unpredictable, and events that are rare but predictable, or 'white swan' ones, with both having extreme tangible and intangible impacts on the global economy (Nicholas 2015; Nicholas, 2008). Over its history, the world economy has witnessed such events which have resulted in either their engagement or disengagement with other countries, both politically and economically. A few of these events, like the financial crisis of 2008, the Covid-19 pandemic, the Russia-Ukraine conflict, and the trade war between the US and China, have resulted in slowing down the phase of globalisation and accelerating the deglobalisation (Eichengreen, 2023; Kumar et al., 2023; Bremmer, 2020). In 2018, USA initiated trade war with China by imposing a 25% tariff on Chinese imports (IMF, 2023). Further, other events like Brexit in 2020, the electoral success of Donald Trump in 2017, and its 'America First' policy were other triggers to implement protectionist measures by countries thus, further hastening the pace of deglobalisation of the world economy (Grunstein, 2022).

These phases of globalisation and deglobalisation had varying impacts on developing and developed countries. In the context of India as a developing economy and the US as a developed economy, it was noted that globalisation positively impacted the Indian economy with increased employment rates, GDP growth, and the rise of the middle class. However, on the other hand, the US economy suffered negatively with reduced employment generation and stagnated GDP growth (Garg and Sushil, 2022). Similarly, the significant positive impact of economic globalisation on the performance of Indian companies in eleven economic sectors was also established (Verma and Srivastava, 2023).

While noting the slower annual global GDP growth at 2.7% between the 9-year timespan of 2009 to 2018 as compared to the period before the great financial crisis, it was also established that there is a tremendous increase in protectionist measures by the USA through reduction in the approval of technology export licenses from 27% in 2016 to -9% in 2020, resulting in increased decoupling phase between two major economies of the world (García-Herrero and Tan, 2020).

While macroeconomic trade policies drive deglobalisation and decoupling, the execution of these is undertaken by multinational companies (MNCs) operating in different geographies through the 'Reverse Internationalization' process. (Gnizy and Shoham, 2017), while defining it as *"a voluntary or forced process of decreasing involvement in international operation"*, noted that the international business literature is still at the nascent stage in defining and analysing these phenomena, and more empirical studies are required to understand the level of reverse internationalisation, its causes- planned or forced, its pace, timings, and the consequences to the organisation in terms of loss of market share, loss of profitability and employment (Gnizy and Shoham, 2017).

As reverse internationalisation is an execution of the strategic intent of an organisation to exit from the current market either completely, through the sale of subsidiaries or operations, or partly through the reduction in its product offerings in the market, further research is needed in this domain to analyze whether the exit was complete or partial, what was the mode of exit, like a complete sell-off, dilution of stakes, what changes were made in the product or market portfolio, and finally, what long term strategies are being adopted by organisations to mitigate the effect of reverse internationalisation.

Where We Are on Deglobalization Curve

There is an abundance of empirical evidence that suggests that the world economy has entered a phase of deglobalisation (Witt et al., 2023; IMF, 2023; Miśkiewicz and Ausloos, 2010) and same could intensify in future (IMF, 2023). However, the debate on the measure of its matrix is still ongoing. While defining Gross Domestic Product (GDP), GDP per capita, annual hours worked, and employment per capita as scales for deglobalisation, it was noted that few countries have entered this phase by the beginning of this century (Miśkiewicz and Ausloos, 2010). Along these lines, by taking GDP as an independent variable and FDI, exports, and imports as dependent variables, it was determined that the deglobalisation process in the USA has already migrated to the next higher and more dynamic phase, resulting from increased trade tariffs and skewed trade agreements (Garg and Sushil, 2022). On the other hand (Witt et al., 2023) argued that the matrix of measurement of deglobalisation should be relative as absolute measures have low construct validity and, hence, proposed trade deglobalisation as a measure of the ratio of global import to global GDP and FDI deglobalisation as a measure of the ratio of global inward FDI to global GDP.

By undertaking an analysis of 45 countries between 1985 and 2019, it is noted that the capital control policies of governments promote reserve accumulation, which has led to their financial deglobalisation, and this correlation is stronger for emerging economies as compared to developed economies (Bergin et al., 2023). Further, the reduction in the exchange of goods between countries, increased trade protection, and reduction in world outward FDI as % GDP from 2.7% in 2018 to 1.2% in 2018 were clear indicators that the world economy is heading towards a deep crest of deglobalisation and slower economic growth (García-Herrero and Tan, 2020). Further evidence of future deglobalization was put forward while projecting a reduction in the volume of world trade from 5.1% in 2022 to 2.4% in 2023(IMF, 2023). trade.

On the energy front, the Oil and gas importing countries suffered the brunt of the Russia-Ukraine conflict, resulting in the decoupling and regionalisation of their energy requirements leading to swift diversification towards green energy alternatives (Eichengreen, 2023). On the other hand, few countries through their policy interventions were able to secure crude oil supplies from Russia at steep discounts as compared to global market rates (Anand et.al., 2023; Sharma, 2023)

Historically and even today, imposing trade barriers is the most populist stance being adopted by countries towards driving their protectionist agenda. This trend of protective trade is evident from the imposition of more than 6,000 trade barriers by developed economies between 2008 and 2016, resulting in a reduction of cross-border trade volumes and the migration of manufacturing and trade from exporting to consuming economies like Vietnam and India (Sharma, 2017). From the foregoing, it is evident that the current global economy has been in a phase of deglobalization since the great financial crisis of 2008, and with the impact of the pandemic, the US-China trade war, trade protection measures being

Table 1. Leading exporters and importers in world trade, 2022 (billion dollars and percentage)

	Exporter Country			Importer Country		
Rank	**Exporters**	**Value ($ Billion)**	**Share (%)**	**Importers**	**Value ($Billion)**	**Share (%)**
1	China	3594	14.4	United States of America	3376	13.2
2	United States of America	2065	8.3	China	2716	10.6
3	Germany	1655	6.6	Germany	1571	6.1
4	Netherlands	966	3.9	Netherlands	899	3.5
5	Japan	747	3.0	Japan	897	3.5
6	Korea, Republic of	684	2.7	United Kingdom	824	3.2
7	Italy	657	2.6	France	818	3.2
8	Belgium	633	2.5	Korea, Republic of	731	2.9
9	France	618	2.5	India	723	2.8
10	Hong Kong, China	610	2.4	Italy	689	2.7
18	India	453	1.8			
	World*	24905	100.0	World (3)	25621	100.0
*Includes significant re-exports or imports for re-export.						

Source: WTO, World Trade Statistical Review 2023

adopted by economies and the decoupling of US-China economies, the depth of the spiral and end point of deglobalization is hard to predict.

The entry of China as a permanent member of the World Trade Organization on 11th December 2001 was a watershed moment not only for China but for world trade. Owing to its low labour cost, investment in the development of railway and road network, rise in technical prowess, government policy and overall cost advantage, it emerged as the world's manufacturing hub, thus, transforming itself as one of the pillars of world trade and economy. Further, the rise of the middle class turned China into a self-consuming global market for products and services, forcing every economy, and every organization to have its 'China strategy' on their table to leverage Chinese advantage for the growth of trade and services. This resulted in rapid growth in the Chinese economy and its GDP (Current USD) touched $6.09 Trillion in 2010 and surpassed Japan as the world's 2nd largest economy (World Bank, 2023).

The rise of China from 2001 onwards greatly affected global trade. As of 2022, China, with net exports of $3594 billion, is the world's leading exporter of goods and services and has a 14.4% share of global exports, which is 26% more than the US and almost eight times that of India. On the other hand, due to the rise of its aspirational middle class, China, with an import bill of $2716 billion, continued to be the second-largest importer of raw materials, natural resources and agricultural products. The trade surplus of $878 billion of the Chinese economy and the trade deficit of $1131 billion of the US economy is testimony to the fact that despite global slowdown and the policy protectionist measures being taken by the governments, the dependency of US and World economy on Chinese goods and services is staggering and would continue to remain so (Table 1). Thus, a clear indicator is that the road ahead is of the co-existence of economic policies by countries taking balanced views on decoupling and deglobalisation.

The World trade data on merchandise export from 2000 and services export from 2005 to 2020 (Figure 1 and Figure 2) was analysed and it was noted that during the golden age of globalisation after

Figure 1. World merchandise export (2000-2020)
Source: World Trade Organization (2023)

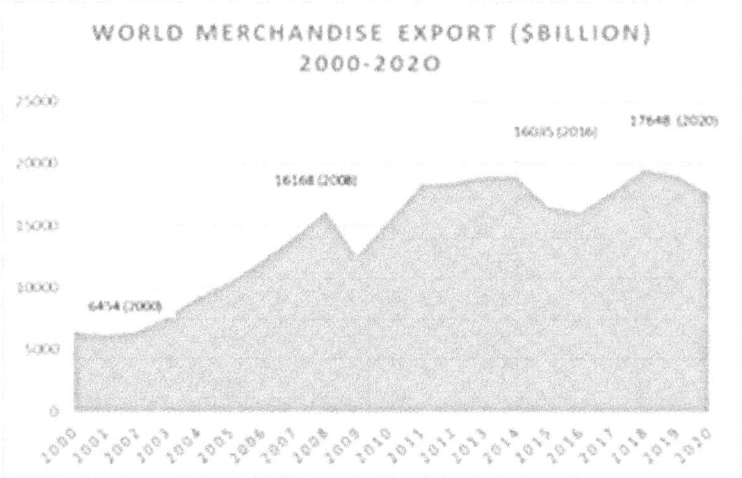

Figure 2. World service export (2005-2021)
Source: World Trade Organization (2023)

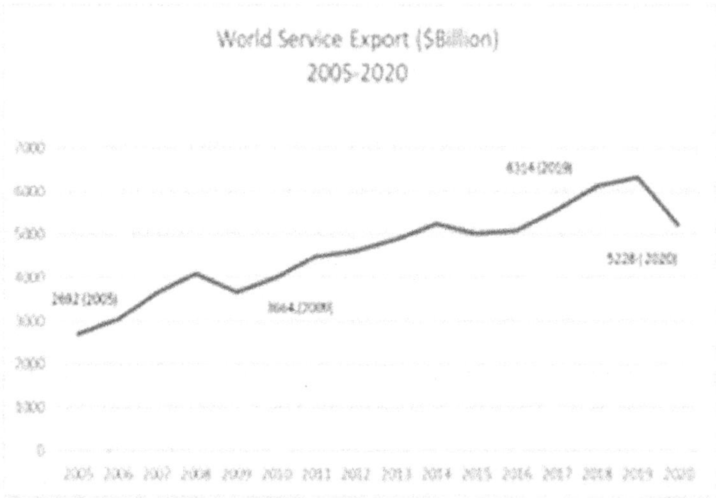

the entry of China into WTO in 2001, the world's merchandise export augmented by two times from $6454 Billion on 2001 to $16168 Billion in 2008 at eight-year CAGR of 12.16%, a year before the great financial crisis. The stagnation of world exports is evident after the great financial crisis, followed by the trade war between the US and China in 2016, the COVID-19 pandemic in 2019 and the ongoing Russia-Ukraine conflict. This has resulted in low and sluggish world trade growth at an 11-year CAGR of 3.13% from 2009-2020, clearly indicating the phase of deglobalisation, decoupling of economies and increase in unliteral trade measures resulting in fragmentation of the world economy. On the other hand, the service exports continued to rise from $2691 Billion in 2005 to $6314 Billion in 2019 at 14 years CAGR of 6.29%.

Along these lines, a threefold increase in global export of digital services from 2005 to 2019 was reported by WTO in its trade report with the potential of surpassing $7 Trillion in the near future (World Trade Organization, 2023). On the other hand, the report also noted considerable increase in protection measures by its member countries from 2 in 2007 to 41 in 2021 and projects that the deglobalisation would lead the world's middle class to be poorer and economies to be less efficient with long term dominance of inflationary pressures (World Trade Organization, 2023). Even before the pandemic, China had emerged as the dominant economic power. With the rise of India as a growing economy, the trade balance is gradually shifting towards Asia, leading to a multipolar world with dynamic development pathways (Dunford and Qi, 2020).

The above discussions led us to three critical enquiries to analyse not only from the point of view of economies but also from the business perspective. First, how would the European Union and other countries respond to the ongoing trade war between the US and China and the current phase of deglobalisation to maintain their economic relevance? Secondly, what would be the contours of the new world order and what role would economic corridors like the Belt and Road Initiative (BRI), India-Middle East-Europe Corridor(IMEC) and others would play in building up a multipolar world? Furthermore, as a result, what strategies do international organisations need to deploy in this multipolar world, which is turning to be economically more regional than global?

3. NEW PILLARS OF INTERNATIONAL COLLABORATION

The World Trade Organization was established in 1995 as an umbrella organisation to facilitate, promote and expand international trade and to provide a level playing field for developing countries (World Trade Organization, 2023). However, since its inception, different rounds of negotiations over agriculture, intellectual property, fishery, investment, etc., have led to stalemate among developed and developing countries wherein developing countries seek fair trade opportunities while developed countries propagate free trade, which as per developing nations are detrimental to their growth aspirations.

This stalemate of seeking consensus among countries initiated during the Doha Development Agenda 2001 is continuing with few successes in agreements on fishery trade. However, overall, the agreements and negotiations are far from their logical conclusion. This has led member nations to look for second-best options by entering into bilateral or multilateral Free Trade Agreements (FTAs), Economic Integration Agreements (EIA) and Regional Trade Agreements (RTAs). Analysis of WTO data on regional trade agreements suggests their rapid pace with 99 'in force' trade agreements in 2002, a year after the Doha Development Agenda, to a staggering 595 'in force' agreements and 361 bilateral and multilateral agreements under the negotiations as on 2023 (Figure 3). On a regional basis, out of these 595 agreements, European countries, are signatories to a maximum of 163 agreements, followed by 103 agreements entered into by East Asian countries, of which Japan, China and South Korea dominate the signatory list. On the other hand, South and North American countries are part of 71 and 50 treaties, respectively (Figure 2).

Countries while signing these free trade treaties are well aware of the future uncertainties that the agreement after enforcement may benefit other country(ies) more than themselves. Analysis of 9 mega free trade agreements entered by the USA and Japan indicates that Japan has a more advantageous position and benefitted more from these FTAs than the USA (Breuss, 2022). However, these FTAs continue to operate as geopolitical measures and balance the growth strategy of countries amid the US-China trade war and lack of agreements at WTO. On similar lines, for developed economies, it was noted that

Figure 3. Growth of reginal trade agreements (1948-2023)
Source: World Trade Organization (2023)

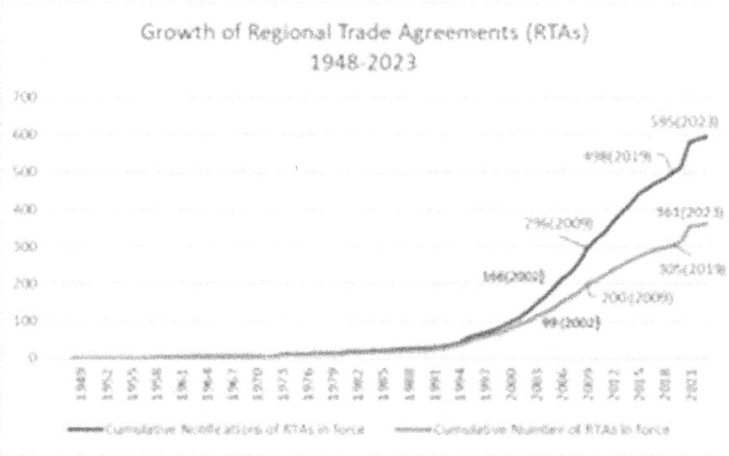

Figure 4. Reginal trade agreements (in force and under negotiations)
Source: World Trade Organization (2023)

since the onset of the great financial crisis in 2008, there had been a gradual increase towards trade regionalisation through entering into bilateral and multilateral Free Trade Agreements (FTAs) and other treaties (Kim et al., 2020; Witt, 2019)

However, since the great financial crisis of 2008, the rate of finalisation of number of agreements has slowed to 19 agreements per year till 2023 from 24 agreements per year from 2002 to 2008 due to prolonged negotiations on some of the contentious issues in these agreements. These issues where countries could not agree often results in delay in these treaties. Given the ineffectiveness of WTO in cutting across global agreements on trade and tariffs, these trade agreements have emerged as the pillars of global cooperation, especially in the era of deglobalisation. A brief analysis of major Free Trade Agreements is presented in Table 2:

Table 2. Key feature of major free trade agreements

Free Trade Agreement	Date of Entry / Coverage	Current Signatories	Purpose / Trade Under FTA
Comprehensive Economic and Trade Agreement (CETA)	21-09-2017 / Goods & Services	27 EU countries and Canada	To eliminate maximum trade tariffs between member countries Euro 66.8 Billion (2019)a Euro 59.2 Billion (2020)a
Comprehensive and Progressive Agreement for Trans-Pacific Partnership (CPTPP)	30-12-2018 / Goods & Services	11 countries - Australia, Brunei Darussalam, Canada, Chile, Japan, Malaysia, Mexica, New Zealand, Peru, Singapore, Vietnam (US withdrew in 2020)	Preferential market access Elimination of 98% duties tariff Trade Data – Not Accessible
EU and Japan Economic Partnership (JEFTA)	01-02-2019 / Good & Services	27 EU countries and Japan	Elimination of tariff and trade barriers, agriculture protection, Liberal public procurement IPR, Investment Expected > Euro 100 Billion
African Continental Free Trade Area (AfCFTA)	May 2019 / Good & Services	African Union 54 countries/ Ratified by 41 countries as of Jan 2022	Removal of traffic and non-tariff barriers, IP protection, FDI Estimate trade: $292 Billion by 2035c
USA – Japan Free Trade Agreement	01-01-2020 / Agriculture and Industrial Goods	US and Japan	Market access for agriculture and industrial goods, Removing trade barriers. Estimate - $7.5 Billion of US exported
Regional Comprehensive Economic Partnership (RCEP)	01-01-2022 / Good & Services	10 members of ASEAN (Brunei, Cambodia, Indonesia, Laos, Malaysia, Philippines, Singapore, Thailand, Vietnam) and 5 FTA partners (Australia, Japan, China, New Zealand, Republic of Korea)*	Eliminates tariff on 91% of goods and services, the World's largest free trade area. Covering 30% of the World population and $12.7 Trillion of World Trade in Goods and services
ASEAN Free Trade Area (AFTA)	01-01-1993 / Goods & Services	East Asia	To lower or eliminate trade tariffs or non-trade barriers through a common effective preferential tariff scheme (CEPT). It comprises separate agreements on trade in Goods, Services and Investments
India-UAE Comprehensive Economic Partnership Agreement (CEPA)	01-05-2022 / Goods & Services	India and UAE	Elimination of 97% of tariff lines for 98% of Indian exports to UAE. Only 9.2% of items remain on the exclusion list. CEPA includes IPR protection, pharma, investment, digital trade, telecom, etc. It estimated bilateral trade of $60 billion
India-Australia Economic Cooperation and Trade Agreement (IndAus-ECTA)	29-12-2022 / Goods & Services	India and Australia	Australia to provide zero-duty access to 100% of tariff lines in phases. Estimated bilateral trade $45 Billion by 2026g
India- Japan Comprehensive Economic Partnership Agreement (CEPA)	15-02-2021 / Good & Services	India and Japan	Removal of 90% of duties on exports from both countries. Bilateral trade - $6.1Billion (2021)h
India-UK Free Trade Agreement	End 2023 (Expected) / Good & Services	India and Japan	Final details to be published

Source:
a: Govt of Canada (2021); Second Canada-EU CETA Joint Committee Meeting
b: EU (2019); EU-Japan trade agreement
c: World Bank (2022); Making most of the African Continent Free Trade Area
d: US ITA (2022); Japan -Country commercial guide
e: RCEP (2022); RCEP agreement comes into force
* India opted out in 2019 due to issues regarding copyright, dumping of agro products and elimination of tariffs
f. PIB (2022) CEPA between India and UAE unveiled
g: MoCA (2022); India-Australia Economic Cooperation and Trade Agreement
h: MoFA (2022); Comprehensive Economic Partnership Agreement between Japan and the Republic of India

Analysis of these collaborative agreements between nations suggests that they all have a common theme of providing mutual trade preferences, reduction of tariffs on a majority of goods and services, the digital flow of information, strong intellectual property protection, cultural exchange, increasing trade between the countries thus, strengthening the trade regionalisation rather than globalisation.

4. THUCYDIDE TRAP AND THE GREAT RESET

The Trap

The ‘ Peloponnesian war’ between Persians and Sparta gave rise to ‘The Thucydides trap’ which is finding relevance in international business and global economy literature with 16 of the case studies undertaken from 16th Century to 2016 leading to all-out war between the nations in 11 of these cases (Allison, 2017).

Fast forward to 2023, the current rivalry between the USA and China is a mirror image and has all the hallmarks of the Thucydides trap, with both countries vying for influence over future technologies of 5G, robotics, quantum computing, Artificial Intelligence and space dominance leading to a loss in mutual trust and decoupling of a global economy which may lead to full-scale deglobalisation and if mismanaged to full-scale cold war leading to more economically devastating consequences than US and USSR cold war saga (Roubini, 2020).

Another theory in play in this economic war of dominance is the Temporal theory of international politics, which puts forward the argument that existing or current economic power, in this case the USA, may cooperate or compete with rising economic power, in this case, China, considering time horizon of ‘When to act’ and ‘How to act’ (Can, 2022). It is argued that after the great financial crisis of 2008, there was an apparent decline in the relative power of the US concerning the rest of the world, and a new world order is emerging with China and India as nodes of this new economy into a multipolar world (Witt, 2019).

The Great Reset

On 10 March 2023, the coronavirus resource centre of John Hopkins University and Medicine stopped collecting data on the COVID-19 pandemic. By then, the World had reported 676,609,955 cases and 6,881,955 deaths, and through more than 4 billion vaccination-administered doses, it has come out of the grip of the worst human suffering in living memory (John Hopkins, 2023). However, the impact of the pandemic is more profound on the livelihood of individuals and the economy as a whole. The world’s global trade, which had already entered the phase of gradual deglobalisation, moved swiftly towards the era of decoupling of economies (Eichengreen, 2023; Kim et al., 2020) and the great reset of the global economy towards deglobalisation was accelerated by the pandemic (Chaudhary and Sharma, 2021).

The trade war between the two major economies of the world, followed by the pandemic-triggered restrictions on trade, investment, immigration, trade protection policies and stressed supply chains have adversely impacted organisations’ international business practices, thus, forcing them to be more focused towards self-resilient and just in time production model (Eichengreen, 2023). For instance, Apple, as part of its decoupling and China+1 strategy, has established its manufacturing hub in India for a range of its iPhones, including the latest version of the iPhone 15+pro. Along similar lines, Samsung has identified

South Asia as its additional manufacturing hub as an alternative to China for its Galaxy series tablets, thus moving towards more economically and politically friendlier manufacturing bases.

The impact of pandemic on the global economic and geopolitical landscape has been unprecedented. While during the 9/11 and 2008 financial crises, the global order dominated by the USA was not challenged, but now post-Covid the challenge from China is loud and clear, and its ascendence in world trade would lead to more intense trade wars in future with central themes of deglobalisation, nationalism and protectism (Bremmer, 2020; Hitt et al., 2021; Zahra, 2021). Further, it is noted that the pendulum of current economic conflict has tilted towards nationalism, resulting in a change in international business practices of multinational enterprises towards diversification in production base, small and efficient supply chain, redundancy in operations and more geographical concentration (Hitt et al., 2021)

From the perspective of organisations, it is evident that the great reset of Covid 19, accelerated deglobalisation, trade wars, populist movements, slowdown in the flow of technology, weak intellectual property protection and hard-to-bridge tariff and non-trade barriers have negatively impacted their business ecosystem, networks and capital flow (Zahra, 2021).

5. "THE ANSWERS"

Shaping of New World Order

The World has been in the midst of a trade war between the USA and China since 2015 which is unlikely to ease in the near future. Every bloc, no matter how much it is collaborating with partners in international trade, always keeps the interest of its industry at the forefront. For example, in the era of the energy transition, countries are moving from gasoline-guzzler vehicles towards environment-friendly Electric Vehicles forcing major economies to provide much-needed tax breaks and incentivise to automotive industry to make them cost-competitive. Similar incentives introduced by China for the industry resulted in glut of low-cost Chinese EVs in the European markets forcing the EU to retaliate by initiating probes into the Chinese subsidy policies for electric vehicles (Hancock, 2023). This led us to our first enquiry into the possible shape of the new economic world order resulting from the trade war between the USA and China and the response of other economic blocs like the EU in these evolving scenarios.

The use of game theory modelling in international trade with complex economic scenarios has been extensively researched with two different schools of thought of achieving equilibrium, one through classic game theory and its extended form of dynamic sequential games (Stones, 2001) and the second one using, theory of moves in which player can move between different outcomes through sequence of moves by looking ahead of the curve i.e. analysing the moves and countermoves of other players (Brams, 1993). Utilizing game theory, payoff prospects in the currency war between the US, EU and China were studied, and it was analysed that protective strategies by countries and negotiations are sufficient deterrents to halt a devaluation war (Askari et al., 2020). On the other hand, (Yin and Hamilton, 2018) argued that the order of play does not impact the end result but should be analysed in tandem with the policy framework similarly, Correa (2001) also suggested the limitation of game theory in international relations as the strategies adopted are interdependent and the payoff could change dynamically.

While the debate on the utilisation of game theory models in the analysis of international trade is continuing, this section of the chapter, for illustration of moves and countermoves by the economic blocs and the impact on payoff utilized the extension form of classical game theory involving three players

Figure 5. USA-China-EU dynamic sequential game model

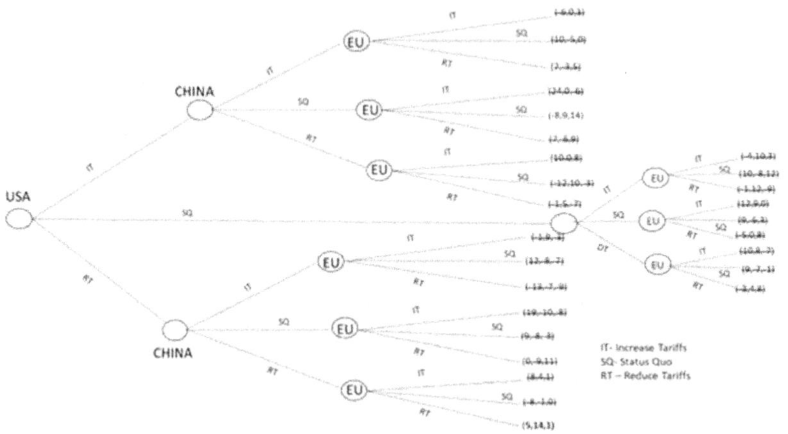

each having three choices (a,b,c) to make. As the trade and war moves taken by the aggressor country and the countermove by the defender cannot taken simultaneously, using an extension form of classical game theory is appropriate as one country will analyse the move of another country before deploying its counter-strategic options. This when applied to the trade war between the USA, China and the European Union, with USA - as an aggressor or initiator of the trade war, China- as a defender with the first counter move and the European Union- as the third impacted trade bloc yielded dynamic game theory model as illustrated in Figure 2 below.

In this dynamic model, every entity or country would attempt to maximise its returns and would decide upon the counter moves by taking into account the strategies adopted by the other country. For example, the US could take three postures in the ongoing trade war first as an aggravator by further Increasing Tariffs (IT), maintaining the Status Quo (SQ) or taking a backstep by Reducing tariffs (RT). Upon these moves, China would analyse and take counter moves of IT, SQ or RT in all three scenarios, leading to 9 dynamic situations. EU, the new entity in the trade war, would then react with its counter moves against these nine dynamic situations, taking the number of possible outcomes at 27, as illustrated in Figure 5. The payoffs (a,b,c) in each of the 27 scenarios are also presented, and each country would try to have a maximum payoff. For example, the US would like to maximise its payoff 'a' at '24' by expecting China to maintain the status quo (SQ) and the EU to increase tariffs (IT) against its move tariff increase (IT), leading to (IT, SQ, IT) as most dominating strategy for a maximum payoff for the US. On the other hand, China would be maximising 'b' at '14' and expects the US and EU to reduce tariffs (RT) and would respond in kind, leading to the most dominating option for China as (RT, RT, RT). Finally, the EU would be maximising 'c' at '14' by maintaining the status quo (SQ) and expect China to do the same against the US move of increasing tariffs (IT), resulting in (IT, IT, SQ) as a strategy with a most favourable outcome for EU.

This takes us to the second enquiry of this chapter: What would be the economic shape of this New World order? As discussed above, the moves and countermoves by the US and China resulted in a division of world trade into multiple economic blocs where countries collaborate and compete at different points in time by entering into trade agreements, and economic treaties and moving out of such treaties as well. Currency war is another weapon in the arsenal of countries in the trade war which can take various devaluation strategies to hurt the interest of opponents. It is anticipated that in the next phase of a trade war, to retain the dominance of the US dollar, in the worst-case scenario, the USA would attempt to restrict China and Russian access to Society for Worldwide Interbank Financial Telecommunications (SWIFT), a critical cog in global finance and to counter this move both these countries have developed their alternate baking platforms of Cross-Border Interbank Payment System (CIPS) and SPFS respectively and have entered into a currency swap with 40 countries (Dunford and Qi, 2020). This would be another pole leading to a new economic world order away from the dominance of the US dollar.

Given the current conflict, an extraordinary attempt is being made by China to realign global geo-politics and economies through its Belt and Road Initiative (BRI). Announced by President Xi in 2013, BRI with 205 agreements signed between 171 countries and organizations, aims to interconnect markets in Central Asia, the Middle East, ASEAN, Africa and Europe (China Org, 2023). It is estimated that the BRI economic corridor would have a significant impact on trade, investment and social conditions of the citizens and world trade is expected to grow by 1.7 to 6.2 per cent and would lift 7.6 million people from poverty (International Bank for Reconstruction and Development,2019). As BRI is developing initiatives, its success will depend upon the future evolving scenarios of the trade-off between the growth of multilateralism on the one hand and the economic growth of the corridor on the other. Four different future scenarios for BRI were developed with strong multilateralism-low economic growth making BRI irrelevant, weak multilateralism-low economic growth scenario confining BRI to Asia, weak multilateralism-high level of economic growth leading to vibrant BRI and lastly, weak multilateralism-high economic growth turning BRI to be an international corridor (Schulhof et al., 2022). However, it is noted that in the zone of deglobalization, the bloc-based regionalism of world trade is on the rise, and countries are increasingly getting sceptical of the rise of these regional economic blocs, especially BRI through which China expects everyone to be part of BRI at its terms and condition with little transparency which countries are resultant to accept as it would means intrusion in their sovereignty (Grunstein, 2022).

It's been ten years since President Xi announced the Belt and Road Initiative to connect three continents with infrastructure connectivity never seen or imagined in the history of mankind. However, BRI is still an evolving initiative and a decade is an extremely short duration to analyse what future course of the project it would take or how it would evolve in future, but whatever form it would take, it would

Figure 6. Belt and road initiative
Source: World Bank, https://www.worldbank.org/en/topic/regional-integration/brief/belt-and-road-initiative

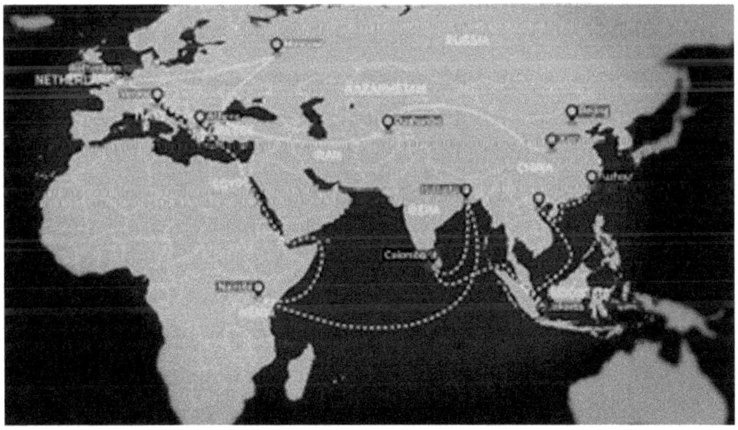

have a major impact on the geopolitics, world trade and global economy resulting in a multipolar world with its core in regionalization rather than globalization.

The establishment of regional economic blocs and economic corridors like Belt and Road Initiative (BRI), India-Middle East-East Europe economic corridor (IMEC) and other economic corridors is on the rise and the onus is on commercial organisations to leverage these collaborative trade treaties, economic corridors and business incentives to develop their international business profile and portfolios. This led us to the third and last enquiry of the chapter on what organisational-level strategies companies are employing or would deploy in this multipolar world, which is increasingly regional than global.

With high dependence of multinational companies on China, this is a tricky question. Organizational with their China-centric policies have already committed part of their commercial operations, especially manufacturing in China, to leverage its low labour cost, strong logistics, access to demand centres, subsidy regime and global market access. However, the rising stakes of the USA and China in an ongoing trade war and government pressures have compelled these companies to set up alternate manufacturing facilities like the newly started chip manufacturing facility of Global Foundries in Singapore (Reuters, 2023). As noted earlier, a strong trend of regionalisation of trade and value chain has resulted in a reverse trend of famous 'offshoring' of operations in the 1990s to 'nearshoring,' i.e. shifting operations nearer to the consumer market. The strong inclination towards 'near shoring' is evident from the analysis of employment data of 45 countries, which indicates a decline in the manufacturing share of China and shifting jobs back to the US and Europe (Lábaj and Majzlíková, 2023).

Another notable emerging trend is 'Friendshoring' defined by Braw (2023) as the *"Relocation of manufacturing, supply chain from China and other geopolitical risky countries to friendlier ones"* for example, moving of production lines of Samsung to Vietnam, a friendly country. But, achieving 'friendshoring' is a difficult task, especially in the short run, as companies struggle to perfect their logistics supply chain and develop new and reliable sub-supplier bases for their new manufacturing facilities to turn them as effective, efficient and productive as their Chinese counterparts (Braw, 2023). However, on the ground, the execution of sovereign decoupling policies is a challenging task for organisations as they possess diverse and multiple product portfolios. Therefore, strategic decisions have to be taken at an organisational level with respect to 'What product(s) to decouple'? and 'to what extent'? A decision matrix of 'Reshorability Vs Strategic Importance' of the product was proposed considering the strategic importance of the product and its ability to move away from the old production centre, utilizing which organisations can take decisions to 'friend shore' - to move the production centre to a friendly country, 'offshore' – to shift the production centre to any other country, 'reshore' – to shift production centre back or closer to the parent country (Witt et al., 2023).

The overall trajectory of a trade war, with rising tariff lines and sanctions, especially on the companies possessing new technologies, is hurting businesses in both the USA and China (Bremmer, 2020). For example, US sanctions on Huwai owing to security risk projected by its 5G equipment has resulted in reduction in company's profits, which shrieked by 65% in the financial year 2022 as compared to 2021, thus leading to business continuity challenges for Huwai (Huawei, 2022; BBC, 2021). Another aspect of rising tariffs on specific products by countries is through imposing anti-dumping duties on certain raw materials and finished products. For instance, India continued its local industry protective policies by imposing anti-dumping duties on imported steel and also imposing land border restrictions on procuring goods and services from China, citing security concerns (PIB, 2020). These restrictions are imposed to make domestic industry more cost competitive and at times, these bear positive results as India witnessed a 76% reduction in the import of solar modules during H12023, in terms of capacity, from China due to the imposition of 40% customs duties on solar PV modules which would result in increasing the domestic capacities and capabilities of the manufacturers in the solar renewable energy segment (Economic Times, 2023).

The Rise of G20

With genesis in 1999, G20- a bloc comprising 19 member countries and the European Union, was established as a platform for finance representatives of member nations to discuss coordinated responses to the financial crisis of 1997-98 and later with the onset of the great financial crisis of 2008 it was level raised to 'Head of State' forum. The expansion of G20 to G20+ with the inclusion of 52 nations African Union (AU), has transformed it into a more inclusive, powerful geopolitical and economic collaborative platform in which the member countries together comprise 85% of global GDP, 75% of the Global trade and 2/3rd of the World's population (G20, 2023).

The 18th summit of G20 under the presidentship of India, may prove to be a turning point for the world to move ahead into the next phase of economic development through forging partnerships and collaboration amongst the member countries. The multi-façade success of the Indian G20 presidency is evident not only in securing complete consensus on the 'New Delhi Leaders Declaration', but also in the summit took a giant leap towards creating it as a comprehensive economic bloc. From the recent declaration, it is expected that G20+ will rise as the mainstay of economic activities between the member countries (Ministry of External Affairs, GoI, 2023) through new trade agreements and the development of economic corridors. For example, during the 18th summit, eight member countries joined hands together to form Global Biofuel Alliance for collaboration in technology, investment and other advancement to enable increased utilisation of biofuels in member countries.

The summit also put forward a big competitor to the ambitious Belt and Road initiative of China with its own India-Middle East- Europe Corridor (IMEC) with USA, India, Saudi Arabia, UAE, France, Germany, Italy and the European Union as signatories with an aim to connecting Europe, Middle East and Asia through railway, advance port infrastructure, undersea fibre cables and other energy infrastructure (Whitehouse, 2023; PIB, 2023). "*It is a big deal*, " as US President Biden put it and this would also be another step toward energy independence of Europe from Russian Oil and gas. IMEC is being projected as a counter-move to China's Belt and Road Initiative but what form or shape it will take in the future is a matter of commitment of countries engaged in this agreement. BRI and IMEC are evolving initiatives focused on economic blocs-based trade movements leading to the multi-polarisation of the global economy and the end of globalisation as we know it!

6. 'DIGI-LOCALIZATION': THE PATH WAY TO DEGLOBALIZATION

The critical nature of digitalization as support pillar for globalization is well understood and documented in literature. However, with the rise of nationalism and multi-polar world trade the paradigm shift in digital strategies of these nations and trade blocs is matter of debate and in future the nature of change may lead to digital trade wars as well. Data protection which may be termed as Digi-localization would be the most critical aspect that would further spur the wave the deglobalization. Today, nations are not only developing war strategies to protect their boundaries but immense focus is to protect their data which is central to their market competitiveness. Today countries at macro level and organizations at micro level seek data to be locally stored, processed and accessed and assessed. This has given rise to strategic data centres especially in developing countries like India. Second area of focus for these nations while moving away from the global economy is the control over key technologies like next generation of Artificial Intelligence for instance in the case of China and US, the two leaders in research and development of generative AI and taking entirely different pathways in this global race. These divergent paths may lead to development of stronger digital boundaries not only between these nations but between the various trade blocs as well. Another aspect of technology protection is the development of digital tools like blockchain to protect this Intellectual property which are facing increasing threats of breaching and creating technological clones in integrated global world. The strength of US Dollar was instrumental in integration of global economies after second world war and the rise of US Dollar has turned it into a primary reserve currency of the world. The nations on the path of deglobalization understand that unless the strong hand of US Dollar is loosened the true nature of trade independence would be impossible to achieve. With trade relations between the two largest economies under constant strains, China is swiftly working towards an independent Digital Currency Electronic Payment (DCEP), an alternate to USD. While Bahamas became first country to issue Sand Dollar- its national digital currency, efforts are ongoing by Russia, Korea and EU to develop their own digital currencies. Not in distant future, we may witness these currencies becoming central to the global trade exchange paving way for financial independence and providing further impetus to the deglobalization pathway.

7. DISCUSSION

This chapter addressed three pertinent questions in international business related to the emergence of the multipolar deglobalized world, which is increasingly getting skewed toward regional economic blocs and how these blocs respond to the ongoing trade war between the US and China. Secondly, will China be able to augment its economic prowess through its Belt and Road Initiative (BRI), and how would the US and other countries counter this dominance through treaties and collaborations? Lastly, on the business front, what growth strategies businesses are and would be deploying to maintain balance in this multipolar world and how they are leveraging these collaborative partnerships for their organizational growth?

It is beyond doubt that the various black and white swan events since the great financial crisis have resulted in a permanent skewness of the global economy towards deglobalization. The ongoing trade war between two major economic powers of the world, the Russian-Ukraine crisis coupled with dynamic trade tactics, treaties, alliances, collaborations and negotiations, is resulting in the consolidation of regional-based economic blocs. In recent business surveys, more than 90% of the organizations opined on the strengthening and further intensification of economic deglobalization, and more than 80% believed that the US and China would decouple extensively in the coming years (Analytica, 2023). Deglobalization is underway, but how deep it will be and how long it will last is difficult to predict.

Mutual trust, transparency and cooperation were the pillars of phases of historic growth of economies in the globalized world, which owing to geopolitical and economic uncertainties, have been replaced by scepticism and competition for global supremacy. Thus, leading the world economy towards a new normal of a decoupled and multipolar world which has shaken the need for the perfect balance between two ends of the spectrum of global trade – globalization and deglobalization. The diminishing relevance of WTO as an enabler towards global fair and free trade and the rising economic dominance of China through protective measures and government incentives is leading to unfair competition in the global market.

Thus, forcing other economies to forge standalone bilateral and regional trade agreements and treaties. The growth of Belt and Road initiative of China, the formation of India-Middle East-Europe Economic corridor during G20 and other economic treaties would further expand the deglobalization agenda and future trade wars and diplomacies would be fought through these corridors itself. These developments would greatly impact the international business profile of not only multinational organizations but also the new technology-driven ventures currently at the initial stages of their growth trajectory.

From the venture perspective, at ground level, the developing and evolving nature of these agreements and treaties, if executed with positive intent, provide great opportunities for the business to grow, prosper and expand their global footprints through these treaties. These opportunities for new businesses are more beneficial, if facilitated by Government representatives, under these treaties resulting in new alliances between the organizations. It is also noted that these agreements sometimes are more skewed towards specific economic sectors with more negotiation prowess and thus, depriving other sectors of the tangible benefits that these treaties or agreements offer such as tariff reduction, preferential treatment and technology exchange. It is also worthwhile to note that the countries today, under the shadow of deglobalization, are seeking not only the trade independence through entering into various treaties and establishment of trade blocs, but also are developing macro level strategies towards technology, financial and digital independence. Thus, moving from global digital economy toward 'Digi-localization'.

One of the contentious issues that businesses face, which makes them sceptical of these treaties, is protecting their Intellectual Property. This is a difficult issue in international trade and has a direct impact on how technology-driven organizations conduct their businesses when they are both creators and traders of IP. While trade treaties and alliances have attempted to strengthen IP protections and respect for IP rights, the same has resulted in varying success in different geographies. Therefore, the dilemma facing technology driven companies is the choice between business growth and apprehension of losing IP theft when these are loosely protected by the treaties. To overcome these, businesses seek partners with shared interests and are developing independent protective mechanisms around their IPs during the penning of collaborative agreements. While there is continuous debate both at the national and business levels on the technology independence which would be driven by strong IP protection both by developing regulatory. While efficacy of WTO is debated in this chapter at length, the Trade Related Aspects of Intellectual Property Rights (TRIPS) agreement for resoling IP related disputes between the nations is excellent platform to build upon by the nations entering into various treaties and trade blocs. With technology providing unique competitive advantage to the nations, it is evident that the future trade deals will have the development of protection and regulatory mechanism for Intellectual Property between the member countries as central theme of negotiations.

In context to the multinational companies, the trade war and resultant deglobalization might have severe consequences on their international business growth strategies and at times, they have to take the hard decision of shifting their operations to friendly countries– a trade-off between growth and survival. But to survive and grow, organizations need to be flexible and agile in their decision-making which at times are marred by uncertainties. The decision-making approach of organizations in light of geopolitical and trade uncertainties and the changing face of global trade dominance is a matter of future research.

REFERENCES

G20. (2023, March 6). *G20- Background Brief.* Retrieved September 25, 2023, from https://www.g20.org/content/dam/gtwenty/about_g20/overview/G20_Background_Brief_06-03-2023.pdf

Allison, G. (2017). The thucydides trap. *Foreign Policy*, *9*(6), 73–80. Advance online publication. doi:10.7551/mitpress/9780262028998.003.0006

Analytica, O. (2023). How are global businesses managing today's political risks? 2023 survey report. *WTW*, 1-34.

Anand, A., Sagar, S. R., & Kumar, C. (2023). How India Is Able to Control Inflation During the Russia-Ukraine War. In Cases on the Resurgence of Emerging Businesses (pp. 237-251). IGI Global. doi:10.4018/978-1-6684-8488-3.ch017

Askari, G., Gordji, M. E., Shabani, S., & Filipe, J. A. (2020). *Game theory and trade tensions between advanced economies*. Academic Press.

BBC. (2021, March 31). *Huawei's business damaged by US sanctions despite success at home.* Retrieved September 21, 2023, from https://www.bbc.com/news/technology-56590001

Bergin, P., Choi, W. J., & Pyun, J. (2023). *Catching Up by 'Deglobalizing': Capital Account Policy and Economic Growth.* doi:10.3386/w30944

Brams, S. J. (1993). Theory of moves. *American Scientist*, *81*(6), 562–570.

Braw, E. (2023). How to "Friendshore". American Enterprise Institute for Public Policy Research.

Breuss, F. (2022). Who wins from an FTA induced revival of world trade? *Journal of Policy Modeling*, *44*(3), 653–674. doi:10.1016/j.jpolmod.2022.05.003

Can, C. M. (2022). Temporal Theory and US-China Relations. *Journal of Strategic Security*, *15*(2), 1–16. doi:10.5038/1944-0472.15.2.1985

Chaudhary, P., & Sharma, K. K. (2021). Effects of Covid-19 on De-globalization. *Globalization, Deglobalization, and New Paradigms in Business*, 133–153. doi:10.1007/978-3-030-81584-4_8

China Org. (2023). *The Belt and Road Initiative.* BRI. Retrieved September 20, 2023, from http://belt.china.org.cn/

Correa, H. (2001). Game theory as an instrument for the analysis of international relations. *Ritsumeikan Annual Review of International Studies*, *14*(2), 187–208.

Dunford, M., & Qi, B. (2020). Global reset: COVID-19, systemic rivalry and the global order. *Research in Globalization*, *2*, 100021. doi:10.1016/j.resglo.2020.100021

Economic Times. (2023, September 14). *India's solar imports from China down nearly 80% by $2 billion in H1 2023: Ember.* Retrieved October 1, 2023, from https://economictimes.indiatimes.com/industry/renewables/indias-solar-imports-from-china-down-nearly-80-by-2-billion-in-h1-2023-ember/articleshow/103667346.cms

Eichengreen, B. (2023). Globalization: Uncoupled or unhinged? *Journal of Policy Modeling*, *45*(4), 685–692. Advance online publication. doi:10.1016/j.jpolmod.2023.02.008

EU. (2019, January 31). *EU-Japan trade agreement enters into force*. Retrieved September 16, 2023, from https://ec.europa.eu/commission/presscorner/detail/en/IP_19_785

García-Herrero, A., & Tan, J. (2020). Deglobalisation in the context of United States-China decoupling. *Policy Contribution*, *21*, 1–16.

Garg, S., & Sushil. (2022). Impact of de-globalization on development: Comparative analysis of an emerging market (India) and a developed country (USA). *Journal of Policy Modeling*, *44*(6), 1179–1197. doi:10.1016/j.jpolmod.2022.10.004

Gnizy, I., & Shoham, A. (2017). Reverse Internationalization: A Review and Suggestions for Future Research. *Advances in Global Marketing*, 59–75. doi:10.1007/978-3-319-61385-7_3

Govt of Canada. (2021, March 25). *Comprehensive Economic and Trade Agreement (CETA) 2nd Meeting of the CETA Joint Committee*. Retrieved September 18, 2023, from https://www.international.gc.ca/trade-commerce/trade-agreements-accords-commerciaux/agr-acc/ceta-aecg/2021-03-25-joint_report-rapport_conjoint.aspx?lang=eng

Grunstein, J. (2022). Globalization, Real and Imagined. *Orbis*, *66*(4), 502–508. doi:10.1016/j.orbis.2022.08.005

Hancock, A. (2023, September 13). *EU to launch anti-subsidy probe into Chinese electric vehicles*. Retrieved September 30, 2023, from https://www.ft.com/content/55ec498d-0959-41ef-8ab9-af06cc45f8e7

Hitt, M. A., Holmes, R. M. Jr, & Arregle, J.-L. (2021). The (COVID-19) pandemic and the new world (dis)order. *Journal of World Business*, *56*(4), 101210. doi:10.1016/j.jwb.2021.101210

Huawei Investments & Holding Co Ltd. (2022). *2022 Annual Report*. Huawei. https://www.huawei.com/en/annual-report/2022

Ian Bremmer. (2020). Coronavirus and the World Order to Come. *Horizons: Journal of International Relations and Sustainable Development, 16*, 14-23.

IMF. (2023). *World Economic Outlook*. IMF.

Johns Hopkins. (2023, March 10). *Johns Hopkins University & Medicine Coronavirus Resource Centre*. Retrieved September 30, 2023, from https://coronavirus.jhu.edu/map.html

Kim, H.-M., Li, P., & Lee, Y. R. (2020). Observations of deglobalization against globalization and impacts on global business. International Trade. *Politics and Development*, *4*(2), 83–103. doi:10.1108/ITPD-05-2020-0067

Kumar, S., Chavan, M., & Pandey, N. (2023). Journal of International Management: A 25-year review using bibliometric analysis. *Journal of International Management*, *29*(1), 100988. doi:10.1016/j.intman.2022.100988

Lábaj, M., & Majzlíková, E. (2023). How nearshoring reshapes global deindustrialization. *Economics Letters*, *230*, 111239. doi:10.1016/j.econlet.2023.111239

Ministry of External Affairs. (2023, September 9). *G20 New Delhi Leaders' Declaration*. Retrieved September 25, 2023, from https://www.mea.gov.in/bilateral-documents.htm?dtl/37084/G20_New_Delhi_Leaders_Declaration

Miśkiewicz, J., & Ausloos, M. (2010). Has the world economy reached its globalization limit? *Physica A*, *389*(4), 797–806. doi:10.1016/j.physa.2009.10.029

MoCA. (2022, June 30). *India-Australia Economic Cooperation and Trade Agreement (INDAUS ECTA) between the Government of the Republic of India and the Government of Australia*. Retrieved September 30, 2023, from https://commerce.gov.in/international-trade/trade-agreements/ind-aus-ecta/

MoFA. (2022, July 22). *Comprehensive Economic Partnership Agreement between Japan and the Republic of India*. Retrieved October 1, 2023, from https://www.mofa.go.jp/region/asia-paci/india/epa201102/index.html

Nicholas, N. (2008). The black swan: The impact of the highly improbable. *Journal of the Management Training Institut*, *36*(3), 56.

Nicholas Taleb, N. (2015). The black swan: The impact of the highly improbable. Victoria, 250, 595-7955.

PIB. (2020, July 23). *Restrictions on Public Procurement from certain countries*. Retrieved September 18, 2023, from https://pib.gov.in/PressReleasePage.aspx?PRID=1640778

PIB. (2022, March 27). *Comprehensive Economic Partnership Agreement (CEPA) between India and the United Arab Emirates (UAE) Unveiled*. Retrieved September 30, 2023, from https://www.pib.gov.in/PressReleaseIframePage.aspx?PRID=1810279

PIB. (2023, September 9). *India-Middle East-Europe Economic Corridor promises to be a beacon of cooperation, innovation, and shared progress*. Retrieved October 2, 2023, from https://pib.gov.in/PressReleaseIframePage.aspx?PRID=1955842

RCEP. (2022, January 1). *RCEP Agreement enters into force*. Retrieved October 2, 2023, from https://rcepsec.org/rcep-agreement-enters-into-force/

Reuters. (2023, September 12). GlobalFoundries opens $4 billion Singapore chip fabrication plant. *Economics Times*. Retrieved September 26, 2023, from https://economictimes.indiatimes.com/tech/technology/globalfoundries-opens-4-billion-singapore-chip-fabrication-plant/articleshow/103606474.cms

Roubini, N. (2020). The Specter of Deglobalization and the Thucydides Trap. Horizons. *Journal of International Relations and Sustainable Development*, *15*, 130–139.

Schulhof, V., van Vuuren, D., & Kirchherr, J. (2022). The Belt and Road Initiative (BRI): What Will it Look Like in the Future? *Technological Forecasting and Social Change*, *175*, 121306. doi:10.1016/j.techfore.2021.121306

Sharma, R. (2017). The Boom Was a Blip: Getting Used to Slow Growth. *Foreign Affairs*, *96*(3), 104–114.

Sharma, S. (2023, July 5). Discounted Russian crude imports saved Indian refiners $7 billion. *Indian Express*. https://indianexpress.com/article/business/commodities/discounted-russian-crude-imports-saved-indian-refiners-7-billion-8751745/

Stone, R. W. (2001). The use and abuse of game theory in international relations: The theory of moves. *The Journal of Conflict Resolution*, *45*(2), 216–244. doi:10.1177/0022002701045002004

US ITA. (2022, November 4). *Japan - Country Commercial Guide*. Retrieved October 2, 2023, from https://www.trade.gov/country-commercial-guides/japan-market-overview

Verma, B., & Srivastava, A. (2023). Impact of different dimensions of globalisation on firms' performance: An unbalanced panel-data study of firms operating in India. *World Review of Entrepreneurship, Management and Sustainable Development*, *19*(3-5), 360–378. doi:10.1504/WREMSD.2023.130618

White House. (2023, September 9). *FACT SHEET: World Leaders Launch a Landmark India-Middle East-Europe Economic Corridor*. Retrieved October 2, 2023, from https://www.whitehouse.gov/briefing-room/statements-releases/2023/09/09/fact-sheet-world-leaders-launch-a-landmark-india-middle-east-europe-economic-corridor/

Witt, M. A. (2019). De-Globalization: Theories, Predictions, and Opportunities for International Business Research. SSRN *Electronic Journal*. doi:10.2139/ssrn.3315247

Witt, M. A., Lewin, A. Y., Li, P. P., & Gaur, A. (2023). Decoupling in international business: Evidence, drivers, impact, and implications for IB research. *Journal of World Business*, *58*(1), 101399. doi:10.1016/j.jwb.2022.101399

World Bank. (2022). *Making the Most of the African Continental Free Trade Area: Leveraging Trade and Foreign Direct Investment to Boost Growth and Reduce Poverty*. World Bank. https://openknowledge.worldbank.org/entities/publication/09f9bbdd-3bf0-5196-879b-b1a9f328b825

World Bank. (n.d.). *World Bank national accounts data, and OECD National Accounts data files*. Retrieved October 5, 2023, from https://data.worldbank.org/indicator/NY.GDP.MKTP.CD?end=2022&locations=CN-US-JP&start=2002

World Trade Organization. (2023a). *Re-globalization for a secure, inclusive and sustainable future*. World Trade Report 2023. https://www.wto.org/english/res_e/publications_e/wtr23_e.htm

World Trade Organization. (2023b). *World Trade Statistical Review 2023*. World Trade Report 2023 https://www.wto.org/english/res_e/booksp_e/wtsr_2023_e.pdf

World Trade Organization. (2023c). Retrieved September 20, 2023, from https://www.wto.org/english/thewto_e/thewto_e.htm

Yin, J. Z., & Hamilton, M. H. (2018). The conundrum of US-China trade relations through game theory modelling. *Journal of Applied Business and Economics*, *20*(8).

Zahra, S. A. (2021). International entrepreneurship in the post Covid world. *Journal of World Business*, *56*(1), 101143. doi:10.1016/j.jwb.2020.101143

Chapter 2
Decentralization in the Digital Age

Swaty Sharma
Lovely Professional University, India

ABSTRACT

This chapter explores how digital technologies are reshaping decentralization in our fast-evolving technological era. It defines decentralization and examines its diverse aspects in politics, economics, and technology. The chapter emphasizes blockchain technology, highlighting its immutability, decentralized consensus, and smart contracts as pivotal elements driving decentralized applications and disrupting traditional paradigms. To provide context, it conducts a thorough literature review, tracing decentralization's historical evolution, exploring its impact on various sectors like decentralized finance, governance, financial inclusion, and system resilience. The chapter also addresses challenges such as regulatory uncertainties, scalability issues, and security concerns, drawing from authoritative sources and academic research. In conclusion, it sets the stage for a deeper exploration of decentralization's applications and impact across sectors, acknowledging both its potential and complexity as a transformative paradigm shift.

1. INTRODUCTION

The digital age has ushered in a profound transformation across various aspects of society, including governance, finance, and communication. At the heart of this transformation lies the concept of decentralization, a paradigm shifts away from traditional centralized structures towards more distributed and inclusive systems ((Tapscott & Tapscott, 2016). The notion of decentralization, while not novel, has taken on new dimensions in the context of contemporary digital technologies. It signifies the dispersal of power, authority, or decision-making from a central entity to a network of smaller units or actors, creating a distributed and inclusive system (Ostrom, 2010). This concept transcends traditional domains, manifesting in political governance, economic transactions, and technological infrastructures. As we journey through this chapter, we will unveil the multifaceted nature of decentralization and its pervasive influence in our digitally-driven world. In this chapter, we will delve into the multifaceted nature of

DOI: 10.4018/979-8-3693-1762-4.ch002

decentralization in the digital age, exploring its definitions, driving forces, and implications. It explains that the digital age has given rise to decentralized systems that challenge traditional centralized structures. It outlines the objectives of the chapter, which include defining decentralization, exploring its types, and examining the role of digital technologies in enabling decentralization.

2. LITERATURE REVIEW: A SCHOLARLY DISCOURSE ON DECENTRALIZATION

To provide a rich context for our exploration, we venture into the scholarly landscape that underpins our understanding of decentralization in the digital age.

Historical Perspectives on Decentralization

The concept of decentralization has historical roots in governance and organization theories (Ostrom, 2010). Early discussions on decentralization in political science and public administration highlight the benefits of dispersing authority and decision-making to local or regional levels. Such decentralization aims to enhance governance efficiency, promote citizen participation, and improve service delivery (Faguet, 2012).

Enlightenment Era and Individual Liberty

The Enlightenment period, spanning the 17th and 18th centuries, was marked by a profound emphasis on individual liberty and autonomy. Thinkers like John Locke, Jean-Jacques Rousseau, and Thomas Jefferson championed the idea that individuals should have the freedom to make decisions that affect their lives (Locke, 1689; Rousseau, 1762; Jefferson, 1776). These philosophers laid the groundwork for the decentralization of power, advocating for governance systems that respected individual rights and localized decision-making.

Local Governance and Decentralized Authority

Throughout history, local governance has been a pillar of decentralization. Ancient city-states like Athens in Greece and the Italian city-states during the Renaissance era are prime examples. These city-states had considerable autonomy and self-governance, making decisions that were best suited to their unique needs (Hansen, 2006; Hale, 2012). In more recent times, Switzerland's system of direct democracy, where decisions are made at the local level through referendums and citizens' assemblies, exemplifies the enduring relevance of decentralized governance (Swiss Government, n.d.).

The Early Internet and Decentralized Information Exchange

Even before the advent of blockchain and cryptocurrencies, the internet itself had a decentralized essence in its early days. The World Wide Web, developed by Tim Berners-Lee in the late 1980s, was designed to be decentralized (Berners-Lee, 1989). Information was distributed across multiple servers, allowing users to share and access data without a central authority. This decentralized architecture ensured that no single entity could control or censor the flow of information. An excellent example is the development of email protocols, which enabled direct communication between individuals without intermediaries (RFC 5321, 2008).

Case Studies

1. Bitcoin and the Cypherpunk Movement:

Bitcoin, the pioneering cryptocurrency introduced by the pseudonymous Satoshi Nakamoto in 2009, can be viewed as a contemporary embodiment of decentralization. Nakamoto's whitepaper, titled "Bitcoin: A Peer-to-Peer Electronic Cash System," sought to create a decentralized digital currency that operates independently of banks and governments (Nakamoto, 2008). Bitcoin's block chain, a decentralized ledger, allows users to transact directly without intermediaries. It has garnered global attention for its potential to disrupt traditional financial systems.

2. Localized Decision-Making in Porto Alegre, Brazil:

The city of Porto Alegre, Brazil, implemented a pioneering experiment in decentralized governance in the late 1980s and early 1990s. The city introduced participatory budgeting, a process where citizens directly decide how to allocate a portion of the municipal budget. This initiative empowered local communities to have a say in public spending and fostered a sense of civic engagement and accountability (Avritzer, 2002).

3. Ethereum and Decentralized Applications (DApps):

Ethereum, a block chain platform launched in 2015 by Vitalik Buterin, expanded the concept of decentralization beyond cryptocurrencies. Ethereum introduced smart contracts, self-executing code that enables the creation of decentralized applications (DApps) (Buterin, 2013). These DApps, running on the Ethereum blockchain, offer various services and functionalities without relying on centralized authorities. This innovation opened the door to a wide range of decentralized applications, including decentralized finance (DeFi) platforms.

In conclusion, decentralization is a concept deeply rooted in history and philosophy. From the Enlightenment's emphasis on individual liberty to localized governance structures and early internet architecture, decentralization has always had a presence. Modern case studies, such as Bitcoin, participatory budgeting in Porto Alegre, and Ethereum's DApps, illustrate how this historical concept is evolving and finding new applications in the digital age. These examples showcase the enduring relevance and transformative potential of decentralization in reshaping various aspects of our society.

Block chain and Decentralization

The advent of blockchain technology has been instrumental in the promotion of decentralization across various domains (Mougayar, 2016). Satoshi Nakamoto's Bitcoin whitepaper (Nakamoto, 2008) introduced the world to blockchain as a decentralized digital currency. Subsequently, blockchain technology has evolved to encompass decentralized applications (DApps), smart contracts, and various other use cases

(Swan, 2015). The literature has extensively explored the technical aspects, security implications, and potential societal impacts of block chain technology (Narayanan et al., 2016).

Economic Decentralization and DeFi

The concept of economic decentralization has gained prominence with the rise of decentralized finance (DeFi) platforms (Gupta & Sengupta, 2020). Researchers have studied the mechanisms, risks, and opportunities associated with DeFi, emphasizing its potential to democratize access to financial services and reduce dependence on traditional intermediaries (Gupta & Sengupta, 2020; Yermack, 2015).

Decentralization and Governance

Scholars have examined the role of decentralization in governance and its implications for participatory democracy (Ribot, 2014). E-governance platforms and block chain-based voting systems have been subjects of research, exploring how they enable citizens to engage in decision-making processes and hold governments accountable (Ribot, 2014).

Challenges and Considerations

The literature review also highlights critical challenges associated with decentralization, including regulatory uncertainty (Casey & Vigna, 2018), scalability issues (Swan, 2015), and security concerns (Tapscott & Tapscott, 2016). Researchers have sought to address these challenges and provide insights into the development and adoption of decentralized systems (Casey & Vigna, 2018).

3. DEFINING DECENTRALIZATION

Decentralization, in its broadest sense, refers to the dispersal of power, authority, or decision-making from a central entity to a multitude of smaller units or actors (Ostrom, 2010). This dispersion can occur across various domains, including political, economic, and technological spheres. In the digital age, decentralization has gained prominence as digital technologies enable novel ways of distributing power and resources.

Types of Decentralization

Political Decentralization: In the realm of governance, political decentralization involves transferring authority from a central government to regional or local authorities (Faguet, 2012). The rise of e-governance platforms and block chain-based voting systems exemplifies this form of decentralization.

Economic Decentralization: Economically, decentralization can manifest through peer-to-peer (P2P) systems, such as decentralized finance (DeFi) platforms, where financial intermediaries are bypassed, and financial transactions occur directly between participants (Gupta & Sengupta, 2020).

Technological Decentralization: This type of decentralization pertains to the distribution of technology resources, such as decentralized data storage networks, where data is stored across a network of nodes rather than on centralized servers (Swan, 2015).

Fundamental Principles of Decentralization

Decentralization is guided by a set of core principles that underpin its philosophy and operation. These principles include autonomy, transparency, resilience, and inclusivity. Understanding these principles is crucial for comprehending the essence of decentralization and its impact on various domains. Here, we delve into each of these principles, supported by examples and case studies.

1. Autonomy:

Definition: Autonomy refers to the ability of individuals or entities to make decisions and act independently within a decentralized system.

Example: Internet Domain Names

The decentralized system for managing internet domain names exemplifies autonomy. Organizations and individuals can register domain names through various domain registrars without centralized control. This decentralization allows autonomy in choosing and managing internet identities.

2. Transparency:

Definition: Transparency implies that the operations and decision-making processes within a decentralized system are open and visible to its participants.

Example: Block chain Transactions

Block chain technology, as seen in cryptocurrencies like Bitcoin, offers a transparent ledger where all transactions are recorded and visible to anyone on the network. Participants can verify transactions without relying on intermediaries, enhancing trust and transparency.

3. Resilience:

Definition: Resilience in decentralization refers to the system's ability to withstand disruptions or attacks by distributing functions across multiple nodes or entities.

Example: BitTorrent

BitTorrent, a decentralized file-sharing protocol, demonstrates resilience. Instead of relying on a single server, BitTorrent distributes file-sharing tasks across multiple peers. This redundancy ensures that even if some peers go offline, the system remains functional.

4. Inclusivity:

Definition: Inclusivity emphasizes that decentralization should enable broad participation, minimizing barriers to entry and ensuring accessibility to a diverse range of participants.

Case Study: Decentralized Autonomous Organizations (DAOs)

DAOs, like "The DAO" on the Ethereum blockchain, exemplify inclusivity. They allow anyone to participate in decision-making and investments within the organization without requiring traditional intermediaries. In 2016, "The DAO" raised over $150 million through crowdfunding, showcasing the inclusivity of decentralized systems.

5. Security:

Definition: Security in decentralization refers to the protection of data and assets from unauthorized access or tampering.

Case Study: Decentralized Exchanges (DEXs)

DEXs, such as Uniswap and SushiSwap, prioritize security by allowing users to trade cryptocurrencies directly from their wallets, eliminating the need for centralized exchanges that can be vulnerable to hacks. These DEXs implement smart contracts and decentralized liquidity pools to enhance security.

6. Efficiency:

Definition: Efficiency entails achieving goals or conducting operations with minimal waste of resources within a decentralized system.

Case Study: Proof of Stake (PoS) Blockchain

PoS blockchains like Cardano aim for energy efficiency by validating transactions and creating new blocks through a consensus mechanism that doesn't require extensive computational power, as seen in Proof of Work (PoW) systems like Bitcoin.

7. Redundancy:

Definition: Redundancy involves having backup or duplicate systems, nodes, or components within a decentralized network to ensure continued functionality.

Example: Tor Network

The Tor network, which provides anonymous internet access, relies on a distributed network of volunteer-run nodes. Redundancy ensures that even if some nodes are compromised, users can still access the internet anonymously through alternative routes.

8. Data Privacy:

Definition: Data privacy emphasizes the protection of individuals' personal information within decentralized systems.

Case Study: Privacy Coins

Privacy-focused cryptocurrencies like Monero and Zcash incorporate advanced cryptographic techniques to obfuscate transaction details, enhancing data privacy for users.

Incorporating these fundamental principles into decentralized systems is essential for creating robust, trustworthy, and resilient networks. These principles are not mutually exclusive and often intersect, contributing to the overall effectiveness and societal impact of decentralized technologies.

4. DIGITAL TECHNOLOGIES AND DECENTRALIZATION

Block Chain Technology

Blockchain technology has emerged as a cornerstone of decentralization in the digital age (Mougayar, 2016). At its core, a blockchain is a distributed ledger that records transactions across a network of computers. Key features of blockchain technology include:

Immutability: Once data is recorded on the blockchain, it is nearly impossible to alter, enhancing trust in the system (Nakamoto, 2008).

Decentralized Consensus: Transactions are verified through consensus mechanisms like Proof of Work (PoW) or Proof of Stake (PoS), eliminating the need for centralized intermediaries (Buterin, 2014).

Smart Contracts: Self-executing contracts coded into the blockchain automate agreements, reducing reliance on legal intermediaries (Szabo, 1997).

Key features of blockchain technology in the context of finance include:

Security and Transparency: Transactions are recorded in a secure, tamper-resistant manner, enhancing transparency and trust.

Elimination of Intermediaries: Traditional financial intermediaries, such as banks and payment processors, can be bypassed, reducing fees and transaction times.

Global Accessibility: Cryptocurrencies and blockchain-based financial products are accessible to anyone with an internet connection, fostering financial inclusion.

Examples and Case Study:

Bitcoin (BTC):

Definition: Bitcoin is the world's first decentralized cryptocurrency, created in 2009 by an individual or group using the pseudonym Satoshi Nakamoto.

Key Features:

Decentralized: Bitcoin operates on a decentralized network of nodes, with no central authority controlling issuance or transactions.

Digital Gold: Bitcoin is often referred to as "digital gold" due to its store of value properties and limited supply (21 million coins).

Global Transactions: Users can send and receive Bitcoin globally without relying on banks or payment processors.

Case Study: Bitcoin's surge in popularity and adoption has led to increased recognition as a store of value and a potential hedge against inflation. High-profile investments by companies like Tesla and Square have further legitimized its role in the traditional financial landscape.

Role of Blockchain Technology in Enabling Decentralization

1. Definition of Blockchain:

Blockchain is a distributed ledger technology that underpins cryptocurrencies like Bitcoin. It consists of a chain of blocks, each containing a record of multiple transactions. These blocks are linked together in a chronological order, creating a secure and tamper-resistant ledger.

2. Decentralization Enabled by Blockchain:

Decentralized Ledger: Blockchain operates as a decentralized ledger distributed across a network of computers (nodes). Each node maintains a copy of the entire blockchain, and no single entity has control over it.

Consensus Mechanisms: Blockchains use consensus mechanisms like Proof of Work (PoW) or Proof of Stake (PoS) to validate and record transactions. These mechanisms rely on network participants rather than a central authority.

Security: Cryptography ensures the security of transactions and prevents unauthorized alterations to the blockchain.

Example: Bitcoin and the Blockchain:

Bitcoin's Blockchain: Bitcoin's blockchain is a public ledger that records all Bitcoin transactions since its inception. Each block contains a group of transactions, and miners compete to validate these transactions by solving complex mathematical puzzles. Once validated, the transactions are added to the blockchain.

Decentralization in Action: Bitcoin's blockchain showcases decentralization in action. It eliminates the need for banks or financial intermediaries to process and verify transactions. Instead, participants in the Bitcoin network collectively validate and maintain the ledger.

Security and Transparency: Bitcoin's blockchain offers security through cryptographic techniques and transparency by making all transactions visible to anyone on the network.

Case Study: Ethereum and Smart Contracts:

Ethereum: Ethereum, another prominent blockchain platform, was launched in 2015 by Vitalik Buterin. It extends the concept of block chain beyond cryptocurrencies.

Smart Contracts: Ethereum introduced smart contracts, self-executing agreements with code that automatically enforces contract terms. Smart contracts have applications in various fields, including legal agreements, supply chain management, and decentralized finance (DeFi).

Decentralized Applications (DApps): Developers can build decentralized applications (DApps) on the Ethereum block chain. These DApps can offer services, conduct transactions, and facilitate interactions without relying on centralized intermediaries.

In conclusion, the birth of cryptocurrencies like Bitcoin and their significance lies in their potential to disrupt traditional financial systems by providing a decentralized, secure, and transparent means of transferring value. Block chain technology, the underlying innovation, plays a pivotal role in enabling this decentralization by providing a tamper-resistant, consensus-driven ledger. Ethereum and its smart contract functionality further illustrate the versatility and potential of block chain technology beyond cryptocurrency.

Decentralized Applications (DApps)

Decentralized applications are software programs that run on blockchain networks (Mougayar, 2016). They leverage smart contracts and decentralized data storage to operate without central control. Examples of DApps include decentralized exchanges (DEXs) and decentralized social networks (Muneeb Ali, 2016).

5. DRIVING FORCES BEHIND DECENTRALIZATION

Trust in Technology

One of the driving forces behind decentralization is the trust engendered by technology (Crosby et al., 2016). In an era of increasing cybersecurity threats and data breaches, decentralized systems offer enhanced security through cryptographic mechanisms and distributed data storage.

Disintermediation

Decentralization often leads to disintermediation, the removal of intermediaries from various processes (Teece, 1986). This can result in cost savings, increased efficiency, and reduced friction in transactions. For instance, DeFi platforms enable users to lend, borrow, and trade assets without banks or financial institutions (Gupta & Sengupta, 2020).

Data Privacy and Ownership

As individuals become more aware of data privacy concerns, decentralized data storage solutions offer greater control over personal information (Swan, 2015). Users can retain ownership of their data and grant or revoke access on their terms.

6. IMPLICATIONS OF DECENTRALIZATION

Inclusive Governance

Decentralization in governance can lead to more inclusive decision-making processes (Ribot, 2014). E-governance platforms enable citizens to participate in policymaking and hold governments accountable.

Financial Inclusion

Decentralized finance (DeFi) has the potential to provide financial services to the unbanked and underbanked populations worldwide (Yermack, 2015). By eliminating the need for traditional financial intermediaries, DeFi platforms can reduce barriers to entry.

Resilience and Redundancy

Decentralized systems are often more resilient in the face of disruptions (Narayanan et al., 2016). Distributed data storage and communication networks can withstand failures at individual nodes without compromising the integrity of the entire system.

7. CHALLENGES AND CONSIDERATIONS

Regulatory Uncertainty

The regulatory landscape for decentralized technologies is still evolving (Casey & Vigna, 2018). Governments grapple with how to regulate decentralized systems, leading to uncertainty and potential legal challenges.

Scalability

While decentralization offers numerous advantages, scalability remains a challenge (Swan, 2015). Some block chain networks struggle to handle large volumes of transactions, resulting in congestion and high fees.

Security Concerns

While block chain technology is inherently secure, vulnerabilities can still be exploited (Tapscott & Tapscott, 2016). Smart contract vulnerabilities and 51% attacks on PoW block chains are notable security risks.

8. THE FUTURE OF DECENTRALIZATION

Scalability Solutions: Scaling solutions, such as layer 2 protocols, sharding, and off-chain scaling, are being developed to address scalability challenges in blockchain networks.

Regulatory Clarity: Governments are working on providing clearer regulations for decentralized technologies. This may include licensing requirements, consumer protection measures, and tax policies.

Improved Security: Developers are continuously working to enhance the security of decentralized systems, including more rigorous auditing of smart contracts and the development of secure coding practices.

User-Friendly Interfaces: Efforts are underway to create more user-friendly interfaces for decentralized applications and services to improve adoption among non-technical users.

Interoperability Standards: Standards for interoperability between different blockchain networks and protocols are being developed to foster a more interconnected and collaborative ecosystem.

Environmental Sustainability: Blockchain projects are exploring more energy-efficient consensus mechanisms, reducing their carbon footprint and addressing environmental concerns.

In conclusion, while decentralization faces its share of challenges, it also holds immense potential to reshape industries and empower individuals. The future of decentralization will likely involve a combination of technological innovations, regulatory developments, and community-driven efforts to overcome these challenges and unlock the full transformative potential of decentralized technologies

9. CONCLUSION

Decentralization in the digital age represents a transformative shift in how power, authority, and resources are distributed. It has the potential to reshape governance, finance, and technology in ways that promote inclusivity and resilience. However, it also poses regulatory, scalability, and security challenges that must be addressed as society continues to adapt to this new paradigm. In the chapters that follow, we will explore specific applications of decentralization and its impact on various sectors in greater detail.

REFERENCES

Buterin, V. (2014). *A Next-Generation Smart Contract and Decentralized Application Platform*. Ethereum White Paper. Retrieved from https://ethereum.org/whitepaper/

Casey, M. J., & Vigna, P. (2018). *The Truth Machine: The Blockchain and the Future of Everything*. St. Martin's Press.

Crosby, M., Pattanayak, P., Verma, S., & Kalyanaraman, V. (2016). Blockchain technology: Beyond bitcoin. *Applied Innovation*, *2*, 6–10.

Faguet, J. P. (2012). Decentralization and Governance. *World Development*, *41*, 67–74.

Gupta, A., & Sengupta, S. (2020). DeFi and the Future of Finance: An Exploratory Study. SSRN Electronic Journal. 10.2139/ssrn.3675187

Mougayar, W. (2016). *The Business Blockchain: Promise, Practice, and Application of the Next Internet Technology*. Wiley.

Muneeb Ali. (2016). Decentralized Apps: What Are They and Could They Take Over? *Forbes*. Retrieved from https://www.forbes.com/sites/muneebali/2016/06/08/decentralized-apps-what-are-they-and-could-they-take-over/

Nakamoto, S. (2008). *Bitcoin: A Peer-to-Peer Electronic Cash System*. Retrieved from https://bitcoin.org/bitcoin.pdf

Narayanan, A., Bonneau, J., Felten, E., Miller, A., & Goldfeder, S. (2016). *Bitcoin and Cryptocurrency Technologies: A Comprehensive Introduction*. Princeton University Press.

Ostrom, E. (2010). Beyond Markets and States: Polycentric Governance of Complex Economic Systems. *The American Economic Review*, *100*(3), 641–672. doi:10.1257/aer.100.3.641

Ribot, J. C. (2014). Cause and Response: Democracy, Rights, and Nature. *Policy Matters*, *21*, 94–105.

Swan, M. (2015). *Blockchain: Blueprint for a New Economy*. O'Reilly Media.

Szabo, N. (1997). Formalizing and Securing Relationships on Public Networks. *First Monday*, *2*(9). Advance online publication. doi:10.5210/fm.v2i9.548

Tapscott, D., & Tapscott, A. (2016). *Blockchain Revolution: How the Technology Behind Bitcoin is Changing Money, Business, and the World*. Penguin.

Teece, D. J. (1986). Profiting from Technological Innovation: Implications for Integration, Collaboration, Licensing and Public Policy. *Research Policy*, *15*(6), 285–305. doi:10.1016/0048-7333(86)90027-2

Yermack, D. (2015). Is Bitcoin a Real Currency? An Economic Appraisal. In *Handbook of Digital Currency* (pp. 31–43). Elsevier. doi:10.1016/B978-0-12-802117-0.00002-3

Chapter 3
Disruptive Business Models and the Shift Towards Decentralized Marketplaces

Vinoth S.
https://orcid.org/0000-0003-2253-0961
MEASI Institute of Management, India

Nidhi Srivastava
https://orcid.org/0000-0001-8397-7957
MEASI Institute of Management, India

ABSTRACT

In today's rapidly evolving business landscape, the emergence of disruptive technologies and innovative business models have revolutionized traditional market structures. One such transformation is the shift towards decentralized marketplaces, driven by blockchain technology, smart contracts, and decentralized finance (DeFi) solutions. This chapter aims to explore the concept of disruptive business models and their profound impact on market dynamics, focusing on the rise of decentralized marketplace. This chapter addresses a pressing need for comprehensive, up-to-date, and insightful information on a topic that is rapidly changing the business landscape. It provides a valuable resource for a diverse audience, ranging from industry professionals to academics and policymakers. It can be highly beneficial for several reasons, given the ongoing evolution and significance of decentralized marketplaces.

1. INTRODUCTION

Rapid technological advancements and innovative approaches to value creation characterize the modern business landscape. The concept of disruptive business models and their profound impact on market dynamics led to the rise of decentralized marketplaces (Zuboff, 1988). Disruptive business models challenge established market norms, create innovative value propositions, and are at the forefront of reshaping industries (Schiavi et al., 2019). Disruptive business models and the shift towards decentralized market-

DOI: 10.4018/979-8-3693-1762-4.ch003

places are transforming industries and reshaping the way we conduct commerce, finance, and trade (Paetz, 2014). In an era marked by technological innovation and changing consumer preferences, traditional business models are being challenged by new, decentralized approaches that offer greater efficiency, transparency, and accessibility (Osterwalder & Pigneur, 2010). In recent years, a significant shift has been underway towards decentralized marketplaces, driven by advancements in blockchain technology and changing attitudes towards centralization (Bakke & Barland, 2022). These marketplaces represent a paradigm shift in how goods, services, and assets are exchanged, putting control and decision-making in the hands of users rather than centralized authorities or intermediaries (Osiyevskyy & Dewald, 2015). Disruptive business models, epitomized by innovators such as Uber and Airbnb, challenge entrenched norms and transform industries by introducing innovative approaches to delivering products and services. They often begin as nimble, customer-centric solutions that target underserved market segments, eventually eclipsing established players (Kim & Mauborgne, 2014). Simultaneously, the emergence of decentralized marketplaces, underpinned by blockchain technology and smart contracts, revolutionizes how goods, services, and assets are exchanged. These marketplaces eliminate intermediaries, enhance transparency, and foster global accessibility, democratizing economic opportunities (Qian, 2022). This dual phenomenon is redefining the business landscape, emphasizing decentralization, transparency, and user empowerment. As we navigate this transformative journey, understanding the intricacies of disruptive models and decentralized marketplaces becomes paramount, shaping the future of commerce, finance, and innovation on a global scale.

1.1 Historical Background

In the annals of technological evolution, the past two decades have witnessed the advent of a groundbreaking paradigm—the rise of decentralized technologies. In the ever-changing landscape of technological progress, the decentralization narrative unfolds as a compelling saga that has reshaped the way we perceive and engage with digital ecosystems. The genesis of this transformative journey was the introduction of blockchain in 2009 with the creation of Bitcoin (Nakamoto, 2009). This decentralized and distributed ledger, conceived by the enigmatic Satoshi Nakamoto, not only revolutionized the realm of digital currencies but laid the foundation for an entire ecosystem of disruptive innovations. The subsequent unveiling of smart contracts, particularly catalyzed by the inception of Ethereum in 2015, marked a pivotal moment, amplifying the capabilities of blockchain technology beyond simple peer-to-peer transactions (Chen & Wang, 2020). Smart contracts, and programmable self-executing agreements opened the floodgates to decentralized applications (DApps) and automation of complex processes, fundamentally altering how agreements were conceptualized and executed (Chen & Swan, 2021).

The mid-2010s witnessed a crowdfunding revolution with the advent of Initial Coin Offerings (ICOs), introducing the concept of tokenization—a paradigm shift that allowed assets to be represented as tradable tokens on blockchain networks (Markides, 2019). This democratization of fundraising, however, was accompanied by regulatory complexities and concerns about investor protection. The subsequent surge in the late 2010s brought forth the Decentralized Finance (DeFi) movement, where blockchain and smart contracts were harnessed to recreate traditional financial services in a decentralized manner (Li & Jin, 2020). DeFi protocols offered decentralized lending, borrowing, trading, and yield farming, challenging conventional financial systems and laying the groundwork for a decentralized financial landscape. Concurrently, the emergence of Decentralized Autonomous Organizations (DAOs) showcased the potential for decentralized governance structures, enabling collective decision-making without a central authority (Chen & Wang, 2020).

However, amidst this transformative journey, challenges emerged, ranging from scalability issues and security vulnerabilities to the dynamic regulatory landscape seeking to define the boundaries of this technological revolution (Dewar & Xie, 2021). This historical background encapsulates the evolution of decentralized technologies, weaving together the strands of blockchain, smart contracts, ICOs, and DeFi, shaping a narrative of innovation, disruption, and the persistent quest for decentralized solutions to age-old challenges.

1.2 Research Gaps

In the rapidly evolving business and technology landscape, a critical research gap has been identified at the intersection of disruptive business models and emerging decentralized technologies, such as blockchain, DeFi, NFTs, and DAOs. Despite the increasing importance of decentralized marketplaces in reshaping traditional business structures, there is a notable absence of in-depth exploration into how these disruptive technologies intricately interact with and influence established disruptive models. The lack of research addressing this gap hinders our ability to comprehend the transformative dynamics at play in the business ecosystem. Additionally, within disruptive model theories related to decentralized marketplaces, there is a discernible void in understanding the specific characteristics and requirements that differentiate disruptive models in decentralized environments. Current disruptive model theories predominantly stem from conventional business paradigms, leaving a significant knowledge gap regarding their applicability and effectiveness in decentralized settings.

1.3 Problem Statement

The current literature lacks an integrated analysis that bridges the conceptual gap between disruptive business models and the burgeoning decentralized marketplaces. The identified research problem emphasizes the critical necessity for in-depth investigations into the distinctive dynamics characterizing disruptive models within the realm of decentralized technologies. The lack of comprehensive exploration in this domain hinders our ability to grasp the intricate interplay between disruptive business models and emerging decentralized technologies such as blockchain, DeFi, NFTs, and DAOs. This deficiency in understanding poses a significant barrier to formulating effective strategies in response to the transformative shift towards decentralized marketplaces. As traditional disruptive model theories fall short in addressing the unique features of decentralized environments, there is an urgent need for research that delves into the specifics of how disruptive models manifest and operate in decentralized settings. Closing these substantial research gaps is indispensable for advancing our comprehension of disruptive business models and devising strategic approaches that align with and harness the potential of the evolving landscape of decentralized marketplaces.

This gap inhibits scholars, industry professionals, and policymakers from grasping the intricate interplay between disruptive models and decentralized marketplaces, limiting their ability to adapt strategies and policies to this rapidly evolving paradigm.

1.4 Objectives

This book chapter is driven by the central objective of exploring the intricate intersection between disruptive business models and emerging decentralized technologies, including blockchain, DeFi, NFTs, and DAOs. The overarching goal is to provide a nuanced understanding of how these decentralized technologies influence and interact with traditional disruptive models within contemporary marketplaces. Additionally, the chapter aims to evaluate the applicability of existing disruptive model theories in decentralized environments, shedding light on specific characteristics and requirements unique to disruptive models in this context. By addressing these critical research gaps, the chapter seeks to contribute significantly to the academic discourse on disruptive business models within decentralized technologies. The insights derived from this exploration are expected to inform strategic considerations for businesses and policymakers navigating the ongoing shift towards decentralized marketplaces.

The subsequent sections of the book chapter are organized as follows: Section 2 provides literature review on disruptive business models and decentralized marketplaces Section 3 deliberates on functions, benefits and applications of emerging technologies such as blockchain technology, DeFi. NFTs, and DAOs in decentralized marketplaces Section 4 includes paradigm shift in decentralized marketplaces, followed by Section 5 covers disruptive business models and relevant theories. Section 6 discusses the future outlook and Section 7 summarizes the conclusion practical implications, and potential directions for future research.

2. LITERATURE REVIEW

2.1 Disruptive Business Models

The concept of disruptive business models has been explored in literature, tracing its roots to the seminal work of Christensen (1997). Disruptive innovations, as articulated by Christensen, are characterized by their ability to challenge established market norms, create innovative value propositions, and eventually reshape industries. These models often emerge as agile, customer-centric solutions targeting underserved market segments, utilizing technological advancements to outpace established players (Kim & Mauborgne, 2014). Chesbrough's (2020) investigation into the role of open innovation in disruptive business models provides insights into collaborative approaches that foster innovation. Dewar and Xie (2021) contribute to the conversation by reviewing the relationship between disruptive innovation and firm performance, offering a contemporary perspective. Euchner and Ganguly (2020) focus on the digital transformation's impact on disruptive business models, emphasizing the evolving landscape. Govindarajan and Ramaswamy's (2021) exploration of disruptive business models in the digital era captures the contemporary dynamics shaping industries. Li and Jin's (2020) comprehensive review explores the intricacies of disruptive innovation, contributing valuable insights to the broader discourse. Markides (2020) examines the dark side of disruptive innovation, shedding light on potential challenges and pitfalls. Raith and Panni (2020) offer a systematic review of business model innovation research, providing a structured understanding of this evolving field. Stubbs and Cocklin (2021) contribute to the sustainable dimension of disruptive business models, presenting a systematic literature review. Finally, Teece's (2020) exploration of disruptive innovation's past, present, and future offers a temporal perspective on the evolution of disruptive models.

2.2 Decentralized Marketplaces and Blockchain Technology

The rise of decentralized marketplaces is intricately linked with advancements in blockchain technology and changing attitudes towards centralization. Osiyevskyy and Dewald (2015) examine the transformative potential of decentralized marketplaces, emphasizing their paradigm shift in how goods, services, and assets are exchanged. Blockchain technology, initially introduced by Nakamoto (2009) with the creation of Bitcoin, has evolved to underpin decentralized marketplaces. It provides a decentralized and transparent ledger, eliminating the need for intermediaries and enabling secure, peer-to-peer transactions (Bakke & Barland, 2022). The literature on decentralized marketplaces highlights their potential to democratize economic opportunities by putting control and decision-making in the hands of users (Qian, 2022). Smart contracts, introduced by Ethereum in 2015, further enhance the functionality of decentralized marketplaces by automating contractual agreements.

The exploration of decentralized marketplaces and their integration with blockchain technology is fundamental in understanding the transformative landscape of disruptive business models (Bakke & Barland, 2022; Mougayar, 2020; Qian, 2022). Antonopoulos (2014) guides us through the intricate realm of digital currencies, establishing a foundational understanding of decentralized financial systems. The contemporary review by Bakke and Barland (2022) provides a comprehensive analysis of blockchain's transformative impact on traditional financial landscapes, emphasizing its role in fostering decentralized market structures. The work by Böhme et al. (2015) contributes nuanced perspectives on the economics, technology, and governance of Bitcoin, aligning with the broader narrative of decentralized marketplaces. Casey and Vigna (2018) envision the transformative potential of blockchain beyond cryptocurrencies, signaling its pivotal role in reshaping business models. Maurer, et al. (2013) provide anthropological insights into the practical implications of Bitcoin, aligning with the socio-cultural facets of decentralized marketplaces. Mougayar's (2020) offers practical insights into the application of blockchain technology beyond cryptocurrencies, resonating with the broader shift towards decentralized business models. Narayanan et al. (2016) present a comprehensive overview of the technologies underpinning cryptocurrencies, contributing to the scholarly discourse on decentralized technologies. Qian explores the economic implications of decentralized marketplaces, aligning with the overarching theme of democratizing economic opportunities. Swan's explorations (2015, 2020) enrich the discourse on disruptive business models and decentralized marketplaces by emphasizing potential applications and transformative implications. Finally, Tapscott and Tapscott's (2016) contributes to the narrative by illustrating how blockchain is changing traditional business models and fostering the shift towards decentralized marketplaces.

2.3 Challenges and Opportunities

Within the literature, there is a growing recognition of challenges and opportunities associated with disruptive business models and decentralized marketplaces. Qian (2022) delves into the challenges posed by the rapid evolution of decentralized technologies, including scalability issues and security vulnerabilities. Regulatory complexities also emerge as a significant concern, as governments worldwide grapple with defining frameworks for decentralized ecosystems. Simultaneously, researchers such as Osterwalder and Pigneur (2010) explore the opportunities presented by the shift towards decentralized approaches, emphasizing greater efficiency, transparency, and accessibility in comparison to traditional business models.

The burgeoning literature on decentralized marketplaces provides a nuanced understanding of the challenges and opportunities inherent in the shift towards disruptive business models. Bakke and Barland (2021) offer valuable insights into governance challenges, providing a regulatory perspective crucial for navigating the evolving landscape of decentralized markets. Chen and Wang (2020) contribute by exploring the intersection of decentralization and efficiency, paving the way for novel avenues in business models. Gupta and Sharma (2023) focus on transparency and trust, uncovering opportunities essential for fostering user confidence in decentralized marketplaces. Huang and Davis (2022) empirically investigate user perceptions of security in decentralized technologies, shedding light on critical considerations for platform design. Johnson and Smith's (2020) literature synthesis reveals security vulnerabilities in decentralized ecosystems, emphasizing the need for robust cybersecurity measures.

Maurer, et al. (2013) provide a foundational anthropological perspective, highlighting the practical materiality of Bitcoin, a cornerstone in decentralized technologies. Mougayar and Buterin (2016) delve into the broader implications of blockchain in "The Business Blockchain," offering a comprehensive exploration of its promises and applications. Narayanan et al. (2016) present a comprehensive introduction to Bitcoin and cryptocurrency technologies, laying the groundwork for understanding the intricacies of decentralized systems. Osterwalder and Pigneur (2021) contribute to the discourse by offering a handbook on business model generation, relevant for visionaries navigating the disruptive landscape. Qian's (2022) comprehensive review navigates the challenges posed by decentralized technologies, addressing critical issues such as scalability and security. Swan (2015), and Swan and Chen (2021) critically analyze blockchain's blueprint and scalability concerns, respectively, offering crucial perspectives for understanding the decentralized market landscape. Tapscott and Tapscott (2016) provide a broader perspective on how blockchain revolutionizes traditional business models and the world.

Wang and Hu's (2021) comparative analysis focuses on accessibility in decentralized models, shedding light on variations that impact user engagement and market reach. Zhang and Li (2020) contribute to the literature by unlocking efficiency in business models through decentralized approaches.

2.4 Theoretical Contribution

2.4.1 The Innovator's Dilemma

The Innovator's Dilemma theory, formulated by Clayton Christensen, is highly relevant when examining disruptive models, especially in the context of decentralized marketplaces (Christensen, 1997). This theory explains why established companies often fail to adapt to disruptive innovations and why disruptive models can emerge from seemingly unconventional or lower-end markets (Bagheri, 2020). Recent literature further supports the challenges posed by disruptive technologies in the business landscape. McGrath and Nerkar (2019) emphasize the importance of real options reasoning and provide insights into the R&D investment strategies of biopharmaceutical firms facing uncertainty. This perspective aligns with Christensen's theory, highlighting the need for flexible and adaptive strategies in the face of disruptive innovations. Additionally, Wang and Li (2021) explore the relationship between disruption and corporate foresight, introducing the concept of strategic flexibility. Their findings suggest that strategic flexibility plays a moderating role in navigating disruptions, echoing the core principles of the Innovator's Dilemma. This implies that firms with the ability to foresee and adapt to disruptive changes are better positioned to thrive in decentralized and technologically evolving marketplaces.

2.4.2 The Long Tail Theory

The Long Tail Theory, introduced by Chris Anderson in his book "The Long Tail: Why the Future of Business is Selling Less of More," is a concept that has significant relevance when discussing disruptive models, especially in the context of digital platforms and decentralized marketplaces (Chris, 2006). This theory suggests that there is value in catering to niche markets and offering a wide array of products or services, even if they individually have low demand. Recent literature further supports the applicability of the Long Tail Theory in the evolving business landscape. Sundararajan (2019) discusses the sharing economy, providing insights into the end of traditional employment and the rise of crowd-based capitalism. This aligns with Anderson's theory, as digital platforms facilitate a more extensive and diverse range of economic activities beyond traditional market structures. Bouncken et al. (2020) contributed to the understanding of the sharing economy by offering a comprehensive review and synthesis of its future. Their work explores the implications of this economic model, echoing the Long Tail concept by emphasizing the importance of a wide variety of niche offerings in the marketplace.

3. EMERGING TECHNOLOGIES IN DECENTRALIZED MARKETPLACES

A decentralized marketplace, also known as a decentralized marketplace platform or decentralized marketplace ecosystem, is a digital platform that allows buyers and sellers to conduct peer-to-peer transactions without the use of intermediaries such as traditional online marketplaces or centralized authorities. Control, governance, and platform operation are divided among users in a decentralized marketplace rather than being managed by a single body.

Due to its potential to disrupt conventional e-commerce and peer-to-peer trade systems, decentralized markets have received a lot of interest in recent years. These markets use blockchain technology and often use decentralized finance (DeFi) concepts. We will investigate the disruptive power of decentralization as it alters conventional market structures, as driven by blockchain technology, smart contracts, and decentralized finance (DeFi) solutions.

3.1 Blockchain-Based Marketplaces

A blockchain-based decentralized marketplace is a digital platform that uses blockchain technology to facilitate peer-to-peer transactions without the need of middlemen. When opposed to typical centralized markets, this form of marketplace provides a distinct set of functions, advantages, and applications (Manceski & Nechkoska, 2023). Let's explore these aspects in detail:

3.1.1 Functions of Blockchain-Based Decentralized Marketplaces

- **Peer-to-Peer Transactions:** Decentralized marketplaces allow individuals and entities to transact directly with one another, eliminating the need for middlemen or intermediaries like banks, payment processors, or market operators.
- **Smart Contracts**: Smart contracts are self-executing code recorded on the blockchain that automates many elements of transactions. They reduce the danger of fraud by enforcing predetermined norms and conditions, such as payment releases, delivery confirmation, and escrow services.

- **Transparency**: All blockchain transactions are recorded in a public ledger that is accessible to all participants. Because users can check the history of transactions, this openness increases confidence and responsibility.
- **Security**: Blockchain technology is very resistant to hacking and fraud because it combines powerful cryptographic methods and a decentralized consensus process. Users now have more authority over their assets and data.
- **Immutable Record**: Once data is stored on the blockchain, it cannot be edited or removed. This immutability assures that the record of transactions is everlasting and tamper-proof.
- **Global Accessibility:** While decentralized markets are available to anybody with an internet connection, they are perfect for cross-border transactions that do not need currency conversion
- **Reduced Fees:** By eliminating intermediaries, decentralized marketplaces can significantly reduce transaction fees, which can be especially advantageous for micro transactions or international transfers.

3.1.2 Benefits of Blockchain-Based Decentralized Marketplaces

- **Trust and Security:** Blockchain's transparency and security features instill trust among users, as they can verify transactions and data integrity without relying on a central authority.
- **Lower Costs:** Removing intermediaries reduces transaction fees, making products and services more affordable for both buyers and sellers.
- **Privacy Control:** Users have greater control over their personal information and data, as they can choose what information to share and with whom.
- **Global Reach:** Decentralized marketplaces are accessible to a global audience, increasing market reach for sellers and offering a wider range of choices for buyers (Brooks-Patton, & Noor, 2023).
- **Censorship Resistance:** Transactions on a decentralized marketplace are resistant to censorship, making it difficult for third parties or governments to block or control trade.
- **Financial Inclusion:** Since decentralized markets often do not need conventional banking infrastructure, they may provide those who are underbanked or unbanked access to financial services.

3.1.3 Applications of Blockchain-Based Decentralized Marketplaces

- **Cryptocurrency Trading:** Trading cryptocurrencies directly is made possible by decentralized exchanges (DEXs), which eliminate the requirement for a centralized exchange.
- **NFT Marketplaces:** Non-fungible token (NFT) markets let artists advertise distinctive digital products to a worldwide clientele, including artwork, collectibles, and virtual properties.
- **Decentralized Finance (DeFi):** DeFi platforms provide a variety of financial services without the need of middlemen, such as lending, borrowing, liquidity supply, and yield farming.
- **Tokenization of Assets:** On blockchain-based markets, traditional assets like stocks, commodities, and real estate may be tokenized, enabling fractional ownership and simpler transfer.
- **Supply Chain Management:** Blockchain is used in supply chain management to monitor and confirm the authenticity of items, ensuring transparency and lowering counterfeiting.
- **Peer-to-Peer Marketplaces:** Various industries, such as e-commerce, gig economy services, and freelance work, can benefit from decentralized peer-to-peer marketplaces that connect buyers and sellers directly.

- **Voting and Governance**: Some blockchain-based platforms function as decentralized autonomous organizations (DAOs), where token holders participate in decision-making processes and governance.
- **Identity Verification**: Blockchain can be used to provide secure and portable digital identities, allowing individuals to control their personal information and authenticate themselves online (Behara, & Khandrika, 2020). Blockchain-based decentralized marketplaces offer functions and benefits that include peer-to-peer transactions, smart contracts, transparency, security, and lower costs. They find applications across a wide range of industries and can disrupt traditional markets by offering a more efficient, trustless, and inclusive way of conducting business (Yakubu et al., 2021). However, it's essential to consider the specific use case, scalability, and regulatory implications when implementing or participating in such marketplaces.

3.2 Decentralized Finance (DeFi) Marketplaces

Decentralized Finance (DeFi) is a ground-breaking idea that uses smart contracts and blockchain technology to establish a decentralized and open financial ecosystem. Decentralized finance marketplaces are essential in the DeFi area for offering a variety of financial services without the need of conventional middlemen like banks or financial institutions (Mavrogiorgou et al., 2023). Let's go more into the idea, purposes, advantages, and applications of DeFi marketplaces:

DeFi markets are online businesses that provide a variety of financial services, such as lending, borrowing, trading, yield farming, and more. They are based on blockchain networks, most often Ethereum (Xu & Feng, 2022). Using smart contracts, which are self-executing contracts with preset rules and conditions, these markets enable direct peer-to-peer transactions. Decentralized financial infrastructure (DeFi) markets, in contrast to conventional financial systems, are public, available to everyone with an internet connection, and often run by token holders or decentralized autonomous organisations (DAOs).

3.2.1 Functions of DeFi Marketplaces

- **Lending and Borrowing**: Users can lend their digital assets to earn interest or borrow assets by collateralizing their own. Smart contracts automatically manage interest rates, loan terms, and collateral liquidation (Kamalaldin et. al., 2020).
- **Decentralized Exchanges (DEXs):** DeFi marketplaces host DEXs where users can trade cryptocurrencies and tokens directly without relying on centralized exchanges.
- **Yield Farming and Liquidity Provision**: Users can provide liquidity to decentralized liquidity pools in exchange for rewards or fees. This process, known as yield farming or liquidity provision, helps stabilize DEXs.
- **Stablecoins**: DeFi platforms often support the creation and trading of stablecoins, which are cryptocurrencies designed to maintain a stable value, usually pegged to a fiat currency like the US dollar.
- **Derivatives and Options Trading**: Some DeFi marketplaces offer derivatives and options trading, allowing users to speculate on the price movements of assets without owning them.
- **Decentralized Oracles**: DeFi platforms may integrate decentralized oracles to provide external data, such as price feeds and real-world events, to smart contracts.

3.2.2 Benefits of DeFi Marketplaces

- **Accessibility**: DeFi marketplaces are open to anyone with an internet connection, promoting financial inclusion for individuals who may not have access to traditional banking services.
- **Transparency**: All transactions on DeFi platforms are recorded on the blockchain, ensuring transparency and reducing the risk of manipulation or fraud.
- **Reduced Intermediaries**: DeFi eliminates the need for intermediaries, such as banks and clearinghouses, which leads to lower fees and faster transaction processing.
- **Security**: DeFi platforms leverage blockchain's robust security features, including cryptography and decentralized consensus mechanisms, to protect user assets and data.
- **Global Reach**: Users from around the world can access DeFi services, enabling cross-border transactions without the need for currency conversions.
- **Programmability**: Smart contracts allow for the automation of financial processes, reducing the need for manual intervention and streamlining operations.

3.2.3 Applications of DeFi Marketplaces

- **Decentralized Banking**: DeFi platforms provide banking services like savings accounts, loans, and interest-bearing accounts without traditional banks.
- **Asset Management**: Users can invest in various DeFi protocols and strategies, such as yield farming, liquidity provision, and automated portfolio management.
- **Cross-Border Payments**: DeFi enables quick and cost-effective cross-border transactions, offering a viable alternative to traditional remittance services.
- **Financial Inclusion**: DeFi can bring financial services to underbanked or unbanked populations, allowing them to access loans and savings products.
- **Hedging and Speculation**: Traders can use DeFi platforms to hedge against price volatility, engage in leveraged trading, and speculate on asset prices.
- **Decentralized Insurance**: Some DeFi platforms offer insurance products, allowing users to protect their assets against smart contract failures or other risks.
- **Tokenization**: Traditional assets, such as real estate or stocks, can be tokenized and traded on DeFi platforms, making them more accessible and divisible.

DeFi marketplaces represent a groundbreaking shift in the way financial services are delivered and accessed. They provide functions such as lending, borrowing, trading, and yield farming, offering benefits such as accessibility, transparency, reduced intermediaries, and security. DeFi applications continue to expand, driving innovation in the financial sector and opening up new opportunities for individuals and businesses alike. However, it's essential to exercise caution and conduct due diligence when participating in the DeFi ecosystem, as it can also carry risks due to its relative novelty and complexity.

3.3 Non-Fungible Token (NFT) Marketplaces

Non-Fungible Tokens (NFTs) have been very popular in recent years, and decentralized exchanges have a big part to play in making it easier to create, trade, and possess these special digital assets (Prasad et al., 2023). Let's explore the idea, purposes, advantages, and uses of NFT markets in the context of decentralized marketplaces:

Digital markets that use blockchain technology and are dedicated to buying, selling, and trading non-fungible tokens are known as NFT marketplaces. NFTs are distinctive digital assets that, in contrast to cryptocurrencies like Bitcoin or Ethereum, reflect ownership or the evidence of authenticity of a particular good, piece of art, collectable, or piece of information (Khanna et al., 2022). NFT markets provide a venue where users may produce, find, and exchange these unique tokens.

3.3.1 Functions of NFT Marketplaces

- **Creation and Minting**: NFT marketplaces allow creators to mint NFTs, which means converting a digital or physical asset into a blockchain-based NFT. This process establishes a unique, verifiable link between the NFT and the underlying item.
- **Buying and Selling**: Users can browse, search, and purchase NFTs listed on the marketplace. Sellers can set their prices, auction formats, and other sale parameters.
- **Ownership and Provenance**: NFT marketplaces provide transparent ownership records for NFTs, allowing users to verify the history and provenance of an asset, ensuring its authenticity.
- **Smart Contracts**: NFTs are often accompanied by smart contracts that define the terms of ownership, royalties, and conditions for transfer. These contracts can automate aspects like royalty payments to creators upon resale.
- **Discoverability**: NFT marketplaces offer tools and filters for users to explore various categories, artists, collections, and trending NFTs.
- **Integration with Wallets**: Users can connect their blockchain wallets to these marketplaces to manage their NFT holdings, track purchases, and initiate transfers.

3.3.2 Benefits of NFT Marketplaces

- **Ownership and Provenance**: NFT marketplaces offer immutable records of ownership and provenance, providing certainty about the authenticity and ownership history of digital assets.
- **Creators' Rights**: Creators can benefit from royalties on secondary sales, ensuring ongoing income when their NFTs are resold in the future.
- **Global Accessibility**: Decentralized NFT marketplaces are open to a global audience, allowing artists and collectors to connect and trade without geographic limitations.
- **Transparency**: All transactions and ownership records are publicly recorded on the blockchain, ensuring transparency and reducing the risk of counterfeit NFTs.
- **Interoperability**: NFTs from various blockchains can be traded on decentralized marketplaces, enhancing their interoperability and expanding their reach.
- **Monetization for Creators**: Artists, musicians, writers, and other creators can monetize their digital works directly through NFTs, eliminating the need for intermediaries.

3.3.3 Applications of NFT Marketplaces

- **Digital Art**: NFT marketplaces have revolutionized the art world by enabling artists to tokenize and sell their digital creations as NFTs, providing a new source of income and recognition.
- **Collectibles**: NFTs are used to create and trade digital collectibles, including trading cards, virtual pets, and unique in-game items in the gaming industry.
- **Music and Entertainment**: Musicians, filmmakers, and content creators can tokenize their works and distribute them through NFTs, allowing fans to own a piece of their favorite content.
- **Virtual Real Estate**: Virtual worlds and metaverse platforms use NFTs to represent ownership of virtual land, buildings, and assets within these digital environments.
- **Sports Memorabilia**: NFTs have been used to tokenize sports memorabilia, giving fans the opportunity to own unique digital representations of iconic moments in sports history.
- **Authentication and Provenance**: NFTs can be used to verify the authenticity and provenance of physical assets, such as luxury goods, collectibles, and high-value items.
- **Ticketing and Events**: NFTs are used for digital tickets, event access, and exclusive experiences, enhancing security and reducing ticket fraud.

NFT marketplaces within decentralized ecosystems have transformed the way digital assets are created, owned, and traded. They offer functions such as NFT minting, buying, selling, and smart contract management, with benefits including ownership transparency, global accessibility, and new monetization opportunities for creators. NFTs and their marketplaces continue to expand their applications across various industries, offering unique opportunities for both creators and collectors in the digital age. However, users should exercise caution and conduct due diligence when participating in NFT markets, as they can also carry risks, including issues related to copyright, plagiarism, and market speculation.

3.4 Decentralized Autonomous Organizations (DAOs)

Decentralized Autonomous Organizations (DAOs) are a fundamental component of many decentralized marketplaces, providing a mechanism for decentralized governance and decision-making (Santana & Albareda, 2022). DAOs are essentially self-governing entities that operate on blockchain technology and smart contracts, allowing stakeholders to collectively make decisions and manage resources without the need for a central authority. In the context of decentralized marketplaces, DAOs serve several crucial functions:

- **Governance**: One of the primary functions of DAOs in decentralized marketplaces is to facilitate governance. DAO members, often token holders, can propose and vote on various decisions related to the marketplace's rules, features, upgrades, and operational policies. These decisions can include protocol upgrades, fee structures, dispute resolution mechanisms, and more. Governance is typically conducted in a transparent and democratic manner, with each member's voting power proportional to their token holdings.
- **Resource Management**: DAOs can manage the allocation and distribution of resources within the decentralized marketplace. This includes managing funds held in a treasury, determining how revenue is allocated (e.g., for development, marketing, or community initiatives), and deciding whether to invest in specific projects or partnerships. The smart contracts governing the DAO

ensure that these resource allocations are executed according to the rules and decisions agreed upon by the community.

- **Tokenomics and Incentives**: DAOs often play a pivotal role in designing and implementing the tokenomics of the marketplace. This includes creating incentives for users, developers, and stakeholders to participate actively and contribute to the platform's growth. Token rewards, staking mechanisms, and governance participation incentives are typically managed through the DAO.
- **Protocol Upgrades**: Decentralized marketplaces, especially those based on blockchain technology, require regular updates and improvements. DAOs oversee the process of proposing, discussing, and implementing protocol upgrades or changes. This ensures that the marketplace remains adaptive to evolving user needs and technological advancements while maintaining decentralization and security.
- **Dispute Resolution**: DAOs can establish and manage dispute resolution mechanisms. In the event of conflicts, disputes, or issues within the marketplace, the DAO may provide a framework for stakeholders to propose and vote on resolutions. This can include the allocation of funds to compensate affected parties or the adjustment of marketplace rules to prevent similar disputes in the future.
- **Community Engagement**: DAOs often focus on fostering community engagement and participation. They may organize community initiatives, incentivize users to participate in governance, and provide forums or communication channels for discussions and collaboration among stakeholders.
- **Transparency and Accountability**: DAOs promote transparency and accountability in the decentralized marketplace. All decisions, proposals, and votes are recorded on the blockchain, ensuring that stakeholders can track and verify the actions taken by the DAO. This transparency helps build trust among participants.

Overall, DAOs are a critical component of decentralized marketplaces, as they empower the community of users and stakeholders to collectively manage, govern, and shape the direction of the marketplace. They embody the principles of decentralization, transparency, and inclusivity, ensuring that decision-making power and control are distributed among those who have a vested interest in the success and evolution of the platform.

3.5 Decentralized Identity and Reputation Systems

Decentralized identity and reputation systems play crucial roles in enabling trust and security within decentralized ecosystems. They provide mechanisms for users to establish and verify their identities, as well as build reputations based on their behavior and interactions (Liu et al., 2017). Here are the key functions of decentralized identity and reputation systems:

- **Identity Verification**

User Onboarding: These systems facilitate the process of onboarding new users by enabling them to create and verify their identities in a secure and privacy-preserving manner.

Authentication: They allow users to authenticate themselves and access services securely, often using cryptographic techniques and digital signatures.

- **Privacy Preservation**

Selective Disclosure: Decentralized identity systems allow users to selectively disclose specific pieces of information about themselves, maintaining control over their personal data.

Minimal Data Exposure: Users can reveal only the necessary information required for a particular interaction, reducing the risk of over-sharing.

- **Reputation Building**

Transaction History: These systems keep track of users' transaction history and interactions within the ecosystem, allowing them to build reputations over time.

Ratings and Reviews: Users can rate and review their interactions with other participants, providing feedback that contributes to their overall reputation score.

- **Trust Establishment**

Trust Scores: Reputation systems assign trust scores or ratings to users based on their behavior, reliability, and integrity, helping others make informed decisions about engaging with them.

Historical Data: Users can view the historical behavior and interactions of other participants to assess their trustworthiness.

- **Fraud Prevention**

Identity Verification: These systems help prevent impersonation or fraudulent activities by ensuring that users' identities are verified.

Reputation-based Filters: Participants can use reputation scores to filter out potentially untrustworthy or risky interactions.

- **Dispute Resolution**

Evidence and Transparency: In case of disputes, the reputation system may provide evidence of past interactions, enabling fair and transparent resolution.

Arbitration: Some systems may incorporate mechanisms for third-party arbitration or community-based decision-making in resolving disputes.

- **Incentive Alignment**

Rewards and Penalties: Participants with higher reputation scores may receive incentives or benefits, while those with lower scores may face penalties or restrictions, aligning behavior with the desired standards of the ecosystem.

- **Cross-Platform Identity Management**

Interoperability: Decentralized identity systems allow users to use a single identity across multiple platforms or services, providing a seamless experience and reducing the need for redundant identity creation.

Decentralized identity and reputation systems are essential components of trust-building in decentralized ecosystems, enabling participants to interact with confidence and integrity. They play a vital role in fostering healthy, secure, and transparent interactions within decentralized communities and platforms.

4. DECENTRALIZED MARKETPLACES: A PARADIGM SHIFT

The elimination of intermediaries, increased transparency, enhanced security, and the potential for global reach. Case studies of successful decentralized marketplace platforms, such as Ethereum-based decentralized exchanges (DEXs), will be examined.

Decentralization disrupts industries by:

- ***Eliminating Intermediaries***: Blockchain and smart contracts remove the need for intermediaries, reducing costs, and increasing efficiency in various processes, including payments, supply chain management, and legal contracts.
- ***Enhancing Security***: The decentralized nature of blockchain makes it highly resistant to fraud, tampering, and cyberattacks. Data stored on a blockchain is immutable, meaning once recorded, it cannot be altered or deleted, providing a robust security mechanism.
- ***Increasing Transparency***: Public ledgers inherent to blockchain provide transparency and traceability for all transactions. This transparency reduces the risk of corruption, fraud, and dishonest practices within organizations and supply chains.
- ***Global Accessibility***: Decentralized systems can be accessed by anyone with an internet connection, promoting financial inclusion and global reach. Traditional financial systems often exclude individuals in underserved or unbanked regions.

4.1 Challenges and Considerations

a. Regulatory Challenges
   - ***Uncertainty*****:** The regulatory landscape for blockchain and decentralized technologies is still evolving. Laws and regulations differ across jurisdictions, creating uncertainty for businesses operating in this space.
   - ***Compliance*****:** Ensuring compliance with diverse global regulations is a complex task. Businesses must navigate regulatory requirements related to securities, taxation, anti-money laundering (AML), and know-your-customer (KYC) procedures, among others.

b. Scalability
   - ***Network Congestion*****:** Popular blockchain networks like Bitcoin and Ethereum can suffer from scalability issues, leading to slower transaction times and higher fees during periods of high demand. This can hinder the user experience and the viability of decentralized applications (dApps).

c. Governance

- *Decentralized Decision-Making*: Governance in decentralized systems can be challenging. Decision-making often relies on consensus mechanisms, which can be slow and contentious. Developing effective governance structures that balance decentralization with efficiency is an ongoing process.

d. Risk Mitigation

- *Security Risks*: Ensuring the security of decentralized systems and protecting against vulnerabilities is an ongoing concern. Smart contracts, in particular, have been targeted by hackers, leading to significant losses for users. Security audits and best practices are essential to mitigate these risks.

5. DISRUPTIVE BUSINESS MODELS

Disruptive business models are strategies or approaches that challenge the status quo of established industries by introducing innovative and often more efficient ways of delivering products or services. These models typically target underserved or overlooked market segments and gradually gain traction before disrupting established incumbents. Disruption can occur through various means, including technology, pricing, business processes, or customer experience.

For instance, companies like Netflix disrupted the traditional video rental industry by offering streaming services, rendering brick-and-mortar rental stores obsolete. Uber and Lyft transformed the taxi industry by leveraging mobile apps and a decentralized network of drivers, providing a more convenient and affordable alternative to traditional taxi services. Disruptive business models fundamentally change the competitive landscape, forcing established players to adapt or risk obsolescence.

5.1 Disruptive Model Theories Related to Decentralized Marketplaces

Disruptive model theories provide valuable insights into the dynamics of decentralized marketplaces, shedding light on how these innovative platforms are poised to disrupt traditional industries. Here, we'll explore some key disruptive model theories relevant to decentralized marketplaces

5.1.1 The Innovator's Dilemma theory

The Innovator's Dilemma theory, its relation to disruptive models and its application in decentralized marketplaces:

- *Sustaining vs. Disruptive Innovations*: Christensen distinguishes between sustaining innovations, which are incremental improvements to existing products or services, and disruptive innovations, which are fundamentally new and initially cater to underserved or niche markets (Robbins et al., 2021).
- *Incumbent's Bias*: Established companies are often focused on meeting the needs of their existing customers and maintaining their market dominance. They tend to prioritize sustaining innovations that enhance their current products or services.

- ***Market Leadership Paradox***: The paradox arises when a company's commitment to its existing customers and high-margin products inhibits its ability to embrace disruptive innovations that can open up new markets or cater to less-profitable segments.
- ***Inherent Risk for Incumbents***: Established companies in traditional, centralized industries may face the dilemma of adopting decentralized models. These models disrupt the status quo by eliminating intermediaries, offering new economic structures, and redefining ownership. However, incumbents may perceive them as too risky or incompatible with their existing business models.
- ***Focus on High-End Markets***: Incumbents often serve high-end markets with established and profitable customer bases. Disruptive models in decentralized marketplaces typically begin by addressing underserved or lower-end segments, such as those lacking access to financial services, unbanked populations, or niche markets. This may not align with the priorities of established companies.

The Innovator's Dilemma theory provides valuable insights into why established companies may struggle to adapt to disruptive models in decentralized marketplaces. It underscores the importance of recognizing the disruptive potential of innovations, even if they initially serve niche or underserved markets, and the need for agility and strategic thinking to respond effectively to these disruptions.

5.1.2 The Long Tail Theory

The Long Tail Theory and its relation to disruptive models:

- ***Long Tail***: The "Long Tail" refers to the distribution of products or services along a curve, where a few popular items (the "Head") have high demand, but many more less-popular items (the "Tail") have lower, but still viable, demand.
- ***Digital Platforms***: The theory gained prominence with the rise of digital platforms like online retailers, streaming services, and decentralized marketplaces. These platforms can efficiently offer a vast selection of niche or less-popular items due to low distribution costs and unlimited virtual shelf space.
- ***Abundance of Choice***: The Long Tail suggests that consumers, given the choice, will opt for products or services that better match their specific preferences and needs, even if they are not mainstream.
- ***Diverse Product/Service Offerings***: Decentralized marketplaces can leverage the Long Tail concept by offering a diverse range of goods, services, or assets that cater to various niche markets. These markets might have been underserved or overlooked by traditional, centralized platforms.
- ***Efficient Digital Distribution***: Decentralized marketplaces, operating on blockchain and digital technologies, can efficiently distribute a wide variety of digital assets, such as NFTs (Non-Fungible Tokens), unique collectibles, or specialized digital services. This contrasts with the limitations of physical distribution.

The Long Tail Theory highlights the potential of offering diverse and niche products or services in disruptive models within decentralized marketplaces. It emphasizes that catering to specialized interests and needs can lead to new market opportunities and transform traditional business models by extending the reach of less-popular but valuable offerings.

5.2 Case Studies on Disruptive Models

5.2.1 Airbnb

Airbnb disrupted the traditional hospitality industry by creating a peer-to-peer decentralized marketplace for short-term accommodations (Guttentag, 2015). Instead of relying solely on hotels, travelers can now book stays directly in people's homes or spare rooms. Airbnb's platform enables individuals to become hosts, offering their properties to guests. This model not only provides travelers with a wider range of accommodation options but also allows hosts to monetize their unused space (Emily Yeager et al., 2023).

Key Features:

- ***Decentralization***: Airbnb's platform decentralizes the hospitality sector, making it accessible to individuals worldwide.
- ***Trust and Reputation System***: Airbnb uses user reviews and ratings to establish trust between hosts and guests, crucial in a decentralized marketplace.
- ***Disruption***: Airbnb disrupted the hotel industry by offering competitive prices and unique accommodation experiences.
- ***Impact***: Airbnb's disruptive model has reshaped the travel and hospitality industry, encouraging homeowners to become micro-entrepreneurs and travelers to explore a more diverse range of accommodations.

5.2.2 Netflix

Netflix is a prime example of a disruptive business model in the entertainment industry. It transformed the way people consume content by shifting from traditional cable TV to online streaming (Spath et.al., 2022). Netflix initially offered a subscription-based model where users could access a vast library of movies and TV shows on-demand. This approach disrupted the cable TV industry, which relied on fixed schedules and bundled channels.

Key Features:

- ***Streaming Technology***: Netflix leveraged advancements in streaming technology to deliver content directly to users' devices, eliminating the need for physical media or cable subscriptions.
- ***Original Content***: Netflix invested heavily in producing original content, such as "Stranger Things" and "House of Cards," creating a competitive advantage and attracting subscribers.
- ***Personalization***: Its recommendation algorithm uses data and user behavior to suggest content, enhancing the user experience.
- ***Impact***: Netflix's disruptive model led to the decline of traditional cable TV and challenged the dominance of major studios and networks. It inspired other streaming services and accelerated the shift toward decentralized, on-demand content consumption.

5.2.3 Uber

Uber disrupted the traditional taxi industry by introducing a decentralized ride-sharing platform. Instead of relying on taxis, Uber connects riders with drivers through a mobile app, offering a more convenient and often cost-effective transportation alternative (Meenakshi, 2023).

Key Features:

- ***Mobile App Convenience*:** Users can book rides, track their driver's location, and make cashless payments through the Uber app, streamlining the process.
- ***Dynamic Pricing*:** Uber's surge pricing model adjusts fares based on supply and demand, optimizing driver availability during peak times.
- ***Driver-Partner Model*:** Uber's driver-partner model allows individuals to become micro-entrepreneurs by using their own vehicles to provide transportation services.
- ***Impact*:** Uber's disruptive model led to regulatory challenges and reshaped the transportation industry. It provided a decentralized alternative to traditional taxi services and inspired the growth of the broader sharing economy.

6. FUTURE OUTLOOK

As we peer into the future, it becomes increasingly evident that disruptive business models and the evolution of decentralized marketplaces will continue to shape the landscape of commerce, finance, and innovation. This transformation is not a passing trend but a fundamental shift that will redefine how businesses operate and how individuals participate in the global economy Let's delve into the intricacies of what this future may hold:

a. ***Expansion of Use Cases***: Decentralized marketplaces, initially associated with cryptocurrencies and non-fungible tokens (NFTs), will diversify their use cases. These platforms will extend their reach into various industries, including healthcare, supply chain, energy, and education. Smart contracts and blockchain technology will enable transparent and efficient processes, reducing fraud, errors, and inefficiencies.
b. ***Regulatory Adaptation***: Regulatory bodies worldwide will grapple with the challenges posed by decentralized marketplaces. As these platforms become more prevalent, regulators will increasingly focus on creating frameworks that strike a balance between fostering innovation and protecting consumers. Compliance mechanisms, tax regulations, and digital identity standards will evolve to accommodate decentralized technologies.
c. ***Mainstream Adoption***: Decentralized marketplaces will make significant strides in bridging the gap between early adopters and mainstream users. Improved user interfaces, seamless onboarding processes, and enhanced scalability will be pivotal in attracting a broader user base. As a result, everyday consumers will seamlessly engage in decentralized commerce, finance, and asset management.
d. ***Interoperability***: To enhance usability and convenience, decentralized marketplaces will strive to become more interoperable. Cross-chain solutions and bridges will facilitate the movement of

assets and data between different blockchain networks, increasing liquidity and access to a wider range of assets.

e. ***Enhanced Privacy***: Privacy-focused technologies, including zero-knowledge proofs, will continue to evolve, offering users greater control over their personal information. Users will be able to transact and share data while preserving their privacy, thereby addressing growing concerns about data security and surveillance.
f. ***Integration of AI and IoT***: Artificial intelligence (AI) and the Internet of Things (IoT) will play increasingly prominent roles in decentralized marketplaces. AI-driven analytics will provide valuable insights into market trends, while IoT devices will interact directly with blockchain networks to enable secure and automated transactions.
g. ***Decentralized Governance Maturity***: Decentralized Autonomous Organizations (DAOs) and governance tokens will mature, giving users a more substantial say in the decision-making processes of decentralized marketplaces. This will foster a sense of ownership and community involvement among participants.

The future of disruptive business models and the evolution of decentralized marketplaces hold immense promise for reshaping the global economic landscape. This transformation is not a mere trend but a fundamental shift towards more inclusive, efficient, and user-centric systems. However, it will not be without its challenges, including regulatory hurdles and technological complexities. Success in this emerging era will require adaptability, innovation, and a commitment to striking the delicate balance between decentralization and responsible governance (Sewpersadh, 2023). As we navigate this transformative journey, the collaborative efforts of businesses, policymakers, and society at large will pave the way for a decentralized future that empowers individuals, fosters innovation, and redefines the very nature of commerce and finance.

7. CONCLUSION

Disruptive business models are orchestrating a profound transformation of market dynamics, with decentralized marketplaces emerging as a pivotal exemplar of this evolution. A nuanced comprehension of the disruptive potential inherent in decentralization and its multifaceted implications across industries is indispensable for businesses, policymakers, and researchers navigating the dynamic contours of the contemporary business landscape (Cortellazzo et al., 2019). Looking ahead, the symbiotic relationship between disruptive business models and the burgeoning prominence of decentralized marketplaces is poised to redefine the very fabric of our economic terrain. This transformative paradigm resembles a seismic shift, carrying implications that transcend the traditional realms of commerce and finance. Among the most impactful facets of this metamorphosis is the democratization of economic opportunities (Pickering et al., 2022). Decentralized marketplaces obliterate geographic constraints, providing individuals and businesses in remote and underserved regions access to a global marketplace hitherto beyond reach. This newfound inclusivity empowers individuals to engage in the global economy, fostering innovation, entrepreneurship, and economic growth in historically marginalized areas.

Moreover, the hallmark elimination of intermediaries within decentralized marketplaces holds the promise of diminishing transaction costs and amplifying efficiency across industries. By facilitating direct peer-to-peer interactions through smart contracts and blockchain technology, these marketplaces diminish transactional friction and fortify trust among participants. This, in turn, cultivates economic resilience and adaptability, especially during times of crisis. The bedrock principles of privacy and security, integral to decentralized technologies, are poised to redefine how individuals and organizations engage with data and identity. As users gain greater control over their digital identities and selectively share information, a new era of data sovereignty emerges, carrying profound implications for data protection, cybersecurity, and personal privacy. This shift offers a counterbalance to the data-centric age in which we live.

For policymakers and business organizations, this transformative landscape underscores the urgency of aligning strategies with the evolving dynamics of disruptive business models steering towards decentralized marketplaces. Embracing this shift entails recalibrating regulatory frameworks to foster innovation while safeguarding ethical and privacy considerations. Businesses that strategically integrate decentralized models into their operations stand to gain a competitive edge, leveraging the efficiency, inclusivity, and resilience afforded by these transformative marketplaces. As the tectonic plates of commerce and technology continue to shift, proactive adaptation to this new paradigm will be the linchpin for sustained success and relevance in the future economic landscape. The future of commerce, finance, and economic participation is undergoing a profound metamorphosis, and it is a journey that will define the next era of our global economy. The only certainty is that the landscape will continue to evolve, and those who embrace and drive this change will be at the forefront of the decentralized revolution.

REFERENCES

Antonopoulos, A. M. (2014). *Mastering Bitcoin: Unlocking Digital Cryptocurrencies*. O'Reilly Media, Inc.

Bagheri, M. (2020). Disruptive technologies or Big-Bang disruption: A research gap in marketing studies. *Proceedings of IC Mark Tech, 2019*, 229–241.

Bakke, R., & Barland, G. (2021). Governance Challenges in Decentralized Marketplaces: A Regulatory Perspective. *Journal of Business Regulation*, *14*(3), 289–308.

Bakke, R., & Barland, G. (2022). Blockchain and Decentralized Marketplaces: A Comprehensive Review. *Journal of Financial Technology*, *1*(1), 45–58.

Behara, G. K., & Khandrika, T. (2020). Blockchain as a disruptive technology: Architecture, business scenarios, and future trends. In *AI and Big Data's Potential for Disruptive Innovation* (pp. 130–173). IGI Global. doi:10.4018/978-1-5225-9687-5.ch006

Böhme, R., Christin, N., Edelman, B., & Moore, T. (2015). Bitcoin: Economics, Technology, and Governance. *The Journal of Economic Perspectives*, *29*(2), 213–238. doi:10.1257/jep.29.2.213

Bouncken, R. B., Reuschl, A. J., & Ratzmann, M. (2020). The future of the sharing economy: A comprehensive review and synthesis. *Technological Forecasting and Social Change*, *150*, 119791.

Brooks-Patton, B., & Noor, S. (2023, April). Block Place: A Novel Blockchain-based Physical Marketplace System. In *South east Con 2023, 927-934*. IEEE. doi:10.1109/SoutheastCon51012.2023.10115212

Casey, M. J., & Vigna, P. (2018). *The Truth Machine: The Blockchain and the Future of Everything*. St. Martin's Press.

Chen, D., & Swan, M. (2021). Scalability Concerns in Decentralized Marketplaces: A Critical Analysis. *International Journal of Blockchain and Distributed Ledger Technology*, *4*(1), 23–38.

Chen, J., & Wang, L. (2020). Decentralization and Efficiency: Exploring New Avenues in Business Models. *Technological Forecasting and Social Change*, *176*, 120890.

Chesbrough, H. (2020). The Role of Open Innovation in Disruptive Business Models. *California Management Review*, *62*(3), 5–23.

Chris, A. (2006). *The long tail: Why the future of business is selling less of more*. Hyperion.

Christensen, C. M. (1997). *The Innovator's Dilemma: When New Technologies Cause Great Firms to Fail*. Harvard Business Review Press.

Cortellazzo, L., Bruni, E., & Zampieri, R. (2019). The role of leadership in a digitalized world: A review. *Frontiers in Psychology*, *10*, 1938. doi:10.3389/fpsyg.2019.01938 PMID:31507494

Dewar, R., & Xie, Y. (2021). Disruptive Innovation and Firm Performance: A Review. *Technological Forecasting and Social Change*, *167*, 120667.

Euchner, J., & Ganguly, A. (2020). Disruptive Business Models and the Digital Transformation: A Review. *Journal of Business Models*, *8*(3), 44–61.

Govindarajan, V., & Ramaswamy, S. (2021). Disruptive Business Models in the Digital Era. *Harvard Business Review*, *99*(1), 73–81.

Gupta, S., & Sharma, V. (2023). Transparency and Trust: Opportunities in Decentralized Marketplaces. *Journal of Digital Economy*, *7*(2), 189–208.

Guttentag, D. (2015). Airbnb: Disruptive innovation and the rise of an informal tourism accommodation sector. *Current Issues in Tourism*, *18*(12), 1192–1217. doi:10.1080/13683500.2013.827159

Huang, L., & Davis, F. D. (2022). User Perceptions of Security in Decentralized Technologies: An Empirical Investigation. *Information Systems Research*, *33*(1), 120–139.

Johnson, A., & Smith, B. (2020). Security Vulnerabilities in Decentralized Ecosystems: A Literature Synthesis. *Journal of Cybersecurity*, *9*(4), 421–438.

Kamalaldin, A., Linde, L., Sjödin, D., & Parida, V. (2020). Transforming provider-customer relationships in digital servitization: A relational view on digitalization. *Industrial Marketing Management*, *89*, 306–325. doi:10.1016/j.indmarman.2020.02.004

Khanna, P., Kumar, S., & Gauba, R., & Aditya. (2022, October). Non-Fungible Tokens' Marketplace: A Secured Blockchain-Based Decentralized Framework for Online Auction. In *International Conference on Computing, Communications, and Cyber-Security* (pp. 841-856). Singapore: Springer Nature Singapore. 10.1007/978-981-99-1479-1_62

Kim, W. C., & Mauborgne, R. (2014). *Blue Ocean Strategy: How to Create Uncontested Market Space and Make Competition Irrelevant*. Harvard Business Review Press.

Li, Y., & Jin, F. (2020). Exploring Disruptive Innovation: A Comprehensive Review. *International Journal of Innovation Management*, *24*(6), 2050052.

Liu, Y., Zhao, Z., Guo, G., Wang, X., Tan, Z., & Wang, S. (2017, August). An identity management system based on blockchain. In *2017 15th Annual Conference on Privacy, Security and Trust (PST)*. IEEE. 10.1109/PST.2017.00016

Manceski, G., & Petrevska Nechkoska, R. (2023). Conceptualisation of Decentralized Blockchain-Based, Open-Source ERP Marketplaces: Disruptive Decentralized Technologies for Co-Creation. In *Facilitation in Complexity: From Creation to Co-creation, from Dreaming to Co-dreaming, from Evolution to Co-evolution* (pp. 175–202). Springer International Publishing. doi:10.1007/978-3-031-11065-8_7

Markides, C. (2019). In Search of Ambidextrous Business Models. *MIT Sloan Management Review*, *61*(4), 22–29.

Markides, C. (2020). The Dark Side of Disruptive Innovation. *MIT Sloan Management Review*, *61*(2), 22–30.

Maurer, B., Nelms, T. C., & Swartz, L. (2013). 'When Perhaps the Real Problem is Money Itself!': The Practical Materiality of Bitcoin. *Social Semiotics*, *23*(2), 261–277. doi:10.1080/10350330.2013.777594

Mavrogiorgou, A., Kiourtis, A., Makridis, G., Kotios, D., Koukos, V., Kyriazis, D., Soldatos, J. K., Fatouros, G., Drakoulis, D., Maló, P., Serrano, M., Isaja, M., Lazcano, R., Vera, J. M., Fournier, F., Limonad, L., Perakis, K., Miltiadou, D., Kranas, P., & Troiano, E. (2023, July). FAME: Federated Decentralized Trusted Data Marketplace for Embedded Finance. In *2023 International Conference on Smart Applications, Communications and Networking (SmartNets)*. IEEE. 10.1109/SmartNets58706.2023.10215814

McGrath, R. G., & Nerkar, A. (2019). Real Options Reasoning and a New Look at the R&D Investment Strategies of Biopharmaceutical Firms Facing Uncertainty. *Organization Science*, *30*(3), 495–515.

Meenakshi, N. (2023). Post-COVID reorientation of the Sharing economy in a hyperconnected world. *Journal of Strategic Marketing*, *31*(2), 446–470. doi:10.1080/0965254X.2021.1928271

Mougayar, W. (2020). *The Business Blockchain: Promise, Practice, and Application of the Next Internet Technology*. John Wiley & Sons.

Mougayar, W., & Buterin, V. (2016). *The Business Blockchain: Promise, Practice, and Application of the Next Internet Technology*. John Wiley & Sons.

Nakamoto, S. (2009). *Bitcoin: A Peer-to-Peer Electronic Cash System*. Retrieved from https://bitcoin.org/bitcoin.pdf

Narayanan, A., Bonneau, J., Felten, E., Miller, A., & Goldfeder, S. (2016). *Bitcoin and Cryptocurrency Technologies: A Comprehensive Introduction*. Princeton University Press.

Osiyevskyy, O., & Dewald, J. (2015). Transformative Potential of Decentralized Marketplaces: A Paradigm Shift in Exchange Processes. *Journal of Business Research*, *68*(7), 1458–1466.

Osterwalder, A., & Pigneur, Y. (2010). *Business model generation: a handbook for visionaries, game changers, and challengers (1)*. John Wiley & Sons.

Osterwalder, A., & Pigneur, Y. (2021). *Business Model Generation: A Handbook for Visionaries, Game Changers, and Challengers*. John Wiley & Sons.

Paetz, P. (2014). A Disruptive Business Model. In *Disruption by Design*. Apress. doi:10.1007/978-1-4302-4633-6_9

Pickering, J., Hickmann, T., Bäckstrand, K., Kalfagianni, A., Bloomfield, M., Mert, A., Ransan-Cooper, H., & Lo, A. Y. (2022). Democratising sustainability transformations: Assessing the transformative potential of democratic practices in environmental governance. *Earth System Governance*, *11*, 100131. doi:10.1016/j.esg.2021.100131

Prasad, C., Rao, B. S., Pujari, J. J., & Hema, C. (2023, August). Developing a Non-Fungible Token-Based Trade Marketplace Platform Using Web 3.0. In *2023 5th International Conference on Inventive Research in Computing Applications (ICIRCA)* (pp. 312-316). IEEE. 10.1109/ICIRCA57980.2023.10220823

Qian, Y. (2022). Navigating the Challenges of Decentralized Technologies: A Comprehensive Review. *Journal of Information Technology*, *37*(2), 215–234.

Raith, M., & Panni, M. F. (2020). Beyond Disruption: A Systematic Review of Business Model Innovation Research. *Journal of Business Research*, *110*, 377–389.

Robbins, P., O'Gorman, C., Huff, A., & Moeslein, K. (2021). Multidexterity—A new metaphor for open innovation. *Journal of Open Innovation*, *7*(1), 99. doi:10.3390/joitmc7010099

Santana, C., & Albareda, L. (2022). Blockchain and the emergence of Decentralized Autonomous Organizations (DAOs): An integrative model and research agenda. *Technological Forecasting and Social Change*, *182*, 121806. Advance online publication. doi:10.1016/j.techfore.2022.121806

Schiavi, G. S., & Behr, A. (2018). Emerging technologies and new business models: A review on disruptive business models. *Innovation & Management Review*, *15*(4), 338–355. doi:10.1108/INMR-03-2018-0013

Schiavi, G. S., Behr, A., & Marcolin, C. B. (2019). Conceptualizing and qualifying disruptive business models. *RAUSP Management Journal*, *54*(3), 269–286. doi:10.1108/RAUSP-09-2018-0075

Sewpersadh, N. S. (2023). Disruptive business value models in the digital era. *Journal of Innovation and Entrepreneurship*, *12*(1), 1–27. doi:10.118613731-022-00252-1 PMID:36686335

Spath, D., Gausemeier, J., Dumitrescu, R., Winter, J., Steglich, S., & Drewel, M. (2022). Digitalisation of Society. In A. Maier, J. Oehmen, & P. E. Vermaas (Eds.), *Handbook of Engineering Systems Design*. Springer. doi:10.1007/978-3-030-81159-4_5

Stubbs, W., & Cocklin, C. (2021). Sustainable Business Models: A Systematic Literature Review. *Organization & Environment*, *34*(3), 288–317.

Sundararajan, A. (2019). *The Sharing Economy: The End of Employment and the Rise of Crowd-Based Capitalism*. The MIT Press.

Swan, M. (2015). *Blockchain: Blueprint for a New Economy*. O'Reilly Media, Inc.

Swan, M. (2020). *Blockchain: Blueprint for a New Economy*. O'Reilly Media.

Swan, M., & Chen, D. (2021). Scalability Concerns in Decentralized Marketplaces: A Critical Analysis. *International Journal of Blockchain and Distributed Ledger Technology*, *4*(1), 23–38.

Tapscott, D., & Tapscott, A. (2016). *Blockchain revolution: how the technology behind bitcoin is changing money, business, and the world*. Penguin.

Teece, D. J. (2020). Disruptive Innovation: Past, Present, and Future. *Long Range Planning*, *53*(6), 101956.

Wang, C., & Hu, Y. (2021). Accessibility in Decentralized Models: A Comparative Analysis. *Journal of Information Systems and Technology Management*, *18*, e202112.

Wang, H., & Li, X. (2021). Disruption and corporate foresight: The moderating role of strategic flexibility. *Technological Forecasting and Social Change*, *162*, 120341.

Williams, L. (2015). *Disrupt: Think the unthinkable to spark transformation in your business*. FT Press.

Xu, J., & Feng, Y. (2022). Reap the harvest on blockchain: A survey of yield farming protocols. *IEEE Transactions on Network and Service Management*, *20*(1), 858–869. doi:10.1109/TNSM.2022.3222815

Yakubu, B. M., Khan, M. I., Javaid, N., & Khan, A. (2021). Blockchain-based secure multi-resource trading model for smart marketplace. *Computing*, *103*(3), 379–400. doi:10.100700607-020-00886-7

Zhang, H., & Li, X. (2020). Decentralized Approaches: Unlocking Efficiency in Business Models. *Journal of Business Efficiency and Innovation*, *1*(1), 56–72.

Zuboff, S. (1988). *In the age of the smart machine: The future of work and power*. Basic Books, Inc.

Chapter 4
Unpacking the Role of Service Quality of AI Tools in Catalyzing Digital Transformation:
A Bibliometric Analysis

Namita Sharma
https://orcid.org/0009-0002-1108-4089
Chitkara Business School, Chitkara University, Punjab, India

Urvashi Tandon
Chitkara Business School, Chitkara University, Punjab, India

ABSTRACT

The aim of this chapter is to do a comprehensive examination of the existing scholarly literature pertaining to artificial intelligence (AI) along with the service quality of AI tools within the framework of digital transformation, managing the effects of deglobalization, and fostering the use of technology. An in-depth study was carried out on a dataset consisting of 156 articles that were taken from the Scopus database. For the purpose of conducting this inquiry, bibliometric methods were utilized, and more specifically, VOSviewer software was utilized, in order to assess the performance analysis and science mapping of the literature that was under investigation. The findings of the study imply that there is a taxonomical representation of current scientific research on the incorporation of artificial intelligence with technological adoption across a variety of fields, with a particular emphasis on online shopping and the level of service quality provided by AI tools.

INTRODUCTION

In the last thirty years, there has been a significant increase in the digitization of the global economy. This has necessitated a shift in operational and managerial practices within businesses, moving away from traditional methods towards a technologically controlled approach. The objective of this shift is to

DOI: 10.4018/979-8-3693-1762-4.ch004

ensure the survival and long-term existence of these businesses. Contemporary technological advancements have facilitated the development of machines that possess the capability to do a wide range of labor-intensive tasks (Singh et al., 2019; Clauberg, 2020). The advancement of artificial intelligence has yielded substantial economic advantages for humanity and has positively impacted several facets of life. Furthermore, it has significantly propelled social progress and ushered society into a new era (Hershbein & Kahn, 2018; Kreps & Neuhauser, 2013; Yin & Qiu, 2021).

Artificial intelligence (AI) refers to a methodology employed to replicate human cognitive abilities through the utilization of a set of algorithms, resulting in the development of a novel computer system capable of executing tasks identical to those performed by people, while concurrently engaging in parallel computing (Al-Adwan, 2020; Bharadiya,2023; Dai et al., 2011). In the field of artificial intelligence, intelligence can be broadly characterized as the capacity to effectively manipulate and convert input into meaningful knowledge, which then guides purposeful actions (Antwi, 2021; Paschen et al., 2019; Cockburn et al, 2018; Aggarwal et al., 2019). To handle persistent profit restrictions, faster strategy cycles, and increased consumer expectations, businesses are increasingly turning to AI technologies that are supported by data analytics. This is happening in order to meet the needs of their customers. The advancements in artificial intelligence (AI) possess the capability to enhance the consumer experience through the augmentation of organizations' understanding of their buying habits and shopping behaviors (Dornberger et al., 2018; Evans, 2019). The deliberate implementation of AI technology at several critical consumer touchpoints has the potential to yield substantial advantages for companies, potentially resulting in heightened levels of customer satisfaction.

One of the most common AI tools used in online shopping is Chatbot. An AI chatbot refers to a conversational system that is powered by artificial intelligence technology, namely in the areas of natural language comprehension, machine learning, and big data analysis (Lee et al., 2021). With the use of AI technology, certain AI chatbots exhibit superior capabilities compared to human workers in areas such as gathering data, memory retention, and computational ability.

The objective of this essay is to offer valuable perspectives on the subsequent research questions pertaining to artificial intelligence:

1. What is the observable trend in the distribution of scholarly literature about Artificial Intelligence (AI) and Service Quality across the years 2014-2023?
2. Which countries, documents, sources, keywords and authors are considered the most noteworthy in the available literature?

The current chapter of this book is structured in the subsequent manner. Section 2 offers a comprehensive analysis of pertinent scholarly works concerning artificial intelligence within the framework of digital transformation and the challenges posed by deglobalization. The third section of this book chapter provides a comprehensive description of the materials and procedures employed in conducting bibliometric analysis. Moreover, Section 4 provides an in-depth examination of data analysis. Section 5 of the document offers a more detailed explanation of the preceding debate, along with an examination of the consequences and limitations.

LITERATURE REVIEW

When compared to the decades that came before, the current era of digitization makes it very necessary for businesses to reduce the amount of time spent waiting and to improve their level of knowledge regarding the environment of the market, which is prone to rapid shifts. According to this point of view, a wide variety of businesses have been adopting emerging technologies in the hopes of achieving greater performance and getting an advantage over their rivals (Balog, 2020; Shalender & Yadav, 2019). Among these advancements, Artificial Intelligence (AI) has been an important factor and has attracted a substantial amount of interest from scholars as well as the business world (Akhtar et al., 2019; Davis et al., 1989; Petropoulos, 2018).

The advancement and spread of artificial intelligence are not independent from other technical advancements; rather, it is ingrained in the overarching process of digitalization (Kitzmann, 2021). Additionally, improvements in processes and storage techniques as well as the growing prevalence of information communication all contribute to the acceleration of digital change.

The ability of an organization to increase the efficacy and efficiency of its services by capitalizing on breakthroughs in information and communication technology and adapting to the changing needs of society as a whole is what is meant by the term "digital transformation" (Arora and Narula, 2018; Bitner, 2000; Calp, 2020). Digital transformation may be achieved by an organization when it is able to capitalize on these improvements in information and communication technology and adapt to the changing demands of society. Additionally, the digital transformation ought to be seen not just as a manifestation of technological innovation or revolution, but also as a process that involves the digitization of human activities and operational procedures. This is because the digital transformation is a process that involves the digitization of human activities and operational procedures. According to Marquardt (2017), the urgency for digital transformation has emerged as a result of changes in demographics, the restricted availability of resources, increased worldwide competition, and the globalization of markets.

According to Garbuio and Lin (2019), an increasing number of businesses are relying more and more on artificial intelligence (AI) to support or even take the place of their ongoing efforts in digital transformation. Because the integration of AI and digital transformation is so tightly entwined, a concerted effort is required in order to completely take advantage of the possibilities that are inherently associated with this association.

The advent of developing digital technologies has had a profound impact on contemporary society and the economy (Fang, 2019; Frank et al., 2018; Hershbein & Kahn, 2018). The evolving landscape of digital technology skills is undergoing transformation. The labour market is subject to dynamic shifts that will inevitably shape the trajectory of employment in the future (Frey & Osborne, 2017; Lim et al.,2021; Sigelman, 2019). According to the findings of a study that was conducted by Lyu and Liu (2021) on the subject of the application of artificial intelligence with digital transformation in the energy sector, it was discovered that large energy companies should take the initiative in embracing new digital technologies so that they can make the most use of the benefits that these technologies have to offer. Importantly, large energy businesses should pay artificial intelligence (AI) special attention and considerably enhance their recruiting of AI professionals among these new digital technologies (Gursoy et al., 2019; Kasilingam, 2020; Sheth et al, 2021; Sheth et al., 2022). This is because AI is expected to play a key role in the future of the energy industry.

According to Grönroos (1984), service quality refers to the evaluation of an organization's ability to meet the expectations of its customers in terms of the quality of service delivered. The SERVQUAL model is employed to quantitatively assess the extent of customer satisfaction with the whole service encounter (Zeithaml & Bitner, 2000). Service quality refers to the extent to which consumers' assessments of a service's performance deviate from their perceived expectations for that service (Parasuraman et al., 1988). Moreover, the dimensions and scale proposed by Parasuraman et al. (1988) in their study are widely regarded as the predominant findings in the field of service quality. Moreover, according to Berry et al. (1988), the quality of service is considered a crucial distinguishing factor and the most potent competitive advantage for service-oriented organizations. The validity of the five-element SERVQUAL scale, which includes assurance, empathy, tangibles, reliability, and responsiveness, has been established in several sectors such as e-commerce and both offline and online contexts (Gefen, 2002; Semeijn et al., 2005).

According to Parasuraman et al. (2005, p. 5), e-service quality refers to the degree to which a website facilitates efficient and effective browsing, purchase, and delivery processes. The absence of face-to-face interaction in e-services may result in heightened concerns over privacy and perceived risk among shoppers, as compared to services offered in traditional brick-and-mortar establishments (Sony, 2020; Bitner et al., 2000; Dabholkar, 1996). Blut et al. (2015) developed a conceptual framework to understand the diverse models of e-service quality proposed by different authors, as well as the multiple variables used to examine the impact of e-service quality on customer satisfaction, repurchase intentions, and word-of-mouth communication. This framework was based on the means-ends-chain theory.

Service robots have the ability to engage with consumers on a large scale while maintaining consistent quality (Wirtz et al., 2018). They also have the capability of automating social interactions at the frontline of service (Van Doorn et al., 2017), and have now become a regular element of service experiences (Mende et al., 2019).

The prevalent and increasing applications of artificial intelligence (AI) in diverse service functions underscore the importance for service providers to thoughtfully deliberate on the utilization of AI in order to efficiently interact with customers in a rational and strategic manner.

THEORETICAL UNDERPINNING

According to Parasuraman et al. (1988), the SERVQUAL scale is a measurement tool that determines how happy an organization's clients are with the standard of the service customer received. The importance-performance matrix that was produced by Martilla and James (1977) was the starting point for the SERVQUAL measurement. Since then, the SERVQUAL model that was constructed by Parasuraman et al. (1985) and the SERVPERF framework that was developed by Cronin and Taylor (1992) have both been incorporated into the SERVQUAL measurement. The SERVQUAL measuring scale was developed by Parasuraman and his colleagues in 1988 for use in a variety of industries, most notably the retail services sector.

QUALITY OF RETAIL SERVICE

Previously conducted investigations (Finn, 1991; Mehta et al., 2000; Wong & Sohal, 2003; Gaur & Aggarwal, 2006; Naik et al., 2010) have provided substantial validation for the SERVQUAL Scale. In the words of Dabholkar et al. (1996), the SERVQUAL scale is more well designed for the service

Figure 1. Quality of services

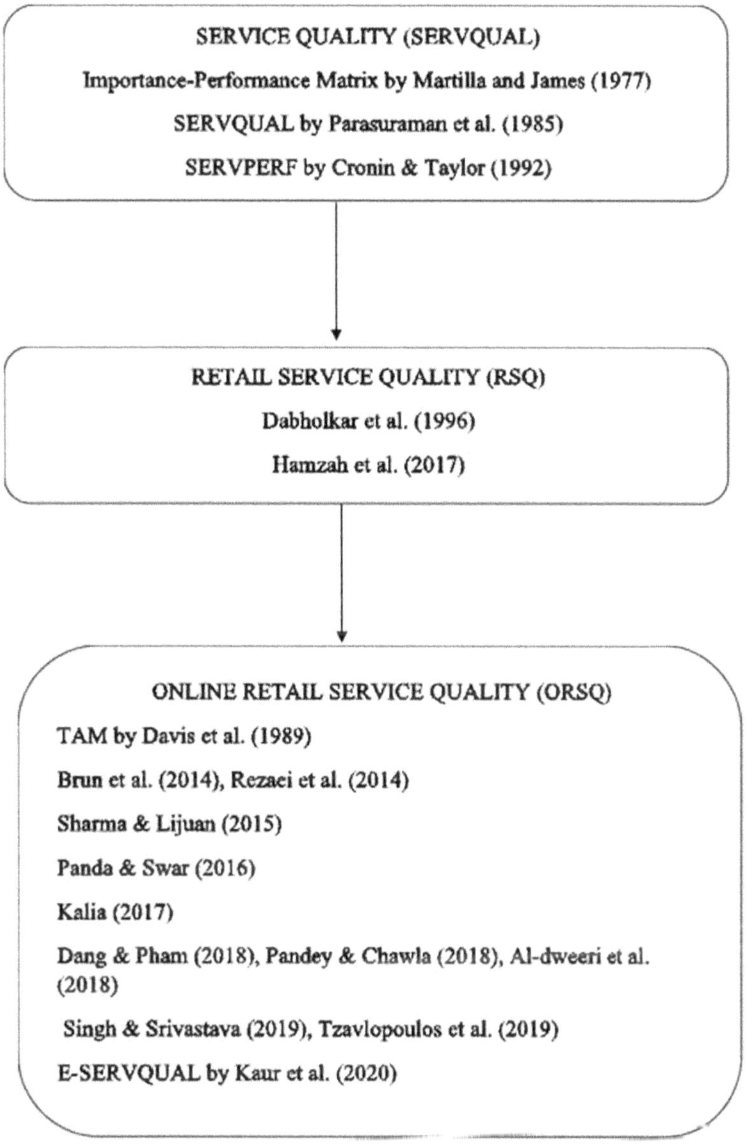

industry than it is to the retail industry. As a result, they developed a new scale for RSQ with a total of five dimensions, which include "physical aspects," "personal interaction," "problem solving," and "reliability," and to use SERVQUAL especially in the offline retail sector. Researchers from Malaysia named Hamzah et al. (2017) conducted a study that looked at RSQ across four dimensions, including security, empathy, reliability, and tangibles. They found that empathy is the most accurate indicator of retail service quality out of the four factors.

THE STANDARD OF SERVICE IN ONLINE RETAILING

The topic of discussion pertains to the quality of service provided by online retail platforms. According to Davis et al. (1989), the TAM dimensions include notions such as "perceived ease of use" and "perceived usefulness" for the purpose of understanding the factors that influence the adoption of technology, such as online or website purchase. Kaur et al. (2020) devised a scale to quantify the aspects of e-SERVQUAL that they recognized as being significant in online buying. These criteria include privacy, information quality, system availability, dependability, usability, efficiency, security, and assurance.

RESEARCH METHODOLOGY

Researchers in the field of management are increasingly using the bibliometric method to examine the relationships between different types of publications, regions, journals, and authors (Wu et al., 2021). Numerous scholarly investigations have utilized bibliometric analysis as a methodological approach to examine institutions and countries (Farrukh et al., 2021; Qamar and Samad, 2022; Wijewickrema, 2023), prominent publications (Wu et al., 2021; Farrukh et al., 2021; Qamar and Samad, 2022; Wijewickrema, 2023; Saha et al., 2020), and significant keywords (Qamar and Samad, 2022; Mahadevan and Joshi, 2021; Ülker et al., 2022). Bibliometric analysis is a method used to assess the body of published academic literature by examining bibliographic data, such as the quantity of citations and publications, thematic patterns, trends, and co-authorship (Farrukh et al., 2021; Ellitan & Richard, 2022).

This study encompasses the examination of performance analysis and science mapping, which are crucial components of bibliometric analysis. Performance analysis involves the assessment of various publications and citations, which serve as indicators of production and impact. Science mapping, however, serves the purpose of visually representing the structure and dynamics of a certain topic (Gao et al., 2021). Hence, the study seeks to achieve its goals through the utilization of bibliometric analysis.

The bibliographic data was extracted from the Scopus database due to its extensive article coverage, surpassing other databases such as "Web of Science," making it one of the most substantial repositories of academic publications (Farrukh et al., 2021). Previous researches have utilized Scopus extensively for bibliometric analysis, as evident by the works of Donthu et al. (2021), Nobanee et al. (2021), Chabowski et al. (2013), and Budd (1988).

To ensure that all relevant literature and resources pertaining to AI in their respective domains were uncovered, multiple search strategies were applied. TITLE-ABS-KEY ("Artificial intelligence" OR "Chatbot" OR "Chatbots" OR "Recommender System" OR "Virtual try on technology" OR "Augmented Reality" OR "Gamification" OR "Image Interactivity technology" AND "Service Quality" AND "E-retail" OR "Online shopping" OR "Trust" OR "Perceived Value" OR "Satisfaction" OR "Loyalty") AND PUBYEAR > 2013 AND PUBYEAR < 2024 AND (LIMIT-TO (PUBSTAGE, "final")) AND (LIMIT-TO (LANGUAGE, "English")) AND (LIMIT-TO (DOCTYPE, "ar") OR LIMIT-TO (DOCTYPE, "cp")) was the string used as query. The title-abstract-keyword column of Scopus database was used to search the abovesaid string. The initial inquiry included a comprehensive examination of 206 academic articles that had been published between the period of 2014 to 2023. The authors of the study applied inclusion and exclusion criteria to the documents. This study exclusively considered publications and conference papers that were published in the English language. Hence, a comprehensive collection of 156 scholarly publications was acquired for the study conducted over the timeframe spanning from 2014 to 2023.

Figure 2. Data retrieval process

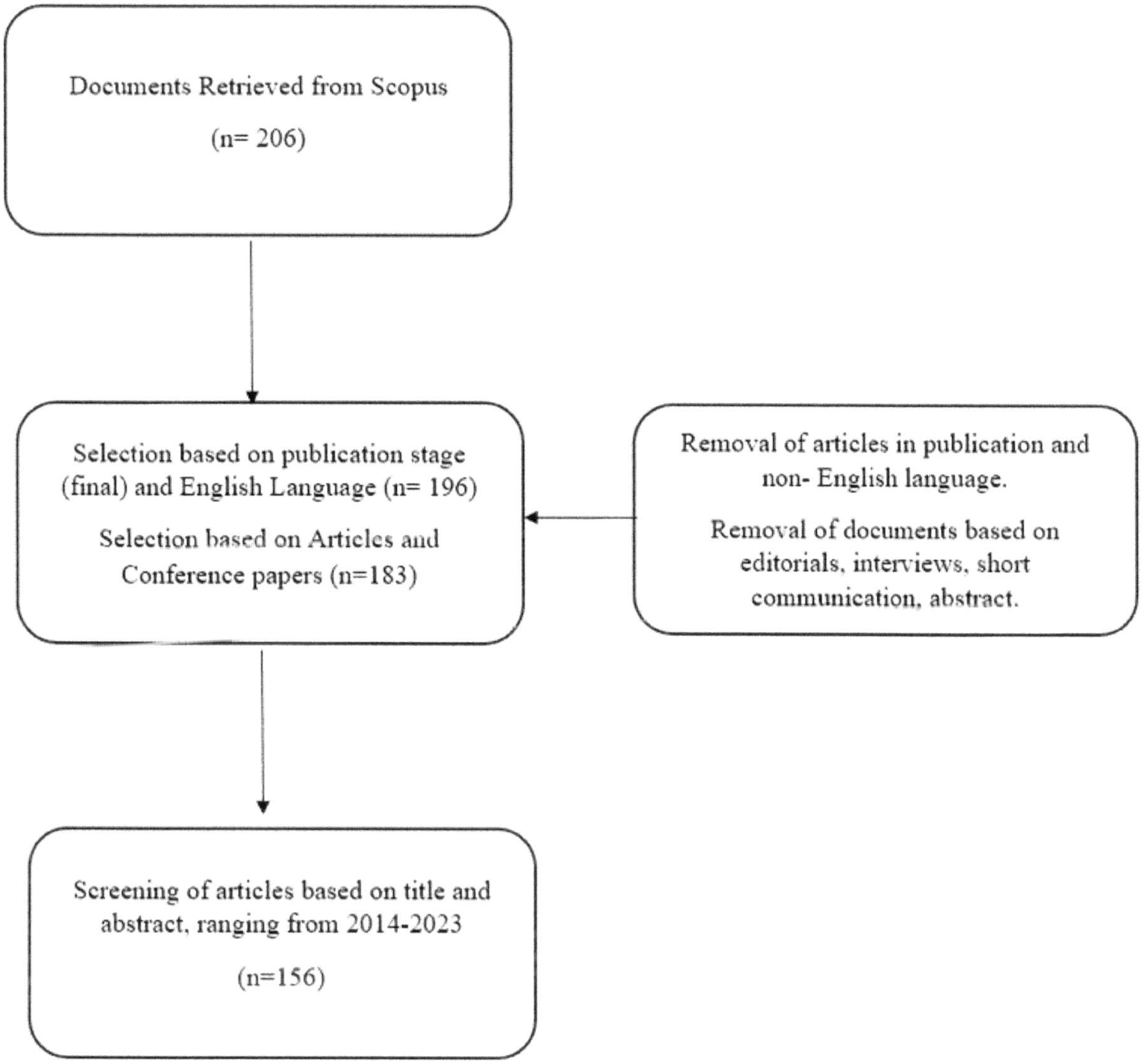

DATA ANALYSIS

Table 1 presents an analysis of the data source acquired from the Scopus database. A thorough compilation of 156 papers was acquired, featuring contributions from 447 individuals situated in 47 countries. These publications were derived from 119 sources, including journals and conferences, covering the time frame from 2014 to 2023.

Table 1. Descriptive analysis

S.No.	Description	Count
1	Total Documents	156
2	Publication Period	2014-2023
3	Articles	75
4	Conference Papers	44
5	Countries	47
6	Sources	119
7	Authors	447

TRENDS IN PUBLICATION

A multitude of scholarly investigations have been conducted on the topics of service quality, contentment, trust, and loyalty prior to the year 2014. Research articles focusing on the intersection of Artificial Intelligence and service quality, pleasure, loyalty, and trust have been increasingly available since 2014. Figure 2 provides significant information into the temporal evolution of study subjects. Several papers in this particular field were published in the year 2014 and continued to experience growth until the year 2016. A marginal decrease can be discerned in the year 2017. The emergence of concerns regarding service quality and AI garnered significant attention and discussion within the public sphere from the year 2017onwards. There is a modest decline in publishing rates observed in 2021, followed by a substantial increase in 2022. The year 2023 is expected to witness a substantial number of publications in the fields of service quality, trust, loyalty, satisfaction, and artificial intelligence.

MOST PROMINENT REGIONS GEOGRAPHICALLY

On a global scale, the concept of artificial intelligence has attained widespread acceptance and is held in exceptionally high esteem. Consequently, the papers are authored by persons hailing from diverse geographic locations. There have been contributions to the field of artificial intelligence made by a total of 47 countries. The top 12 geographical regions that have had the most significant impact on this study are listed in Table 2. When only looking at publications, it is easy to see that Asian countries are in the lead. This is especially apparent when one takes a myopic view of the situation. On the basis of the number of documents released, Asian nations including China, India, and South Korea are leading the top positions in the fields of service quality and artificial intelligence. These nations have 35 publications, 19 publications, and 17 publications, respectively. In contrast to what was stated earlier, it is

Figure 3. Trends of publication in the domain of artificial intelligence

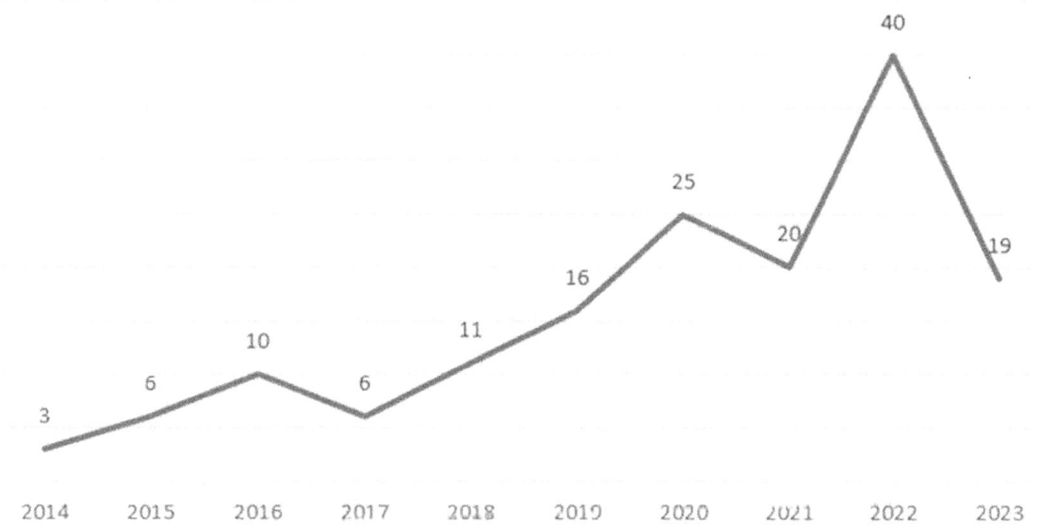

clear from the citations that are at hand that the United States of America is in first place in the league, with a grand total of 417 citations.

Table 2. The most productive countries

Rank	Country	Continent	Documents	Citations
1	China	Asia	35	638
2	India	Asia	19	176
3	South Korea	Asia	17	143
4	United States	North America	15	357
5	Australia	Australia	12	219
6	Taiwan	Asia	10	98
7	Indonesia	Asia	9	73
8	Malaysia	Asia	8	86
9	Germany	Europe	7	49
10	United Kingdom	Europe	6	363
11	Turkey	Asia	5	16
12	Vietnam	Asia	5	110

Table 3. Co-authorship of countries

From	To	Link Strength
China	South Korea	2
China	United States	6
China	United Kingdom	1
South Korea	United States	2
Australia	China	2
Australia	Vietnam	2
Australia	India	1
Australia	United Kingdom	1
India	Vietnam	1
Germany	United Kingdom	1
Indonesia	Malaysia	2
Malaysia	Taiwan	1
Malaysia	United Kingdom	1

Leading Countries' Co-Authorship

The co-authorship of the various regions of the world is represented in tabular form in Table 3, which is provided for your convenience. The analysis of patterns of co-authorship that span numerous countries may yield substantial information regarding the ties that individual authors have with a range of national associations if the right questions are asked. The United States and China have worked together on the vast bulk of the research that has been published on artificial intelligence (AI) and service quality in relation to trust, customer loyalty, and satisfaction with services. This research has been published in a variety of academic journals.

Bibliographic Coupling of Countries

Figure 3 presents a graphic representation depicting the countries that are actively engaged in research pertaining to the field of Artificial Intelligence. Bibliographic coupling analysis was performed on nations that have published a minimum of five documents, with the requirement that a country must have received at least three citations. Out of a total of 47 countries, there are a total of 12 that fit the criteria. The size of the circle in the network diagram represents the amount of contribution made by that country; the larger the circle, the more significant the contribution will be. Each circle in the network diagram represents a different nation. The creation of four distinct clusters was the end result of a process known as bibliographic coupling. The red cluster is comprised of countries such as China, South Korea, Turkey, and the United States, with China taking the lead. Conversely, the green cluster exhibits a predominant

Figure 4. Bibliographic coupling of countries

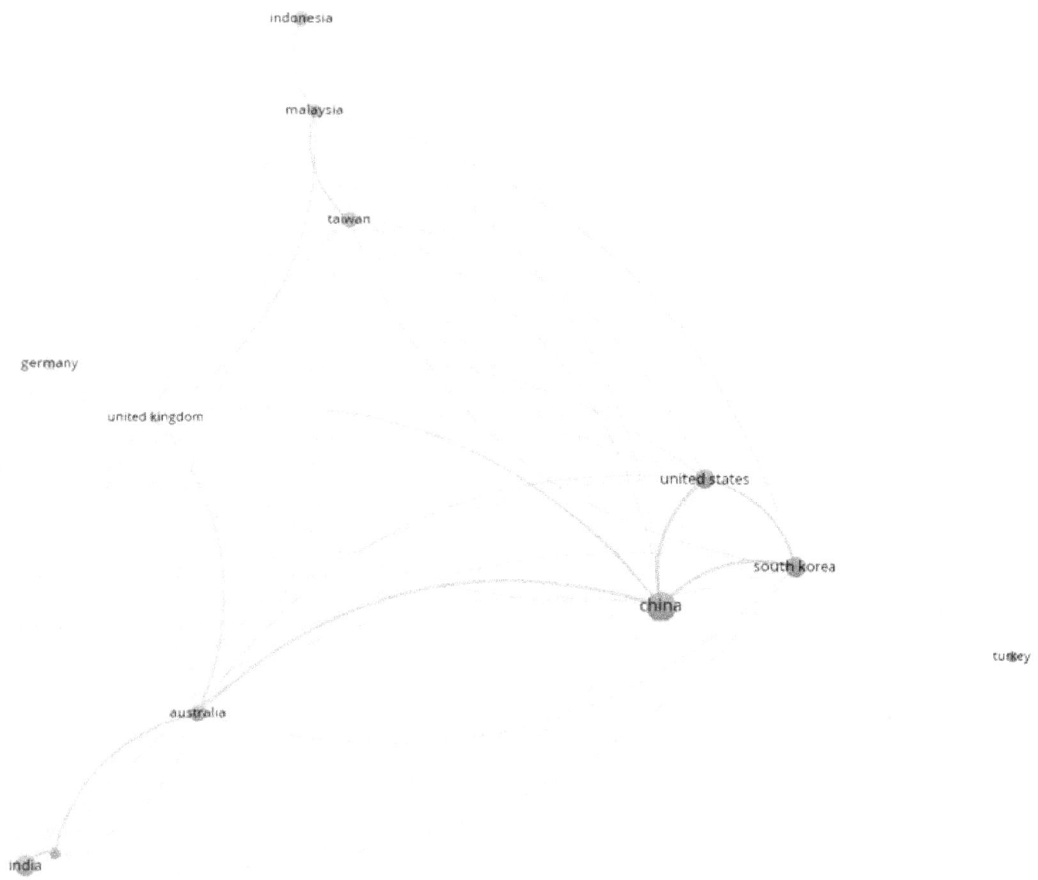

presence of Indonesia, alongside the inclusion of Malaysia and Taiwan. The blue cluster is composed of Australia, India, and Vietnam, whereas the yellow cluster is jointly occupied by Germany and the United Kingdom.

Leading Journals

The leading academic journals in the fields of artificial intelligence and service quality are outlined in Table 4, which provides a ranking of these publications based on the number of documents they have published. The following are the criteria that were used for the selection of sources based on the number of citations: a minimum requirement of 2 documents from a source and a minimum requirement of 5 citations for a source. Only 13 of the totals of 119 sources that were looked at satisfy the criteria that was provided. From this subgroup, ten of the most important sources were chosen for more investigation. Two documents in the field of telematics and informatics received a combined total of 227 citations, propelling the field to the top of the list of subjects most frequently cited. This academic journal is responsible for the dissemination of publications that pertain to the domains of engineering, computer science, and the social sciences. When viewed through a narrow lens, it is possible to see quite clearly

Table 4. Top journals on the basis of citations

S.No.	Source	No. of Publications	Citations
1	Telematics and Informatics	2	227
2	Sustainability	7	151
3	Journal of Hospitality Marketing and Management	3	119
4	Journal of Retailing and Consumer Services	5	66
5	Procedia Computer Science	3	47
6	Technology in Society	2	43
7	ACM International Conference Proceeding Series	7	21
8	Advances in Intelligent Systems and Computing	3	15
9	Proceedings of the Annual Hawaii International Conference on System Sciences	2	15
10	Proceedings - 20th ieee/acis International Conference on Software Engineering, Artificial Intelligence, Networking and Parallel/Distributed Computing, 2019	2	7

that Sustainability has a greater number of publications published to its credit, namely seven, although its citation score is 151.

Prominent Authors

Table 5 presents a comprehensive overview of the most influential authors in the domain of artificial intelligence and online commerce, as ascertained through an analysis of the citations received by their published works. The criteria for selecting authors based on citations were as follows: a minimum threshold of 2 documents authored by an individual and a minimum threshold of 5 citations received by an individual. Out of the total number of 447 authors, it was found that 20 authors satisfied the given criteria. For further analysis, the top 10 writers with the highest number of citations were chosen from this particular subset. Yun J. has been regarded as the most significant author. In addition, Prentice C., Lu Y., and Malik A. also demonstrated exceptional performance. Typically, scholars engaged in the study of artificial intelligence pertaining to service quality, trust, perceived value, satisfaction, or online purchasing predominantly hail from institutions situated in South Asian nations.

Co-Occurrence of Author Keywords

The utilization of a co-occurrence network of keywords is a scholarly methodology that offers link-based insights into the interconnections of various sub-fields of research, with the aim of identifying prominent research themes (Ülker et al., 2022). Keyword co-occurrence analysis, when employed proficiently, has the potential to unveil noteworthy patterns within textual data, hence facilitating comprehension of the

Table 5. Top cited authors based on citations

S.No.	Author	Documents	Citations
1	Yun J.	2	196
2	Prentice C.	4	169
3	Lu Y.	2	37
4	Malik A.	2	35
5	Nguyen T.M.	2	35
6	Lam H.Y.	3	20
7	Li Y.	2	20
8	Tang V.	2	14
9	Suhartanto D.	2	11
10	Syarief M.E.	2	11

Figure 5. Co-occurrence frequency of author keywords

interrelationships among different concepts and words (Arora et al., 2022). Out of the total of 550 keywords utilized for analysis, only 7 of them were found to meet the minimum threshold of 05 occurrences.

The analysis focused on the network visualization of keyword co-occurrence, as shown in Figure 4. This visualization exhibited nodes and links that represented the relationships between words. The nodes function as placeholders for the keywords, with their size reflecting the frequency of occurrence of the keywords in the documents. A greater node size suggests a higher frequency of presence, while a smaller size indicates a lower frequency.

The seven recognized keywords were further categorized into following two clusters:

Cluster 1: Artificial Intelligence, Customer Experience, Customer Satisfaction, Service Quality.

Cluster 2: Chatbot, Customer Service, Trust

The utilization of artificial intelligence (AI) technology inside digital transformation endeavors has demonstrated considerable efficacy in enhancing corporate performance and accessibility (Calp, 2020; Chen et al., 2023; Dahiya, 2017; Park et al., 2021; Pelau et al, 2021). In essence, the utilization of AI approaches will enable managers to enhance the effectiveness and precision of their decision-making processes.

IMPLICATIONS

The research presented examines the integration of artificial intelligence (AI) in the context of online purchase and its effects on service quality, thereby highlighting several social and ethical considerations. Based on the facts supplied, several considerations can be identified.

1. Effects on the Economy and the Labor Force

The growing adoption of AI technology, particularly chatbots and automation, within enterprises has raised concerns regarding the possibility of job displacement. Artificial intelligence (AI) technologies, as employed in many online purchasing platforms, have the potential to supplant certain human positions, either resulting in unemployment or necessitating a transformation in the skill sets demanded within the labor market.

The ethical question at hand pertains to the imperative of facilitating a fair and equitable transition for workers who are impacted by automation. This entails the provision of ample chances for retraining and upskilling, enabling them to effectively navigate and adapt to the evolving employment landscape.

2. Security and Privacy of Data

The study discusses the utilization of artificial intelligence (AI) on digital platforms, whereby there is a frequent occurrence of gathering and examining substantial volumes of client data. The ethical considerations encompass the implementation of strong data privacy protocols in order to safeguard sensitive consumer information.

It is imperative for businesses to uphold transparency regarding the utilization of customer data by AI systems, acquire informed consent, and establish robust security measures to mitigate the risks of data breaches and unauthorized exploitation.

3. Concerns on Unfairness in AI Systems

Artificial intelligence (AI) algorithms, such as those utilized in chatbots and recommender systems, have the potential to unintentionally sustain biases that are inherent in the data used for training purposes. This phenomenon has the potential to result in inequitable treatment towards specific demographic cohorts.

Regularly conducting audits and proactively addressing biases inside AI systems is of utmost importance for businesses, as it serves the purpose of promoting justice and mitigating the risk of discriminatory outcomes. Ethical considerations encompass the imperative of ensuring transparency in algorithmic decision-making processes and effectively mitigating prejudice present in training datasets.

4. The Influence on Customer Relationships

AI technology, such as chatbots, have the potential to improve productivity and customer service. However, it is important to consider the potential negative consequence of reduced human connection. Maintaining excellent customer connections necessitates the establishment of a delicate equilibrium between automated interactions and human engagement.

Ethical considerations encompass the imperative to maintain transparency with clients regarding the utilization of artificial intelligence, establishing channels for human intervention when necessary, and guaranteeing that interactions driven by AI are in accordance with the values held by customers.

5. Inequalities in the World's Economies

This study examines the worldwide dissemination of articles pertaining to artificial intelligence (AI), identifying specific countries that have emerged as the forefront in this field. The ethical considerations encompass the need to confront the potential economic inequities that may arise between nations leading in AI development and others that may fall behind.

It is imperative to exert endeavors in order to assure the equitable distribution of the advantages stemming from artificial intelligence (AI). Furthermore, fostering international collaboration is highly recommended as a means to overcome the existing technology gap.

6. Availability and Acceptance

The study discusses the influence of artificial intelligence (AI) on the labor market and emphasizes the necessity for firms to strengthen their recruitment efforts in the field of AI specialists. The imperative of guaranteeing accessibility and inclusion in the development and implementation of artificial intelligence (AI) cannot be overstated.

Ethical considerations encompass the imperative to ensure the accessibility of AI technology to a wide range of communities, taking into account several factors including linguistic diversity, cultural differences, and socioeconomic disparities.

In conclusion, it is imperative for businesses to exercise caution and deliberation when incorporating artificial intelligence (AI) into their operations, as they must carefully contemplate the social and ethical ramifications associated with these technologies in order to guarantee their responsible and sustainable utilization. The need of striking a balance between the prospective advantages and ethical considerations is crucial in ensuring the enduring efficacy and favorable societal consequences of artificial intelligence (AI) implementations.

DISCUSSION

The field of artificial intelligence, in conjunction with service quality, is a burgeoning area of study within the academic community.

The main aim of this research was to explore the incorporation of artificial intelligence into online purchasing and its impact on service quality. This investigation builds upon existing literature and employs bibliometric analysis as a research methodology. This study makes a scholarly contribution to the existing literature on artificial intelligence in the context of online purchasing and service quality. It achieves this by utilizing bibliometric analysis and content analysis techniques, which enhance the rigorousness of methodology and extensiveness of literature in the field of AI research. The objective of this study is to ascertain the notable authors, geographic areas, and publications within the field of artificial intelligence and online commerce.

The study conducted a co-authorship analysis and identified the three authors with the highest level of influence. Yun J., Prentice C., and Lu Y. have the highest degree of connectivity with other writers. The analysis of bibliographic coupling among countries indicated that China had the highest number of publications, followed by the United States and South Korea.

Furthermore, it is worth noting that researchers from China and the United States exhibit a significant number of articles that have been jointly authored. Prentice C exhibited the greatest number of publications, surpassing all other authors. Lam H.Y. followed closely behind in terms of publication count. However, Yun J. emerged as the most often referenced author among the aforementioned individuals. The field of Telematics and Informatics has garnered a cumulative total of 227 citations, hence elevating its position to the highest rank among subjects that are most frequently mentioned.

CONCLUSION

The findings of this research carry value not only for researchers but also for organizations in their efforts to improve strategy planning. Although there has been some recent and gradual growth in interest in the field of artificial intelligence and service quality, there is still a significant need for additional research into the topic. The descriptive analysis makes it abundantly evident that there have not been very many studies published on the topic of artificial intelligence and service quality. This study presents a glimpse of the existing state of knowledge and makes a proposal for an in-depth investigation of the relevant literature that specifically highlights areas where more research is needed. The findings of this study have both theoretical and applied value because of their connection to the process of commercializing AI's improved service quality. Therefore, despite the fact that artificial intelligence has the ability to cut down on labor and expenses, businesses still need to exercise caution when choosing the right kind of AI tool in order to improve their customer experience.

LIMITATIONS

Initially, it is important to note that the research undertaken is thorough in nature, although it should be acknowledged that it is not exhaustive. In addition, the Scopus database was utilized in this study. However, it is suggested that future research endeavors consider incorporating additional pertinent databases, such as Web of Science, in order to conduct a comprehensive analysis. Furthermore, it is possible that the filtering technique employed to export the information may have inadvertently omitted certain relevant studies. Additionally, future studies could employ a quantitative content analysis methodology to systematically and theoretically analyze Artificial Intelligence.

REFERENCES

Agrawal, A., Gans, J. S., & Goldfarb, A. (2019). Exploring the impact of artificial intelligence: Prediction versus judgment. *Information Economics and Policy*, *47*, 1–6. doi:10.1016/j.infoecopol.2019.05.001

Akhtar, P., Frynas, J. G., Mellahi, K., & Ullah, S. (2019). Big data-savvy teams' skills, big data-driven actions and business performance. *British Journal of Management*, *30*(2), 252–271. doi:10.1111/1467-8551.12333

Al-Adwan, A. S., Kokash, H., Adwan, A. A., Alhorani, A., & Yaseen, H. (2020). Building customer loyalty in online shopping: The role of online trust, online satisfaction and electronic word of mouth. *International Journal of Electronic Marketing and Retailing*, *11*(3), 278–306. doi:10.1504/IJEMR.2020.108132

Ameen, N., Tarhini, A., Reppel, A., & Anand, A. (2021). Customer experiences in the age of artificial intelligence. *Computers in Human Behavior*, *114*, 106548. doi:10.1016/j.chb.2020.106548 PMID:32905175

Antwi, S. (2021). "I just like this e-Retailer": Understanding online consumers repurchase intention from relationship quality perspective. *Journal of Retailing and Consumer Services*, *61*, 102568. doi:10.1016/j.jretconser.2021.102568

Arora, M., Prakash, A., Dixit, S., Mittal, A., & Singh, S. (2022). A critical review of HR analytics: visualization and bibliometric analysis approach. *Information Discovery and Delivery*.

Arora, P., & Narula, S. (2018). Linkages between service quality, customer satisfaction and customer loyalty: A literature review. *Journal of Marketing Management*, *17*(4), 30.

Ashfaq, M., Yun, J., Yu, S., & Loureiro, S. M. C. (2020). I, Chatbot: Modeling the determinants of users' satisfaction and continuance intention of AI-powered service agents. *Telematics and Informatics*, *54*, 101473. doi:10.1016/j.tele.2020.101473

Balog, K. (2020). The concept and competitiveness of agile organization in the fourth industrial revolution's drift. *Strategic Management*, *25*(3), 14–27. doi:10.5937/StraMan2003014B

Bharadiya, J. P. (2023). A Comparative Study of Business Intelligence and Artificial Intelligence with Big Data Analytics. *American Journal of Artificial Intelligence*, *7*(1), 24.

Bitner, M. J., Brown, S. W., & Meuter, M. L. (2000). Technology infusion in service encounters. *Journal of the Academy of Marketing Science*, *28*(1), 138–149. doi:10.1177/0092070300281013

Budd, J. M. (1988). A bibliometric analysis of higher education literature. *Research in Higher Education, 28*(2), 180–190. doi:10.1007/BF00992890

Calp, M. H. (2020). The role of artificial intelligence within the scope of digital transformation in enterprises. In *Advanced MIS and digital transformation for increased creativity and innovation in business* (pp. 122–146). IGI Global. doi:10.4018/978-1-5225-9550-2.ch006

Chabowski, B. R., Samiee, S., & Hult, G. T. M. (2013). A bibliometric analysis of the global branding literature and a research agenda. *Journal of International Business Studies, 44*(6), 622–634. doi:10.1057/jibs.2013.20

Chen, Q., Lu, Y., Gong, Y., & Xiong, J. (2023). Can AI chatbots help retain customers? Impact of AI service quality on customer loyalty. *Internet Research.*

Clauberg, R. (2020). Challenges of digitalization and artificial intelligence for modern economies, societies and management. *RUDN Journal of Economics, 28*(3), 556–567. doi:10.22363/2313-2329-2020-28-3-556-567

Cockburn, I. M., Henderson, R., & Stern, S. (2018). The impact of artificial intelligence on innovation: An exploratory analysis. In *The economics of artificial intelligence: An agenda* (pp. 115–146). University of Chicago Press.

Dahiya, M. (2017). A tool of conversation: Chatbot. *International Journal on Computer Science and Engineering, 5*(5), 158–161.

Dai, H., Haried, P., & Salam, A. F. (2011). Antecedents of online service quality, commitment and loyalty. *Journal of Computer Information Systems, 52*(2), 1–11.

Davis, F. D., Bagozzi, R. P., & Warshaw, P. R. (1989). User acceptance of computer technology: A comparison of two theoretical models. *Management Science, 35*(8), 982–1003. doi:10.1287/mnsc.35.8.982

Donthu, N., Kumar, S., Mukherjee, D., Pandey, N., & Lim, W. M. (2021). How to conduct a bibliometric analysis: An overview and guidelines. *Journal of Business Research, 133*, 285–296. doi:10.1016/j.jbusres.2021.04.070

Dornberger, R., Inglese, T., Korkut, S., & Zhong, V. J. (2018). Digitalization: Yesterday, today and tomorrow. *Business Information Systems and Technology 4.0: New Trends in the Age of Digital Change*, 1-11.

Ellitan, L., & Richard, A. (2022). The influence of online shopping experience, customer satisfaction and adjusted satisfaction on online repurchase intention to Tokopedia consumers in Surabaya. *Budapest International Research and Critics Institute-Journal (BIRCI-Journal), 5*(2), 16504-16516.

Evans, M. (2019). *Build A 5-star customer experience with artificial intelligence.* https://www.forbes.com/sites/allbusiness/2019/02/17/customer-experience-artificialintelligence/#1a30ebd415bd

Fang, Y. H. (2019). An app a day keeps a customer connected: Explicating loyalty to brands and branded applications through the lens of affordance and service-dominant logic. *Information & Management, 56*(3), 377–391. doi:10.1016/j.im.2018.07.011

Farrukh, M., Raza, A., Meng, F., & Wu, Y. (2022). CMS at 13: A retrospective of the journey. *Chinese Management Studies*, *16*(1), 119–139. doi:10.1108/CMS-07-2020-0291

Festinger, L. (1962). Cognitive dissonance. *Scientific American*, *207*(4), 93–106. doi:10.1038cientificamerican1062-93 PMID:13892642

Frank, M. R., Sun, L., Cebrian, M., Youn, H., & Rahwan, I. (2018). Small cities face greater impact from automation. *Journal of the Royal Society, Interface*, *15*(139), 20170946. doi:10.1098/rsif.2017.0946 PMID:29436514

Frey, C. B., & Osborne, M. A. (2017). The future of employment: How susceptible are jobs to computerisation? *Technological Forecasting and Social Change*, *114*, 254–280. doi:10.1016/j.techfore.2016.08.019

Gao, P., Meng, F., Mata, M. N., Martins, J. M., Iqbal, S., Correia, A. B., & Farrukh, M. (2021). Trends and future research in electronic marketing: A bibliometric analysis of twenty years. *Journal of Theoretical and Applied Electronic Commerce Research*, *16*(5), 1667–1679. doi:10.3390/jtaer16050094

Gursoy, D., Chi, O. H., Lu, L., & Nunkoo, R. (2019). Consumers acceptance of artificially intelligent (AI) device use in service delivery. *International Journal of Information Management*, *49*, 157–169. doi:10.1016/j.ijinfomgt.2019.03.008

Hershbein, B., & Kahn, L. B. (2018). Do recessions accelerate routine-biased technological change? Evidence from vacancy postings. *The American Economic Review*, *108*(7), 1737–1772. doi:10.1257/aer.20161570

Huang, M. H., & Rust, R. T. (2018). Artificial intelligence in service. *Journal of Service Research*, *21*(2), 155–172. doi:10.1177/1094670517752459

Kasilingam, D. L. (2020). Understanding the attitude and intention to use smartphone chatbots for shopping. *Technology in Society*, *62*, 101280. doi:10.1016/j.techsoc.2020.101280

Kaur, B., Kaur, J., Pandey, S. K., & Joshi, S. (2020). E-service Quality: Development and Validation of the Scale. *Global Business Review*.

Kitzmann, H., Yatsenko, V., & Launer, M. (2021). *Artificial intelligence and wisdom*. Academic Press.

Kreps, G. L., & Neuhauser, L. (2013). Artificial intelligence and immediacy: Designing health communication to personally engage consumers and providers. *Patient Education and Counseling*, *92*(2), 205–210. doi:10.1016/j.pec.2013.04.014 PMID:23683341

Lim, W. M., Gupta, S., Aggarwal, A., Paul, J., & Sadhna, P. (2021). How do digital natives perceive and react toward online advertising? Implications for SMEs. *Journal of Strategic Marketing*, 1–35. doi:10.1080/0965254X.2021.1941204

Lv, Z., & Xie, S. (2022). Artificial intelligence in the digital twins: State of the art, challenges, and future research topics. *Digital Twin*, *1*, 12. doi:10.12688/digitaltwin.17524.2

Lyu, W., & Liu, J. (2021). Artificial Intelligence and emerging digital technologies in the energy sector. *Applied Energy*, *303*, 117615. doi:10.1016/j.apenergy.2021.117615

Mahadevan, K., & Joshi, S. (2022). Omnichannel retailing: A bibliometric and network visualization analysis. *Benchmarking*, *29*(4), 1113–1136. doi:10.1108/BIJ-12-2020-0622

Makridakis, S. (2017). The forthcoming Artificial Intelligence (AI) revolution: Its impact on *society and firms. *Futures*, *90*, 46–60. doi:10.1016/j.futures.2017.03.006

Marquardt, K. (2017). Smart services–characteristics, challenges, opportunities and business models. In *Proceedings of the International Conference on Business Excellence* (Vol. 11, No. 1, pp. 789-801). 10.1515/picbe-2017-0084

Mende, M., Scott, M. L., van Doorn, J., Grewal, D., & Shanks, I. (2019). Service robots rising: How humanoid robots influence service experiences and elicit compensatory consumer responses. *JMR, Journal of Marketing Research*, *56*(4), 535–556. doi:10.1177/0022243718822827

Nadikattu, R. R. (2020). Implementation of new ways of artificial intelligence in sports. *Journal of Xidian University*, *14*(5), 5983–5997.

Nobanee, H., Al Hamadi, F. Y., Abdulaziz, F. A., Abukarsh, L. S., Alqahtani, A. F., AlSubaey, S. K., & Almansoori, H. A. (2021). A bibliometric analysis of sustainability and risk management. *Sustainability (Basel)*, *13*(6), 3277. doi:10.3390u13063277

Park, S. S., Tung, C. D., & Lee, H. (2021). The adoption of AI service robots: A comparison between credence and experience service settings. *Psychology & Marketing*, *38*(4), 691-703.

Paschen, J., Kietzmann, J., & Kietzmann, T. C. (2019). Artificial intelligence (AI) and its implications for market knowledge in B2B marketing. *Journal of Business and Industrial Marketing*, *34*(7), 1410–1419. doi:10.1108/JBIM-10-2018-0295

Pelau, C., Dabija, D. C., & Ene, I. (2021). What makes an AI device human-like? The role of interaction quality, empathy and perceived psychological anthropomorphic characteristics in the acceptance of artificial intelligence in the service industry. *Computers in Human Behavior*, *122*, 106855. doi:10.1016/j.chb.2021.106855

Petropoulos, G. (2018). The impact of artificial intelligence on employment. *Praise for Work in the Digital Age*, *119*, 121.

Qamar, Y., & Samad, T. A. (2022). Human resource analytics: A review and bibliometric analysis. *Personnel Review*, *51*(1), 251–283. doi:10.1108/PR-04-2020-0247

Saha, V., Mani, V., & Goyal, P. (2020). Emerging trends in the literature of value co-creation: A bibliometric analysis. *Benchmarking*, *27*(3), 981–1002. doi:10.1108/BIJ-07-2019-0342

Shalender, K., & Yadav, R. K. (2019). Strategic flexibility, manager personality, and firm performance: The case of Indian Automobile Industry. *Global Journal of Flexible Systems Managment*, *20*(1), 77–90. doi:10.100740171-018-0204-x

Sheth, A., Unnikrishnan, S., Bhasin, M., & Raj, A. (2021). *How India Shops Online 2021: A post pandemic view of online shopping*. https://www.bain.com/insights/how-india-shops-online-2021/

Sheth, A., Unnikrishnan, S., Bhasin, M., & Raj, A. (2022). *How India Shops Online 2022: An insight into the e-retail landscape and the emerging trends shaping the market.* https://www.bain.com/insights/how-india-shops-online-2022-report/

Sigelman, M., Bittle, S., Markow, W., & Francis, B. (2019). *The hybrid job economy: How new skills are rewriting the dna of the job market.* Burning Glass Technologies.

Singh, J., Flaherty, K., Sohi, R. S., Deeter-Schmelz, D., Habel, J., Le Meunier-FitzHugh, K., Malshe, A., Mullins, R., & Onyemah, V. (2019). Sales profession and professionals in the age of digitization and artificial intelligence technologies: Concepts, priorities, and questions. *Journal of Personal Selling & Sales Management, 39*(1), 2–22. doi:10.1080/08853134.2018.1557525

Soni, N., Sharma, E. K., Singh, N., & Kapoor, A. (2020). Artificial intelligence in business: From research and innovation to market deployment. *Procedia Computer Science, 167*, 2200–2210. doi:10.1016/j.procs.2020.03.272

Soni, V. D. (2020). Emerging roles of artificial intelligence in ecommerce. *International Journal of Trend in Scientific Research and Development, 4*(5), 223–225.

Tizhoosh, H. R., & Pantanowitz, L. (2018). Artificial intelligence and digital pathology: Challenges and opportunities. *Journal of Pathology Informatics, 9*(1), 38. doi:10.4103/jpi.jpi_53_18 PMID:30607305

Ülker, P., Ülker, M., & Karamustafa, K. (2023). Bibliometric analysis of bibliometric studies in the field of tourism and hospitality. *Journal of Hospitality and Tourism Insights, 6*(2), 797–818. doi:10.1108/JHTI-10-2021-0291

Van Doorn, J., Mende, M., Noble, S. M., Hulland, J., Ostrom, A. L., Grewal, D., & Petersen, J. A. (2017). Domo arigato Mr. Roboto: Emergence of automated social presence in organizational frontlines and customers' service experiences. *Journal of Service Research, 20*(1), 43–58. doi:10.1177/1094670516679272

Wijewickrema, M. (2023). A bibliometric study on library and information science and information systems literature during 2010–2019. *Library Hi Tech, 41*(2), 595–621. doi:10.1108/LHT-06-2021-0198

Wirtz, J., Patterson, P. G., Kunz, W. H., Gruber, T., Lu, V. N., Paluch, S., & Martins, A. (2018). Brave new world: Service robots in the frontline. *Journal of Service Management, 29*(5), 907–931. doi:10.1108/JOSM-04-2018-0119

Wu, Y., Farrukh, M., Raza, A., Meng, F., & Alam, I. (2021). Framing the evolution of the corporate social responsibility and environmental management journal. *Corporate Social Responsibility and Environmental Management, 28*(4), 1397–1411. doi:10.1002/csr.2127

Yin, J., & Qiu, X. (2021). AI technology and online purchase intention: Structural equation model based on perceived value. *Sustainability (Basel), 13*(10), 5671. doi:10.3390u13105671

Chapter 5
The Digital Transformation:
Crafting Customer Engagement Strategies for Success

Gajalakshmi N. S. Yadav
https://orcid.org/0009-0001-7890-753X
Christ University, India

R. Seranmadevi
https://orcid.org/0000-0002-4559-4100
Christ University, India

ABSTRACT

In the business world, the digital transformation has ushered in an era of unprecedented change. The explosive growth of digital technology over the past three decades has profoundly changed how businesses operate, compete, and engage with their customers. Businesses across industries have been forced to navigate this constantly changing environment due to the adoption of cloud computing, data analytics, the growth of e-commerce, and the rise of artificial intelligence. Organizations are forced to reconsider their strategies, operations, and consumer engagement models as a result of this significant instability. This chapter discusses the role of artificial intelligence in aiding customised and personalized marketing strategies, acknowledging the diverse preferences and behaviour of consumers. It shares insights on branding and customer management in an AI-driven environment. It also emphasises on online and technology-mediated customer engagement.

1. INTRODUCTION

Digital transformation is critical in enhancing customer experience. Being in sync with the latest technology helps brands align with customer requirements, leading to increased customer satisfaction and experience. Digital transformation in business creates new channels of communication. Emails, mobile applications, chatbots, and social media are modern digital communication channels activated in transi-

DOI: 10.4018/979-8-3693-1762-4.ch005

tioning companies. Digital technologies are acting as a game changer in the business. They are not only enhancing revenues but also playing a crucial role in building loyal customers.

Digital technologies are without any doubt becoming more and more prevalent in our daily lives (Colbert, Yee, & George, 2016). Since the introduction of the Internet at the beginning of the new millennium, the range of options that digital technologies provide and, consequently, their influence on people, businesses, and society, have increased. The nature of digital technologies has changed over time. Due to factors including growing computational power, lower computing costs, the availability of enormous amounts of data, and the development of machine learning algorithms and models, artificial intelligence (AI) in marketing is rapidly becoming more and more significant. AI is widely used in many different marketing areas.

Expert digital leaders also stress the importance of tackling digital transformation jointly using exploratory methods and precise, even measurable, targets. These leaders stand out due to their emphasis on delivering measurable results and introducing and disseminating innovative techniques within the organization.

2. BACKGROUND TO THE STUDY

The CE (customer engagement) can be defined as "A consumer's positively valenced brand-related cognitive, emotional, and behavioral activity during or related to focal consumer/brand interactions". Islam and Rahman (2016) have also similarly defined CE as "the customer's willingness to actively engage and communicate with the focal object (e.g., brand, organization, community, website, or organizational activity), [which] varies in magnitude (high/low) and direction (positive/negative) based on the type of customer interaction with various touch-points (physical/virtual)." CE has acquired a central position in the recent marketing literature.

Customers now have nearly complete brand transparency as a result of the advent of the digital era. Customers are so empowered in this increasingly networked world that everything can be replaced for them, including functions, services, content, and products. This shift is being driven by the emergence of interactive Web 2.0 as well as technical developments like artificial intelligence (AI), big data, augmented reality (AR), and the Internet of Things (IoT). These innovations have become an essential part of consumers' lives, coupled with various social networking platforms. Today's tech-savvy consumers spend a significant amount of time online due to the increasingly interactive nature of the environment. To engage consumers with their businesses, marketers are consequently investing heavily in offering seamless digital experiences and quick, personalized solutions.

In the last several years, the online CE landscape has expanded quickly, and consumers have moved from being "passive information receivers" to active "value co-creators" who participate in the creation of new products and services. This participation of consumers would bring a paradigm shift in business operations shortly.

In the next paragraphs, we try to look into the importance of digital transformation in the current scenario of the business.

2.1 Importance of Digital Transformation in Today's Business Landscape

1. **Competitive edge**: Gaining a competitive edge is possible for businesses that embrace digital transformation. They can react to market changes more quickly, develop more quickly, and provide stronger consumer experiences, which puts them ahead of less digitally advanced competition.
2. **Enhanced Customer Experience**: In today's increasingly digital environment, customers need quick, convenient, and seamless interactions with brands. By streamlining customer journeys, delivering specialized goods, and enhancing customer support via digital channels, digital transformation enables businesses to satisfy these expectations.
3. **Productivity and Efficiency**: Automation and digital tools simplify internal operations, lowering the effort and errors associated with manual labor. As a result, businesses have improved operational effectiveness, lower expenses, and higher overall productivity.
4. **Data-Driven Decision-Making**: The digital transformation generates massive amounts of data that can be utilized for developing insight. Businesses can employ artificial intelligence and data analytics to create more data-driven decisions that will improve their strategy, marketing initiatives, and product development.
5. **Global Reach**: Due to digital technologies, companies may reach a wider audience. Without making substantial expenditures in physical infrastructure, e-commerce, online advertising, and digital marketing strategies assist businesses in expanding into new markets and client groups.
6. **Innovation and agility**: Digital transformation encourages experimentation and quick iterations, which drives creativity. Businesses that are agile and innovative may quickly test new ideas, adjust to shifting market conditions, and pivot as needed.
7. **Cyber security and risk management**: They are crucial as we become increasingly dependent on digital systems. To guard against data breaches and cyber threats, digital transformation entails investments in cyber security measures and risk management techniques.
8. **Supply Chain Optimisation**: Digital technologies improve supply chain visibility and control. As a result, there may be fewer interruptions to the supply chain, better inventory control, and enhanced logistics, all of which can increase the dependability of goods and services.
9. **Environmental Sustainability**: By lowering the need for physical resources and enabling more effective resource management, digital transformation can help sustainability objectives. For instance, the demand for physical office space and transportation can be reduced with the help of remote work and digital collaboration tools, which also lowers carbon emissions.
10. **Adaptation to Market Trends**: Market trends and customer preferences are subject to quick change. Businesses may adapt their strategy in response to these developments thanks to digital transformation, maintaining their long-term relevance and profitability.
11. **Compliance & Regulation**: Especially in areas like data privacy and cyber security, digital transformation helps businesses maintain compliance with ever-evolving regulations. To prevent legal and reputational problems, compliance with these rules is essential.
12. **Crisis Resilience**: The COVID-19 pandemic revealed the value of digital capabilities for maintaining business continuity. Strong digital infrastructures allowed businesses to adapt to distant work, sustain relationships with customers, and weather crisis-related disruptions.

3. OBJECTIVES OF THE STUDY

3.1. To explore the role of Artificial Intelligence in aiding customized and personalized marketing strategies.
3.2. To share insights on branding and customer management in an AI-driven environment.
3.3. To explain the potential of digital technologies and manage the transformation.
3.4. To assist practitioners in understanding the challenges in crafting customer engagement strategies using Artificial Intelligence and also future directions for researchers.

This study is novel in its contribution by increasing awareness of the fascinating field of CE in the rapidly changing digital environment by analyzing and synthesizing some noteworthy research themes and dimensions.

4. ROLE OF ARTIFICIAL INTELLIGENCE IN PERSONALIZED ENGAGEMENT MARKETING

To create robots that behave as though they are intelligent, John McCarthy, one of the founders of AI, was the first to define the phrase artificial intelligence (McCarthy, 1955). Artificial intelligence (AI) is defined as a variety of intelligent human behaviors that can be achieved artificially by a machine, system, or network, including perception, memory, emotion, judgment, reasoning, proof, recognition, understanding, communication, conception, thought, learning, forgetting, creating, etc. (Deyi et al. 2017).

Digital advertising and digital tactics, in general, can be optimized in a variety of ways, but solutions utilizing artificial intelligence in marketing can go much further in terms of in-depth data analysis on a broad scale. To have a better offer and solution, this technology enables the utilization of hidden Internet users' data in keyword searches, social network profiles, and other online data (Tjepkema, 2018). Marketing experts now can feed consumer profiles in response to this wonderful data. Artificial intelligence-based solutions offer a comprehensive perspective of online users and potential clients, enabling the correct message to be sent at the right moment and to the right person. The key is to gather information from each user's interaction.

The ability of artificial intelligence to organize and analyze massive amounts of content and consequently find trends is its key feature. With this strategy, marketers can continue to engage users in live, ongoing conversations or events online. Users' purchasing decisions are directly influenced by prompt communication with them. Using social media and other digital platforms strategically is also made possible by artificial intelligence (Tjepkema, 2018).

The Eldorado of futuristic marketing is ultimately artificial intelligence. Today, we must deal with the enormous marketing prospects provided by artificial intelligence. Utilising this technological advancement will therefore enable a certain progress in this discipline.

4.1 Applications of Artificial Intelligence in Social Media

Social networks collect so much data that it would be nearly difficult for a human to sort, analyze, or even make use of it. This is why using artificial intelligence effectively on social media is crucial.

a) Chatbots

On various platforms, including email applications, websites, and mobile applications, the chatbot is an artificial intelligence program that can sustain a conversation or discussion with a user using natural language (Dagnon, 2018; Frankenfield, 2018). Chatbot responses are highly advanced and extremely promising in human-machine interaction. On the technical front, however, chatbots are a simple evolution of a question-and-answer system built on natural language processing (Frankenfield, 2018). Chatbot applications make interactions between humans and machines less artificial, which enhances the user experience.

Chatbots can direct customers to brands and items in instant messaging programs or even accompany them while they navigate a website, creating a highly customized user experience with the brand. On the website, chatbots are also employed to initiate interactive conversations with users and provide assistance and follow-up. To help users through the conversion process, they are also included in order pages and contact pages (Frankenfield, 2018).

It is important to understand that marketing encompasses more than just bringing in new clients. The brand must also interact with Internet users. This work is well-suited for chatbots, which can also follow and analyze customer shopping history. With this overview of Internet user behavior, marketers may alter and retarget their digital campaigns whenever they see fit to take the recommendations given by the data gathered and boost conversion rates.

Despite significant technological advancements, chatbots cannot, however, take the place of people. Their sole purpose is to automate essential tasks so that marketing teams can concentrate on more creative work. Additionally, chatbots require frequent maintenance and updating.

Chatbots are customer service solutions that can handle minor issues. An intelligent chatbot will be able to determine when to transfer control to a person. As a result, chatbots are not stand-alone solutions but rather landing pages with a clear and flexible purpose (Dagnon, 2018).

b. Predictive Analytics

The term "predictive analytics" describes the application of statistics and machine learning to behavior analysis and prediction. Nevertheless, because we all follow patterns like waking up in the morning, cleaning our teeth, taking a shower, dressing, and eating breakfast, humans are fairly predictable. By using this feature of prediction, marketing experts can foresee future events and modify their marketing strategies accordingly (Stelzner, 2018).

Machines have developed the ability to make these predictions more precise since we are predictable and have general knowledge. They can be more effective and efficient, for instance, by understanding when the marketing department needs to do more live Facebook or spend less on advertising. Furthermore, marketing services that can predict can save a lot of time and money.

c. AI-Generated Content

Artificial intelligence creates content using rules, but for it to create a narrative around the facts it needs datasets like a match summary. For instance, creating reports can take a while. Artificial intelligence, on the other hand, can assist businesses in saving time and resources and encourage staff to concentrate on their most important duties.

The challenge is enormous since computers cannot respond on their own, even though the substance of AI appears to be expanding. Human assistance is desperately needed. Even if we incorporate these elements into its algorithms, artificial intelligence is not aware of human emotions, therefore a machine will not understand what we perceive as amusing (Kreimer, 2018).

The next section reveals insights on how to leverage Generative AI in building Winning Customer Experiences.

5. DESIGNING WINNING CUSTOMER EXPERIENCES WITH GENERATIVE AI

The focus must be on the customer and not on the technology.

1. Focus on the learning
   - **Recognition** - The initial phase in a customer journey is the recognition of a customer's need
   - **Request** - In the second phase of the customer journey, these user needs are translated into a Request.
   - **Respond** - Finally, the firm needs to respond to the customer.
   - **Repeat** - To transform a series of experiences into a deeper customer relationship
   - **Resilience** - continuous renewal of competitive advantages
2. Using the technology to complement a firm's capabilities and not as a substitute

Businesses must keep in mind that technology by itself cannot give firms a competitive edge, particularly not when it is widely accessible. The important challenge is how a company may use it in a way that makes it valuable and increases the willingness of its customers to pay for it, while simultaneously making it difficult for competitors to replicate it.

6. BRANDING AND CUSTOMER MANAGEMENT IN AN AI-DRIVEN ENVIRONMENT

6.1 Artificial Neural Networks and Brand Choice

The introduction of barcodes and scanners made it simple for retail establishments to monitor the purchasing patterns of their customers. The products that they have bought are described in detail in the customer data. These days, supermarkets and hypermarkets employ loyalty cards to track individual consumers' preferences and anticipate their future purchasing behavior. This forecast aids in the retail sector's profit maximization. According to Kaya et al. (2010), it is a nonlinear statistical model designed for the human brain that can be educated and programmed to recognize patterns in data. Artificial neural network applications are utilized in the market to optimize corporate operations. Research suggests that the implementation of basic artificial neural networks can improve the precision of consumer preferences, predict their assessment of a brand, and prevent the use of fake data.

6.2 User Generated Content (UGC): Brand Sustainability

Customers posting their branding hashtags enhances the company's return on investment, according to the MIT Sloan Management Review, which states that trade and commerce are urged to reorganize social web network business strategies of brands (Hoffman & Fodor, 2010). Researchers have looked into the motivations behind user promotion and framing of brand-aligned content posts on social media platforms to determine the social interaction and personal identity of users or customers (Mayrhofer et al., 2020). A lot of people post images of themselves with display brands to social media to promote their brand recognition and image (Muntinga et al., 2011). According to Sung et al. (2018), these frequent posts are regarded as brand selfies and have raised user post-exposure on social media. Moreover, content analysis encourages social media to offer social media integration, personal identification, and social collaboration of the customers or users to help the brand sustainability (Smith et al., 2012). The last fifty years have seen a rising and changing theme in sustainability. The ability of the brands to hold onto or improve their marketplace rankings through the benefits of combining data with empathy with rival brands serves as validation for this brand's sustainability. Because of this, an important indicator of a brand's sustainability is its average growth rate (Schultz & Block, 2015).

6.3 Brand Intelligence Analytics

In the AI-driven world of today, several web applications are made to use textual data analysis to assess the importance, positioning, and image of brands (Colladon, 2020). To quantify brand visibility, an app is available that computes the Semantic Brand Score (SBS) (Colladon, 2018). With the help of big data, managers may monitor the positive or negative associations that consumers have with businesses by reviewing developments in consumer trends and attitudes (Colladon, 2020).

The SBS is a novel idea that uses consumer data analytics to uncover brand significance (Colladon, 2020). It is designed to determine the relative importance of several brands by taking into account dynamic longitudinal patterns and using data from various web sources from various angles. This metric is suitable for data analysis on various languages and cultures. According to Keller (2016), these SBS principles mostly assisted well-known brands in building brand equity through brand identification and awareness.

A brand could be a politician or someone, or it could be an idea that uses a few popular terms to build brand value. The SBS is used, among other things, to estimate share market trends, assess competitors, and assess the dynamics of changeover when a new brand replaces an established one (Colladon, 2018). Furthermore, three dimensions make up the SBS: generality, variety, and connection. Generality gauges how frequently a brand name is used, which contributes to brand recall and awareness. Diversity pertains to the assortment of texts associated with a brand, such as lexical embedding. The brand's capacity to create links between different texts is aided by connectivity (Colladon et al., 2020). Finally, further capability in SBS for brand enrichment and diversification of adjustments in the prevalence, diversity, and connectedness that might characterize the new brand metrics could be taken into account in future research.

6.4 Digital Innovations and Brand Excellence

Technology-enabled digital breakthroughs produced a value chain for sustainable development. One of the sustainable development goals can be accomplished by reducing waste through the widespread use of digital technologies, even in the manufacturing sector (Nagel, 2019). Digital innovations encompass various domains such as big data, digital transaction platforms, online and mobile advertising, online privacy, online reviews, and the significance of retail analytics. Additional digital advances support firms in their marketing strategies through online advertising, and consumers receive algorithmic recommendations from a variety of platforms, including Google, Amazon, Spotify, and many more (Ratchford, 2019).

Digital innovations and brand excellence are multifaceted phenomena that exist across all domains of the gig or digital economy. To gain a competitive edge, it combines blockchain technology, cryptocurrencies, and artificial intelligence. To attain the companies' brand excellence, this phenomenon works in concert with data sharing, collaborations, platforms, and acquisitions. To maximize brand quality, digital technologies are incorporated into several industries, including the retail, automotive, and agri-food sectors (OECD, 2020). In addition to these, future research directions suggest conducting studies on digital innovation specifically in the service, software, and new start-up sectors for organizational excellence and brand management from the perspective of business optimization.

Automation makes it possible to provide clients with personalized goods and services based on their brand preferences on online discussion boards and social media. Engagements with customers improve the image of the company. Additionally, AI helps businesses identify customer data through voice recognition, social media interaction, gestures, and facial expressions. Emotional appeal also enables businesses to comprehend the preferences of their customers.

AI also helps the business with segmentation, targeting, and postponement to comprehend the diversity of consumer tastes and preferences. It also suggests purchasing decisions. AI applications in strategic planning raise brand value while increasing customer satisfaction. AI support helps businesses develop new products by finding gaps, customizing, and analyzing social media posts that are trending.

Marketers can use dynamic competitive pricing to help brands compare with competitors by utilizing big data. Additionally, by using robots and drones to support intense distribution channels, marketers enable AI to improve customer service and foster brand loyalty. According to Campbell et al. (2020), artificial intelligence has offered a significant opportunity for integrated marketing communication to improve firms' visibility in front of consumers through the use of contextual ad targeting, retargeting, keyword bidding, and product customization.

Furthermore, both short- and long-term AI predictions boost brand and customer loyalty. In summary, artificial intelligence (AI) helps with personalized products by generating exceptional brand experiences and values, offering more offers at dynamic prices, delivering high-quality services, and selecting the appropriate advertisements to elicit positive feelings from consumers. According to Kumar et al. (2019), long-term company prediction improves customer relationship management, retention, and the growth of brand values and customer equity. Predictive analytics can be used by businesses to develop marketing trends such as interactive bots, recommendations, and real-time optimizations to establish their brand value in the international marketplace (Ma & Sun, 2020).

7. ONLINE AND TECHNOLOGY-MEDIATED CUSTOMER ENGAGEMENT

7.1 Demystifying the Potential of Digital Technologies

More global, collaborative, and open activities are made possible by digital technologies (Bogers, Chesbrough, & Moedas, 2018). Although the promise of digital technologies is widely anticipated, determining the extent and significance of the digital economy presents a difficult task (Brynjolfsson & Collis, 2019). In the digital age, businesses operate in a setting where ongoing connectedness both facilitates and necessitates increased consumer and collaborator engagement and where access to resources frequently takes the place of ownership (McGrath, 2020). Moreover, not every company, procedure, or business model needs a digital transformation, and many executives especially those working for publicly traded firms have little interest in fundamentally altering their firms (Andriole, 2017).

Building on distinctive features like its distributed database and irreversible records, the blockchain has the potential to lower transaction costs, make it easier to access outside resources, or allow for the efficient exchange and management of property rights (Felin & Lakhani, 2018; Pedersen, Risius, & Beck, 2019; Tapscott & Tapscott, 2017). The cost-effective management of IT infrastructure that cloud solutions provide makes them a good substitute for internal IT provision in small and medium-sized businesses (SMEs) (Lacity & Reynolds, 2014; McAfee, 2011). Data analytics and data-enabled learning become a competitive advantage for many businesses as a result of the increasing availability of data and information (Barton & Court, 2012; Hagiu & Wright, 2020). Businesses can generate real-time data with contextual information because of the Internet of Things (IoT), which is backed by intelligent, linked products and devices (Gandhi & Gervet, 2016). As a result, there are more opportunities for cross-functional collaboration, improved coordination of complex processes, and new kinds of linkages and interactions (Porter & Heppelmann, 2014, 2015). AI amplifies this impact even more, providing businesses with previously unavailable opportunities to automate procedures, gain cognitive insights, and develop cognitive enhancement (Watson, 2017; Tarafadar, Beath, & Ross, 2019).

The majority of AI-related initiatives, at least up to this point, have focused on implementing AI within the existing business (Brock & von Wangenheim, 2019), raising concerns about the possibly exaggerated expectations about the technology's capabilities that emerge (Gerbert & Spira, 2019), despite the technology's predicted revolutionary and transformative impact (Iansiti & Lakhani, 2020; Brock & von Wangenheim, 2019). Businesses must adapt with connected strategies to enhance the customer experience as consumers use mobile technology to stay connected all the time (Siggelkow & Terwiesch, 2019; Stieglitz & Brockmann, 2012). Additionally, social media platforms serve as new sources of corporate value while also competing with existing technologies for users' time and attention as tools for social networking and accessing digital information (Culnan, McGough, & Zubilaga, 2010; Kane, 2015).

7.2 Managing the Transformation

Digital technologies are no different from other technologies in terms of acceptance and dissemination; this is a process that is unique to the technology and the environment and is not always simple. For example, developing a sense of justice, accountability, and transparency as well as a realistic understanding of the technology's potential have emerged as critical elements in fostering technology adoption to improve the usage of AI (Davenport, 2019). Additionally, this may call for investments in leaders' and employees' education related to AI as well as senior leadership participation in communicating the necessity of working with AI (Fountaine, McCarthy, & Saleh, 2019).

In general, digital transformation is an iterative process that frequently calls for quick adaption using a participatory approach (Hansen, Kraemmergaard, & Mathiassen, 2011; Smith & Watson, 2019). Moreover, the procedure is frequently more incremental since businesses first concentrate on a small number of well-chosen initiatives before expanding (Davenport & Ronanki, 2018). A successful digital transformation requires six interconnected phases, according to Leonardi (2020): the leadership sells the transformation, employees adopt it, they choose how to use it, data alters employee behavior, local performance improves, and local performance is in line with corporate objectives. Fundamentally, digital transformation is a business transformation with an emphasis on improving operational efficiency and customer experience (Weill & Woerner, 2018). Although the components of digital transformation are not universally understood, most studies identify operational procedures, customer experience, business models, performance management, the workplace, mindset and skills, and a company's IT function as important components (El Sawy, Kraemmergaard, Amsinck, & Lerbech Vincent, 2016; Gurbaxani & Dunkle, 2019; Matt, Hess, Benlian, & Wiesbock, 2016; Westerman, Bonnet, & McAfee, 2014). A well-defined plan, along with an operational foundation and organizational culture, are essential for effectively navigating a digital transition (Iansiti & Lakhani, 2020; Sebastian et al., 2017). Because of the rapid pace of change, businesses must remain vigilant and responsive, even to subtle cues from their surroundings (Venkatraman, 2019).

Businesses that concentrate on servicing new consumer groups rather than just continuing to serve current customers or concentrating on cost-cutting are among those that have successfully responded to digital disruption (Bughin & van Zeebroeck, 2017). Applying the digital lens to the company's current product and service offering is one way to approach digital initiatives (Iansiti & Lakhani, 2014). Digital-physical mashups, which enable a company's customers to take advantage of the benefits of both the digital and physical spaces—such as rich product information, online reviews, and price comparison—are a common feature of successful digitization initiatives (Rigby, 2011; Rigby, 2014). Examples of these advantages include event/experience, testing/trying on, and personal assistance. Hybrid product offerings, which integrate new and existing technologies, frequently assist businesses in understanding and effectively managing the uncertainties associated with new technology (Furr & Snow, 2015).

To develop new value-added services, the company must know who in its ecosystem it needs to involve outside of its boundaries (Vaia, Carmel, Trautsch, DeLone, & Menichetti, 2012). The most challenging aspect of digital transformation for businesses is frequently the necessary internal cultural shift towards interdisciplinary collaboration in a quick-thinking, flexible, and experimentation-friendly atmosphere (Ibarra, 2019; Westerman, Soule, & Eswaran, 2019). In particular, there should be no interruptions in the difficult partnership between business and IT (Dremel, Herterich, Wulf, Waizmann, & Brenner, 2017). A template and business-driven strategy with a matrix organization, strict supplier steering, and cascaded planning proved effective in the context of industrialized digital transformation (Winkler & Kettunen, 2018). Additionally, thinking in code and making code the norm within the company, along with regular standardization and easy-to-use analytical tools, are likely to boost positive outcomes and facilitate the digital transformation process (Barton & Court, 2012; Walter, 2019; Wixom, Yen, & Relich, 2013). From an operational standpoint, businesses may be able to use digital technologies to lessen the inherent complexity of their offering of goods and services, which could allow them to increase the variety of what they have to offer without heading over the sweet spot for complexity (Mocker, Weill, & Woerner, 2014).

7.3 Managerial Implications for Strategy and Business Models

Strong leadership, an agile and scalable core, and a distinct focus on either a customer interaction strategy or a digitalized solutions strategy are essential components of a successful digital business strategy (Ross, Sebastian, & Beath, 2017). A digital business strategy needs constant navigation of the dynamic and evolving digital landscape, as demonstrated by DBS Bank in Singapore. It leverages the wealth of information to create new value for customers (Sia, Soh, & Weill, 2016).

Grover, Kohli, and Ramanlal (2018) advise managers to do a comprehensive sociotechnical, strategic assessment for each digitization project to ascertain the firm's preparedness for digital products as well as any possible customer consequences. Businesses must immediately overcome the hurdles posed by legacy IT infrastructure if they are to succeed in the digital environment, as everything their rivals do with data is scalable, defendable, and replicable (Wessel, Levie, & Siegel, 2016). Algorithms powered by AI offer a lot of potential for helping businesses streamline their digital strategies. For instance, Alibaba uses an algorithmic self-tuning technique to figure out what works and modify and form its strategy (Reeves, Zeng, & Venjara, 2015). However, managers cannot rely solely on algorithms, even while technologies may offer valuable insights. Rather, managers must continue to concentrate on how they apply the knowledge obtained from the data provided to the technology by their strategic goals and by taking the data's long-term consequences into account (Luca, Kleinberg, & Mullainathan, 2016).

Despite all of the big data's potential, very few businesses can fully utilise the data that is already embedded in their operating systems (Parmar, Mackenzie, Cohn, & Gann, 2014). Businesses must adapt their decision-making culture to take advantage of data; senior managers must embrace evidence-based decision-making (Ross, Beath, & Quaadgras, 2013); they must also combine data management approaches to create new roles that are analytics-focused and establish policies for handling the challenges that come with digital transformation (Davenport, 2013). There needs to be coordination among leaders at all levels and an emphasis on the human aspects of analytics (Davenport, 2014). When it comes to finding a balance between defensive data management (such as security and governance) and offensive data management (such as predictive analytics), businesses should think about creating a cogent strategy and determining the true value of the data they discard (DalleMule & Davenport, 2017). According to Short and Todd (2017), data value is the sum of three different types of value: the value of an asset or stock, the value of an activity, and the value that is anticipated or future. Companies should focus on data flows rather than stocks and use data scientists in addition to product and process developers rather than data analysts if they want to take advantage of big data (Davenport, Barth, & Bean, 2012).

Businesses typically use one of three strategies to monetize their data: selling information offerings to both new and existing markets, enhancing internal corporate processes and decision-making, and enclosing information around key products and services (Wixom & Ross, 2017). Watson, Boudreau, Li, and Levis (2010), for instance, demonstrate how UPS reduced mileage, reduced pollutants, and increased safety, and maintenance expenses by collecting data from its trucks' proprietary firmware.

Managers should evaluate whether the model would enable the company to take advantage of scale and aggregation effects, provide a better buyer or seller experience, and address market uncertainties by examining the continuum between a pure reseller and a pure multisided platform (Hagiu & Wright, 2013). Hagiu and Rothman (2016) address issues encountered in creating and growing online markets, including managing disintermediation, maintaining trust, establishing business models, and interacting with regulators at an early enough stage. They illustrate these points with examples from eBay, Lending Club, and Airbnb.

Network effects, clustering, danger of disintermediation, vulnerability to multi-homing, and bridging to different networks are the five fundamental aspects of networks that platform managers depend on to succeed (Zhu & Lansiti, 2019). Incumbent platforms may pursue legal action to try to enforce existing regulations and capitalize on their strengths to compete with emerging platforms (Edelman & Geradin, 2016). The other issue is that a disproportionately large and increasing amount of the value being created is being captured by a small number of digital superpowers (Iansiti & Lakhani, 2017b). By emphasizing what they do better than their rivals, successful platform dethroners can create a potent kind of distinction (Suarez & Kirtley, 2012).

To introduce AI to the market, start-ups can employ a variety of business model paradigms, including supported intelligence (such as image scan analysis), augmented intelligence (such as precision medicine), and autonomous intelligence (such as doctorless hospitals) (Garbuio & Lin, 2019). According to Mandviwalla and Watson (2014), Businesses can use social media to raise financing by using insights and interacting with the factors that influence the production of human, social, and symbolic capital; data-driven analysis and decision-making; and knowledge generation and dissemination to produce and advance human and organizational capital.

8. CHALLENGES IN CRAFTING CUSTOMER ENGAGEMENT STRATEGIES USING ARTIFICIAL INTELLIGENCE

a) Data privacy and security concerns

Consumers' concerns around data security and privacy are growing. AI primarily depends on data, and improper or careless use of consumer data can have negative effects on trust and even legal repercussions. It is vital to strike the right balance between privacy and personalization.

b) Bias and fairness

Unintentionally maintaining biases found in training data can result in unfair or discriminatory outcomes for AI systems. To prevent perpetuating preexisting prejudices, it is crucial to make sure AI systems are trained on a wide range of representative datasets.

c) Transparency and explainability

Deep learning models in particular are frequently viewed as "black boxes" that are challenging to understand. Customers may become concerned if there is a lack of explainability and openness. Gaining the trust of customers requires developing transparent models and offering justifications for judgments made by artificial intelligence.

d) Integration with human touch

Even though AI can automate a lot of operations, customers frequently still prefer human interaction in some situations. Maintaining a pleasant customer experience requires finding the ideal blend between automated and human-led engagement.

e) Adoption and acceptance

Consumers may be distrustful of new technology, such as interactions driven by AI. Gaining customers' acceptance and confidence in AI technologies might be complicated. Businesses must explain the worth and advantages of AI without offending customers who stick with traditional channels of communication.

f) Continuous learning and adaptation

Over time, customer preferences and behaviors change. For AI models to remain relevant, they must be periodically trained and modified. It is a constant challenge to update AI systems to meet evolving customer requirements.

g) Scalability

The number of client interactions rises along with the growth of organizations. One of the biggest challenges is making sure AI systems can scale successfully to handle a lot of users and transactions without losing efficiency.

h) Technology integration

Many organizations already have established procedures as well as technologies in place. It can be difficult to integrate AI with these systems in a seamless manner. It is necessary to solve compatibility problems and the requirement for strong integration solutions.

i) Regulatory compliance

The legal environment of AI is changing. Companies must adhere to privacy and data protection laws, which can differ depending on their place of business. One of the challenges in integrating AI into consumer engagement is navigating these complex regulatory settings.

j) Cost and resource constraints

AI solution implementation can require a lot of resources, both monetary and human. The implementation of AI may provide difficulties for smaller firms because of budgetary constraints and a lack of AI technology proficiency.

To successfully navigate these obstacles, a comprehensive strategy is needed, including careful planning, ethical considerations, and continual monitoring and adaptation to make sure that AI-driven customer engagement strategies meet customer expectations as well as corporate objectives.

9. FUTURE DIRECTIONS FOR RESEARCHERS

Researchers might look into how interactive communication with customers across several touchpoints might increase their engagement and enable them to co-create value with the application of Artificial Intelligence. Furthermore, further research is required to determine how businesses might enhance CE to boost profitability through the formulation of effective CE strategies with the help of Artificial Intelligence. Thus, another interesting topic that needs academic research is how to handle customer retention to increase the share of wallet and encourage customers to pay a premium price through the application of digital innovations. In light of this, further study is encouraged to examine whether and how a company's Corporate Social Responsibility efforts contribute to consumer engagement by promoting favorable perceptions of the company by reflecting their activities through Social media platforms. Finally, it is necessary to conduct an empirical investigation into the directionality and strength of CE as well as its conceptual connections to more contemporary psychological conceptions like brand love and brand hatred by integrating digital transformation tools.

CONCLUSION

In general, digital technology enables businesses to forge closer bonds with their customers (Siggelkow & Terwiesch, 2019) and to strategically craft new product offerings, such as social coupons (Kumar & Rajan, 2012). Businesses are competing more and more in customer journeys with context-based interaction, automation, personalization, and continuous innovation (Edelman, 2015). To help clients navigate through online options, businesses must develop choice engines and foster empathy and emotional relationships with them (Agarwal & Weill, 2012; Thaler & Tucker, 2013). Beyond merely answering simple questions, virtual assistants can offer focused, advanced support (Nili, Barros, & Tate, 2019). Additionally, businesses should think about marketing to AI platforms, since AI assistants are starting to replace customers' trusted advisors (Dawar & Bendle, 2018).

Moreover, analytics and data have assumed a central role in these efforts. By utilizing artificial intelligence, big data, and predictive analytics, companies can gain insights that facilitate more intelligent decision-making and customized customer experiences. This ultimately results in a stronger brand reputation, more committed customers, and steady growth. Businesses need to be flexible and agile as the digital world changes, eager to adopt new trends and technology. The concept of customer interaction is fundamentally versatile, as it is formed by both changing market circumstances and customer feedback. As such, continual assessment and strategy modification is essential.

In the current digital era, characterized by intense competition and elevated consumer expectations, companies that place a high priority on developing efficacious customer engagement tactics will not only endure but also prosper. The secret to success is a blend of data, technology, and a customer-centric mindset. As we proceed, we must keep in mind that the road to digital transformation is a continuous one and that the destination is characterized by a strong and long-lasting relationship with the consumer, fuelled by creativity, keen observation, and an uncompromising dedication to quality.

REFERENCES

Agarwal, R., & Weill, P. (2012). The benefits of combining data with empathy. *MIT Sloan Management Review*, *54*(1), 35.

Andriole, S. J. (2017). Five myths about digital transformation. *MIT Sloan Management Review*, *58*(3), 20–22.

Barton, D., & Court, D. (2012). Making advanced analytics work for you. A practical guide to capitalizing on big data. *Harvard Business Review*, *90*(10), 79–83. PMID:23074867

Benabdelouahed, R., & Dakouan, D. (2020). The Use of Artificial Intelligence in Social Media: Opportunities and Perspectives. *Expert Journal of Marketing*, *8*(1), 82–87.

Bogers, M., Chesbrough, H., & Moedas, C. (2018). Open innovation: Research, practices, and policies. *California Management Review*, *60*(2), 5–16. doi:10.1177/0008125617745086

Brock, J. K.-U., & von Wangenheim, F. (2019). Demystifying AI: What digital transformation leaders can teach you about realistic artificial intelligence. *California Management Review*, *61*(4), 110–134. doi:10.1177/1536504219865226

Brynjolfsson, E., & Collis, A. (2019). How should we measure the digital economy? *Harvard Business Review*, *97*(6), 140–148.

Campbell, C., Sands, S., Ferraro, C., Tsao, H. Y. J., & Mavrommatis, A. (2020). From data to action: How marketers can leverage AI. *Business Horizons*, *63*(2), 227–243. doi:10.1016/j.bushor.2019.12.002

Colbert, A., Yee, N., & George, G. (2016). From the editors: The digital workforce and the workplace of the future. Academy of Management Journal, 59(3), 731–739.

) Colladon, A. F. (2018). The semantic brand score. *Journal of Business Research, 88*, 150–160. doi:. jbusres.2018.03.026. doi:10.1016/j

Colladon, A. F. (2020). Forecasting election results by studying brand importance in online news. *International Journal of Forecasting*, *36*(2), 414–427. doi:10.1016/j.ijforecast.2019.05.013

Culnan, M. J., McHugh, P. J., & Zubilaga, J. I. (2010). How large U.S. companies can use Twitter and other social media to gain business value. *MIS Quarterly Executive, 9*(4), 243–259.

Dagnon, S. (2018). *Using Chatbots for Social Media Marketing*. Available at: https://mavsocial.com/chatbots-social-media-marketing/

Davenport, T. H. (2019). Can we solve AI's "trust problem"? *MIT Sloan Management Review*, *60*(2), 1–5.

Dawar, N., & Bendle, N. (2018). Marketing in the age of Alexa. *Harvard Business Review*, *96*(3), 80–86.

Dremel, C., Herterich, M., Wulf, J., Waizmann, J.-C., & Brenner, W. (2017). How AUDI AG established big data analytics in its digital transformation. *MIS Quarterly Executive*, *16*(2), 81–100.

El Sawy, O. A., Kraemmergaard, P., Amsinck, H., & Lerbech Vinther, A. (2016). How LEGO built the foundations and enterprise capabilities for digital leadership. *MIS Quarterly Executive*, *15*(2), 141–166.

Felin, T., & Lakhani, K. (2018). What problems will you solve with blockchain? *MIT Sloan Management Review*, *60*(1), 32–38.

Fountaine, T., McCarthy, B., & Saleh, T. (2019). Building the AI-powered organization. *Harvard Business Review*, *97*(4), 62–73.

FrankenfieldJ. (2018). *Chatbot*. Available at: https://www.investopedia.com/terms/c/chatbot.asp

Gandhi, S., & Gervet, E. (2016). Now that your products can talk, what will they tell you? *MIT Sloan Management Review*, *57*(3), 49–50.

Gerbert, P., & Spira, M. (2019). Learning to love the AI bubble. *MIT Sloan Management Review*, *60*(4), 8–10.

Gurbaxani, V., & Dunkle, D. (2019). Gearing up for successful digital transformation. *MIS Quarterly Executive*, *18*(3), 209–220. doi:10.17705/2msqe.00017

Hagiu, A., & Wright, J. (2020). When data creates a competitive advantage. *Harvard Business Review, 98*(1), 94–101.

Hansen, A. M., Kraemmergaard, P., & Mathiassen, L. (2011). Rapid adaptation in digital transformation: A participatory process for engaging IS and business leaders. *MIS Quarterly Executive, 10*(4), 175–185.

Hoffman, D. L., & Fodor, M. (2010). Can you measure the ROI of your social media marketing? *MIT Sloan Management Review, 52*(1), 41.

Hollebeek, L. D., Glynn, M. S., & Brodie, R. J. (2014). Consumer brand engagement in social media: Conceptualization, scale development and validation. *Journal of Interactive Marketing, 28*(2), 149–165. doi:10.1016/j.intmar.2013.12.002

Iansiti, M., & Lakhani, K. R. (2020). Competing in the age of AI. *Harvard Business Review, 98*(1), 60–67.

Kane, G. C. (2015). Enterprise social media: Current capabilities and future possibilities. *MIS Quarterly Executive, 14*(1), 1–16.

Kaya, T., Aktas, E., Topçu, İ., & Ulengin, B. (2010). Modeling toothpaste brand choice: An empirical comparison of artificial neural networks and multinomial probit model. *International Journal of Computational Intelligence Systems, 3*(5), 674–687. doi:10.2991/ijcis.2010.3.5.15

Kreimer, I. (2018). *How to Get Started with AI-Powered Content Marketing*. Available at: https://www.singlegrain.com/artificial-intelligence/how-to-get-started-with-ai-powered-content-marketing/

Kumar, V., Rajan, B., Venkatesan, R., & Lecinski, J. (2019). Understanding the role of artificial intelligence in personalized engagement marketing. *California Management Review, 61*(4), 135–155. doi:10.1177/0008125619859317

Lacity, M. C., & Reynolds, P. (2014). Cloud services practices for small and medium-sized enterprises. *MIS Quarterly Executive, 13*(1), 31–44.

Ma, L., & Sun, B. (2020). Machine learning and AI in marketing–Connecting computing power to human insights. *International Journal of Research in Marketing, 37*(3), 481–504. doi:10.1016/j.ijresmar.2020.04.005

Matt, C., Hess, T., Benlian, A., & Wiesbock, F. (2016). Options for formulating a digital transformation strategy. *MIS Quarterly Executive, 15*(2), 123–139.

Mayrhofer, M., Matthes, J., Einwiller, S., & Naderer, B. (2020). User-generated content presenting brands on social media increases young adults' purchase intention. *International Journal of Advertising, 39*(1), 166–186. doi:10.1080/02650487.2019.1596447

McAfee, A. (2011). What every CEO needs to know about the cloud. *Harvard Business Review, 89*(11), 124–132.

Mocker, M., Weill, P., & Woerner, S. L. (2014). Revisiting complexity in the digital age. *MIT Sloan Management Review, 55*(4), 73–81.

Muntinga, D. G., Moorman, M., & Smit, E. G. (2011). Introducing COBRAs: Exploring motivations for brand-related social media use. *International Journal of Advertising, 30*(1), 13–46. doi:10.2501/IJA-30-1-013-046

Nagel, M. (2019). *Exploring digital innovations: Mapping 3D printing within the textile and sportswear industry*. https://www.divaportal.org/smash/get/diva2:1369648/FULLTEXT01.pdf

Nili, A., Barros, A., & Tate, M. (2019). The public sector can teach us a lot about digitizing customer service. *MIT Sloan Management Review, 60*(2), 84–87.

OECD. (2020). *The Digital of Science, Technology and Innovation: Key Developments and Policies*. https://www.oecd.org/science/inno/the-digitalisation-of-science-technology-and-innovation-b9e4a2c0-en.htm

Pedersen, A. B., Risius, M., & Beck, R. (2019). A ten-step decision path to determine when to use blockchain technologies. *MIS Quarterly Executive, 18*(2), 1–17. doi:10.17705/2msqe.00010

Porter, M. E., & Heppelmann, J. E. (2014). How smart, connected products are transforming competition. *Harvard Business Review, 92*(11), 64–88.

Porter, M. E., & Heppelmann, J. E. (2015). How smart, connected products are transforming companies. *Harvard Business Review, 93*(10), 96–116.

Rasool, A., Shah, F. A., & Islam, J. U. (2020). Customer engagement in the digital age: A review and research agenda. *Current Opinion in Psychology, 36*, 96–100. doi:10.1016/j.copsyc.2020.05.003 PMID:32599394

Ratchford, B. T. (2019). The impact of digital innovations on marketing and consumers. *Marketing in a Digital World, 16*, 35–61. doi:10.1108/S1548-643520190000016005

Ross, J. W., Sebastian, I. M., & Beath, C. M. (2017). How to develop a great digital strategy. *MIT Sloan Management Review, 58*(2), 7–9.

Schneider, S., & Kokshagina, O. (2021). Digital transformation: What we have learned (thus far) and what is next. *Creativity and Innovation Management, 30*(2), 384–411. doi:10.1111/caim.12414

Schultz, D. E., & Block, M. P. (2015). Beyond brand loyalty: Brand sustainability. *Journal of Marketing Communications, 21*(5), 340–355. doi:10.1080/13527266.2013.821227

Smith, A. N., Fischer, E., & Yongjian, C. (2012). How does brand-related user-generated content differ across YouTube, Facebook, and Twitter? *Journal of Interactive Marketing, 26*(2), 102–113. doi:10.1016/j.intmar.2012.01.002

Stelzner, M. (2018). *Predictive Analytics: How Marketers Can Improve Future Activities*. Available at: https://www.socialmediaexaminer.com/predictive-analytics-how-marketers-can-improve-future activities-Chris-Penn/

Stieglitz, S., & Brockmann, T. (2012). Increasing organizational performance by transforming into a mobile enterprise. *MIS Quarterly Executive, 11*(4), 189–204.

Suarez, F. F., & Kirtley, J. (2012). Dethroning an established platform. *MIT Sloan Management Review, 53*(4), 35–41.

Sung, Y., Kim, E., & Choi, S. M. (2018). Me and brands: Understanding brand-selfie posters on social media. *International Journal of Advertising*, *37*(1), 14–28. doi:10.1080/02650487.2017.1368859

Tapscott, D., & Tapscott, A. (2017). How blockchain will change organizations. *MIT Sloan Management Review*, *58*(2), 10–13.

Tarafadar, M., Beath, C., & Ross, J. W. (2019). Using AI to enhance business operations. *MIT Sloan Management Review*, *60*(4), 37–44.

Tjepkema, L. (2018). *Why AI is Vital for Marketing with Lindsay Tjepkema*. Available at: https://www.magnificent.com/magnificent-stuff/why-ai-is-vital-for-marketing-with-lindsay-tjepkema

Venkatraman, V. (2019). How to read and respond to weak digital signals. *MIT Sloan Management Review*, *60*(3), 47–52.

Watson, H. J. (2017). Preparing for the cognitive generation of decision support. *MIS Quarterly Executive*, *16*(3), 153–169.

Westerman, G., Bonnet, D., & McAfee, A. (2014). The nine elements of digital transformation. *MIT Sloan Management Review*, *55*(3), 1–6.

Winkler, T. J., & Kettunen, P. (2018). Five principles of industrialized transformation for successfully building an operational backbone. *MIS Quarterly Executive*, *17*(2), 121–138.

KEY TERMS AND DEFINITIONS

Artificial Intelligence: Refers to the replication of human intellect in machines, such as computers. It entails creating software and computer systems that are capable of carrying out operations that normally demand human intelligence. Problem-solving, learning, reasoning, understanding of language used in conversation, pattern recognition, and decision-making are some of these responsibilities.

Artificial Neural Networks: They are an essential feature of artificial intelligence and machine learning, especially in the deep learning discipline.

Big Data: Large and complicated datasets that are difficult to handle, process, or analyze with conventional data processing methods are referred to as "big data. These datasets are characterized by their volume, velocity, variety, and often, their veracity.

Business Models: A business model is a structure that describes how an organization generates, provides, and obtains value. It explains the fundamental components and strategies that a company uses to run, generate financial resources, and sustain its operations.

Chatbots: Other names for chatbots include "virtual assistants" and "conversational agents." They can communicate with people by text or speech, offering information, responding to queries, carrying out tasks, and even having casual conversations.

Digital Transformation: It is the process of utilizing digital technologies to alter different parts of an organization or business in a meaningful and frequently fundamental way. It entails using digital tools and technologies to rethink and reengineer how a business runs, serves its clients and generates value.

Predictive Analytics: In the field of data analytics, predictive analytics employs machine learning and statistical algorithms to find trends and forecast future results or events.

Semantic Brand Score: It entails analyzing how a brand is mentioned, discussed, and seen in a variety of online sources, including social media, news stories, blogs, and customer reviews, using machine learning and natural language processing (NLP) techniques.

Chapter 6
DeFi's Transformative Influence on the Global Financial Landscape

V. Saravanakrishnan
https://orcid.org/0000-0002-4015-6128
Christ University, India

M. Nandhini
Karpagam Academy of Higher Education, Coimbatore, India

P. Palanivelu
Karpagam Academy of Higher Education, Coimbatore, India

ABSTRACT

The rise of decentralized finance (DeFi) has fundamentally reshaped the financial industry, challenging traditional banking systems and opening up a world of possibilities in global finance. This chapter explores the multifaceted impact of DeFi on the global economic landscape, addressing critical themes through a series of subtitles. DeFi is disrupting traditional banking models by offering alternative financial services directly on blockchain networks, such as lending, borrowing, and trading. One of the remarkable achievements of DeFi is its ability to provide financial services to previously underserved and unbanked populations. Tokenization is a crucial aspect of DeFi, enabling the representation of real world assets as digital tokens on the blockchain. DeFi offers numerous advantages but poses security challenges, including smart contract vulnerabilities and hacks. This chapter provides an overview of the major themes and implications of DeFi's influence on finance, highlighting its opportunities and challenges.

1. INTRODUCTION

The world of finance is undergoing a seismic shift, driven by the emergence of a revolutionary concept known as Decentralized Finance, or DeFi (Zetzsche et al., 2020). This chapter explores DeFi's transformative influence on the global financial landscape, providing essential context and defining key concepts.

DOI: 10.4018/979-8-3693-1762-4.ch006

1.1 Defining DeFi: A Revolution in Finance

DeFi is a blockchain-based financial ecosystem that aims to democratize and transform the way financial services are conducted. It uses the principles of decentralization, transparency, and immutability to build open-source platforms and financial protocols (Schär, 2021).

The conventional structure of global finance comprises banks, insurance firms, and brokers. DeFi takes a different approach. According to Wieandt & Heppding (2023), it uses blockchain technology Instead of relying on centralized intermediaries to establish an independent financial system.

Instead of intermediaries, DeFi uses smart contracts, which are self-executing codes on a blockchain (Koulu, 2016). These allow users to perform various financial transactions, such as asset management and lending, without requiring a traditional bank account. Through a decentralized system, users can lower their costs and improve their accessibility (Chen & Bellavitis, 2020).

1.2 The Historical Evolution of DeFi

The development of blockchain technology and cryptocurrency has heavily influenced the evolution of DeFi. This historical perspective covers the various phases of its evolution.

The early cryptocurrency era from 2009 to 2013 witnessed the groundbreaking emergence of Bitcoin, introduced by the pseudonymous Satoshi Nakamoto. Bitcoin, functioning as a peer-to-peer electronic cash system, marked the inception of decentralized digital currency. In 2011, the landscape expanded with the introduction of alternative cryptocurrencies, or altcoins, such as Litecoin and Namecoin, exploring varied use cases and consensus mechanisms (Harvey et al., 2021).

The subsequent period of 2013 to 2014 saw the advent of smart contracts with the creation of Ethereum by Vitalik Buterin in 2015. Ethereum's blockchain platform enabled the execution of programmable contracts, paving the way for decentralized applications (DApps) and laying the foundational infrastructure for the decentralized finance (DeFi) ecosystem (Kitzler et al., 2023).

Moving into 2015 to 2017, the early DeFi protocols emerged with projects like Augur, focusing on prediction markets, and MakerDAO, introducing a decentralized stablecoin. Simultaneously, the year 2017 marked the ICO boom, a popular fundraising method allowing DeFi projects to secure capital for development (Popescu, 2022).

The years 2017 to 2019 witnessed the rise of decentralized exchanges (DEXs) like EtherDelta and IDEX, pioneering trustless trading. However, challenges arose, as security issues and hacks exposed vulnerabilities within DeFi protocols, underscoring the need for robust security measures.

The year 2020 marked a significant DeFi boom, with the prominence of protocols like Compound Finance, Aave, and Yearn Finance, offering lending, borrowing, and yield farming opportunities. Uniswap introduced the concept of Automated Market Makers (AMMs), transforming the landscape of decentralized exchanges (Piñeiro-Chousa et al., 2022).

Yield farming and liquidity mining gained traction in 2020, allowing users to earn rewards by providing liquidity to DeFi protocols. Simultaneously, the introduction of governance tokens incentivized user participation in various projects (Piñeiro-Chousa et al., 2022).

Navigating the challenges of 2020 and 2021, DeFi faced risks such as flash loan exploits and increased regulatory scrutiny, prompting concerns about legal status and compliance (Momtaz, 2022).

In response to scalability issues, the period of 2021 to 2022 witnessed the integration of Layer 2 scaling solutions like Optimistic Ethereum and Arbitrum. These solutions aimed to address scalability concerns, reduce transaction fees, and enhance overall efficiency. Additionally, efforts were made towards achieving cross-chain compatibility for improved functionality and interoperability (Ou et al., 2022)

As of 2022 to 2023, DeFi continues its trajectory of growth and maturation. Integration of oracles, providing improved price feeds and data for accurate information within DeFi protocols, has become a focus. Furthermore, there is an emergence of multi-chain DeFi projects exploring interoperability across multiple blockchains to enhance scalability and overall functionality. The ongoing evolution reflects the dynamic nature of the DeFi space, continually adapting to technological advancements and market demands (Ou et al., 2022).

1.3 The Emergence of Decentralized Finance

According to Wronka (2021), the decentralized financial system, DeFi, was created due to the cryptocurrency and blockchain movement. Bitcoin, which was first introduced in 2009, is considered to be digital gold. It paved the way for the concept of cryptocurrencies that operate on a decentralized ledger.

Ethereum was founded in 2015. It paved the way for decentralized applications (DApps) by enabling developers to create smart contracts on the blockchain. These dApps are the foundation for various DeFi services, such as lending platforms and exchanges (Popescu, 2022). The rise of DeFi applications and protocols attracted a wide range of individuals looking to explore the financial frontier. The ecosystem quickly expanded, attracting the attention of both traditional finance and cryptocurrency (Umar et al., 2022).

According to recent statistics by CoinGecko, the total value locked (TVL) in DeFi protocols has witnessed an exponential surge, reaching over $100 billion by the end of 2022, compared to just $1 billion in early 2020. This dramatic increase in TVL signifies the growing confidence and adoption of DeFi within the global financial ecosystem (CoinGecko, 2022).

2. DISRUPTING TRADITIONAL BANKING

2.1 DeFi's Challenge to Traditional Banking

Decentralized Finance (DeFi) represents a paradigm shift in finance, profoundly challenging the traditional banking model. Centralized intermediaries characterize traditional banking systems, including banks, credit unions, and financial institutions. These intermediaries facilitate financial transactions, maintain records, and provide various services. However, this centralized model has limitations, including high fees, limited accessibility, and a lack of transparency.

The concept of DeFi marks a fundamental shift in how financial transactions are conducted. Instead of relying on central intermediaries, it uses blockchain technology to establish a trustless ecosystem that eliminates the need for banks and other financial institutions (Chen & Bellavitis, 2020). Unlike traditional banking, DeFi gives users more control over their finances. It eliminates the need to rely on banks and provides them various financial services such as loans and trading. Harvey et al. (2021) state that despite the disruptive nature of DeFi, traditional banks are still adapting to the changes brought about by the technology. Some are exploring the possibility of partnering with DeFi platforms, while others monitor the situation to see its impact on their operations.

Critical aspects of DeFi's decentralized nature include:

- Through peer-to-peer transactions, users can make transactions without intermediaries.
- Transactions are kept on a transparent public ledger, which anyone can audit.
- DeFi's decentralized nature helps reduce the likelihood of a single failure point.

The advantages of DeFi over traditional banks are:

Control: The users can manage their financial activities and assets. They no longer need to rely on banks to make financial transactions. (Wronka, 2021), (Dhanambhore, 2023)

Transparency: DeFi's platforms maintain auditability and transparency, which helps reduce the risk of fraud and fosters trust. (Eikmanns et al., 2023)

Efficiency: The efficiency of DeFi's platforms is well known. Its smart contracts can perform transactions accurately and quickly, which helps prevent human error. (Dhanambhore, 2023)

Cost-Effective: Unlike traditional banks, DeFi's platforms do not charge various fees. These include foreign exchange, wire transfer, and account maintenance charges. (White, 2023)

Global Reach: DeFi's global reach enables its platforms to be accessed by users worldwide. This makes financial transactions and services more accessible. (Auer et al., 2023)

According to Cheng et al. (2021), Despite the various threats that DeFi poses to traditional banks, some are still optimistic about the potential of this disruption. Several banks are currently evaluating their options in response to the DeFi challenge.

- Some banks are exploring partnering with DeFi platforms to provide their customers with new financial services (Navaretti et al., 2018).
- As banks look to improve their operations and reduce their costs through blockchain technology, many are already implementing it into their operations (Osmani et al., 2020)
- In response to the rise of cryptocurrencies and the increasing number of DeFi users, several central banks are exploring the possibility of issuing digital currencies (Lee et al., 2021).

The actions of traditional banks toward DeFi vary depending on the regulatory environment and the region. Some institutions are more proactive about incorporating blockchain technology, while others remain watchful and cautious.

2.2 Banking Without Banks: The DeFi Ecosystem

The concept of decentralization is at the heart of DeFi's platform. Unlike traditional banks, which rely on central authorities to manage their operations, DeFi utilizes blockchain networks. These networks are composed of multiple independent nodes constantly recording and verifying transactions. This eliminates the need for a central authority to oversee the system (Jung, 2023).

- Through peer-to-peer transactions, users can efficiently conduct transactions without an intermediary (Lenz, 2016).
- Transactions are documented on a publicly accessible ledger, which anyone can audit (Cai, 2019).
- DeFi's decentralized nature ensures its systems are secure (Ozili, 2022).

Smart Contracts as Financial Intermediaries: Examining the role of smart contracts in replacing banks.

The foundation of DeFi is smart contracts. These are self-executing contractual agreements that are written directly into the code. They can be executed automatically whenever certain conditions are met (Waltl et al., 2018).

- Smart contracts can be used to facilitate peer-to-peer borrowing and lending. They can also automate the management of collateral and interest rates (Zhang et al., 2023).
- Smart contracts are used in decentralized exchange systems (DEXs) to allow investors to conduct transactions without relying on third parties (Lohr et al., 2022)
- Stake your assets in DeFi's protocols and earn rewards through smart contracts (Makarov & Schoar, 2022)

In DeFi, intelligent contracts reduce the possibility of manipulation and fraud. They are immutable and transparent, which makes them an ideal solution for reducing risk.

Through DeFi, users can easily manage their assets and financial activities. Traditional banking can limit their ability to control their finances.

With DeFi, users can own their assets, which can minimize the risk of losing them due to failures of third parties. Moreover, users can easily access various financial services, such as loans and trading, all without requiring banks or other intermediaries. Its decentralized nature enables users from different parts of the world to participate in the platform (Bellavitis et al., 2022)

2.3 The Evolution of Banking in a DeFi World

The financial world has undergone significant changes over the past couple of centuries. From decentralized peer-to-peer systems to more centralized banking organizations, the chapters explore the history of Banking in a DeFi World. It covers how DeFi is helping the industry transform.

2.4 From Centralization to Decentralization: Tracing the Historical Evolution of Financial Services

The origins of financial services can go back to ancient cultures when people conducted transactions through peer-to-peer networks. These communities could transact through local networks (Herbane, 2010).

The emergence of Centralized Financial Inclusion: Mahadeva (2008) explored how trade and society expanded, and various centralized institutions, such as banks and lending companies, emerged. These helped economic growth by offering credit services and standardized processes. Unfortunately, these institutions also introduced barriers to entry and inefficiencies. The rise of digital technology led to the transformation of financial services. For instance, peer-to-peer platforms, blockchain technology, and crowdfunding have helped bring back direct and decentralized models.

2.5 DeFi's Impact on Banking Evolution: Analyzing How DeFi Is Accelerating This Evolution

Werner et al. (2022) explored that DeFi aims to challenge the conventional financial model through decentralization and disintermediation. It eliminates the need for banks and brokers by allowing users to conduct financial transactions directly. The goal of DeFi is to encourage innovation within the financial industry. As its projects disrupt the traditional banking model, significant banks must develop new services and solutions. The decentralized nature of DeFi enables it to be accessed by users worldwide, fostering financial inclusion and globalization (Eikmanns et al., 2023). In underdeveloped regions, it provides a lifeline to individuals previously excluded from the financial sector.

3. INNOVATIVE FINANCIAL PRODUCTS

The rapid emergence and evolution of DeFi have led to the development of new financial products designed to challenge the conventional finance model (Darlin et al., 2022). This chapter discusses the various innovations brought to bear on this ecosystem, such as yield farming and flash loans. It also explores the advantages of these products.

The decentralized finance (DeFi) concept is a revolutionary approach to financial services. It involves using blockchain technology and smart contracts to transform traditional funding methods. At its core, it aims to provide users with global accessibility and transparency.

3.1 Yield Farming: Maximizing Returns in DeFi

One of the most common innovations in DeFi is yield farming, which allows users to earn passive money by supplying liquidity to the platform (Darlin et al., 2022). This involves taking advantage of the decentralized finance ecosystem's exchange rate. Various yield farming strategies, such as simple staking and more complex ones, can be used, and yield farmers are exposed to multiple risks (Darlin et al., 2022).

In addition to being a profit-making mechanism, yield farming has become vital to the DeFi ecosystem. It powers various applications, such as lending platforms, exchanges, and governance tokens. Through the platform, users can earn rewards by participating in the decisions related to the DeFi ecosystem (Saha et al., 2022).

Yield farming provides users with various advantages, such as supporting projects and generating passive income. It also allows them to maintain their assets while taking advantage of the competitive yields (Phalan et al., 2016).

3.2 Flash Loans: On-Demand Liquidity Without Collateral

A revolutionary innovation within the financial ecosystem of DeFi, flash loans allow users to obtain large amounts of assets without requiring collateral. This feature has caught widespread attention and is utilized extensively in the marketplace. Unlike traditional loans, which require collateral to be placed, users of flash loans do not have to lock up any assets. They can easily carry out various financial operations through these loans (Wang et al., 2021).

The various applications of flash loans on DeFi have demonstrated their versatility. They are a valuable asset to the financial services industry.

One of the most common uses of flash loans is to profit from the differences between the prices of different trading pairs or exchanges. Traders can use this type of loan to take advantage of the multiple arbitrage opportunities that are available to them. Leveraged loan transactions, or flash loans, are used to liquidate the debt on financial services platforms (Qin et al., 2021). They can be used to fund assets currently on the verge of being liquidated, preserving the operations of the DeFi platform. To refinance existing debts, users can use flash loans on DeFi (Phalan et al., 2016). They can get funds to pay off old loans at lower interest rates and get better terms. For instance, a user with a high-interest-rate loan might get funds from flash loans to pay off their initial loan and get a better deal (Qin et al., 2021). With the integration of flash loans into complex trading procedures, traders can now complete the process of borrowing assets from one transaction block, eliminating the need to separate the operations involved (Saha et al., 2022).

3.3 Lending Protocols: Decentralized Borrowing and Lending

Several protocols like Aave, MakerDAO, and Compound power the decentralized finance ecosystem. These allow users to lend and borrow assets in a secure and permission-free manner.

Paliwal (2022) users can easily borrow assets from various liquidity pools through the Compound protocol. It uses an algorithm and transparent approach to set interest rates, which are determined according to the market's dynamics. With the introduction of flash loans, Aave has become one of the most innovative platforms in the decentralized finance ecosystem. These loans allow users to borrow money without collateral easily. Its decentralized nature and lending mechanisms have allowed users to achieve efficiency and flexibility. MakerDAO, a pioneering platform in the decentralized finance industry, created a stablecoin DAI (Van der Merwe, 2021). Users can easily create and secure DAI as collateral through its unique governance model. Its decentralized nature and fee system further demonstrate the platform's commitment to the principles of DeFi lending (Paliwal, 2022).

3.4 StableCoin

People can borrow stablecoins to earn interest or fund their daily needs. This liquidity source within the DeFi network enables users to conduct various financial transactions without exposing them to the volatility of other cryptocurrencies. In the lending sector, stablecoins such as DAI, USDC, and USDT are becoming more critical. These digital currencies have a relatively stable value typically linked to real-world assets. They play an essential role in maintaining the stability of the cryptocurrency market (Saengchote et al., 2022).

Through the lending process, stablecoins can be offered to users by their lenders. The interest rates on these cryptocurrencies are typically higher than those offered by traditional bank accounts. In addition, the transparency and global accessibility of DeFi platforms make it an ideal choice for users (Popescu, 2022). People can lend and borrow stablecoins to hedge their assets or take advantage of opportunities that arise without selling them. Stablecoin loans are a dependable funding source for exchange, yield farming, liquidity management, and DeFi trading (Saengchote et al., 2022).

3.5 Automated Market Makers (AMMs)

AMMs have become one of the most significant developments in the decentralized finance industry. These market-making tools have revolutionized how assets are traded. This section will discuss the various advantages that AMMs have to offer to DeFi users.

AMMs have become a vital part of the financial technology industry as a substitute for traditional order book structures. They are based on the concept of algorithmic pricing and provide liquidity. A liquidity pool is an AMM structure where multiple assets are placed. The users of these pools are then referred to as Liquidity provision, responsible for providing liquidity. They are also paid fees for facilitating transactions in the collection. AMMs typically utilize a constant function formula to ensure that the pricing mechanism of a pair of trading pairs remains consistent. This ensures that one side of the market increases its liquidity while the other decreases. AMMs utilize algorithms to determine the value of assets. Traders then use these to execute transactions. The pricing is adjusted according to the demand and supply in the pool (Mohan, 2022).

AMMs have significantly impacted the financial landscape and the trading activity of DeFi. They eliminate the need for centralized exchanges and intermediaries and allow users to conduct transactions without requiring a central hub. Unlike traditional markets with limited trading hours, AMMs allow DeFi users to trade assets 24 hours daily. AMMs make trading more accessible by simplifying the process. Instead of creating complicated order books or navigating through complex trading interfaces, traders can trade assets through these platforms. AMMs democratized the provision of liquidity. Users can become LPs and earn fees from transactions within the liquidity pool (Bartoletti et al., 2022).

The advantages and impact of AMMs on the DeFi market are the focus of case studies presented by two prominent AMMs, UniSwap and SushiSwap. Due to its popularity in the DeFi market, UniSwap has been regarded as one of the first significant AMMs. It enables users to trade various assets and earn fees. Its efficient trading methods and user-friendly interface are some of the factors that have contributed to its success. As an AMM and decentralized exchange, SushiSwap enhances the AMM framework by introducing various features, such as yield farming. It has also incentivized LPs with the SUSHI token, contributing to its rapid popularity (Jensen et al., 2021).

4. NAVIGATING TRENDS IN DECENTRALIZED FINANCE

4.1 Decentralized Exchanges (DEXs) Dominance

Decentralized exchanges (DEXs) have risen to prominence within the DeFi space, exemplified by platforms like Uniswap. Uniswap, employing the Automated Market Maker (AMM) model, has become a leading decentralized trading venue. Its innovation of allowing users to trade directly from their wallets without the need for traditional order books has revolutionized the way users engage in decentralized trading. Uniswap's success has inspired a wave of similar DEXs, such as SushiSwap and PancakeSwap, showcasing the growing dominance of decentralized trading platforms. These DEXs have provided users with increased financial sovereignty and addressed liquidity challenges traditionally associated with centralized exchanges.

4.2 Liquidity Mining and Yield Farming

The trend of liquidity mining and yield farming gained substantial traction, with platforms like Yearn Finance leading the way. Yearn Finance introduced a yield aggregation strategy, enabling users to automatically seek the highest yield across various lending protocols. The concept of yield farming incentivizes users to provide liquidity to DeFi protocols in exchange for governance tokens or other rewards. Projects like Curve Finance and Compound further contributed to the popularity of these mechanisms. For instance, Curve Finance's stablecoin pools allowed users to earn fees and CRV governance tokens, showcasing the symbiotic relationship between liquidity providers and DeFi protocols.

4.3 NFTs and DeFi Integration

The intersection of Non-Fungible Tokens (NFTs) and DeFi gained momentum, with projects like Aavegotchi leading the way. Aavegotchi combines decentralized finance with unique NFTs, allowing users to stake their NFTs as collateral to earn interest or borrow against them. This integration provides a novel use case for NFTs beyond the realm of digital art, creating a synergistic relationship between decentralized finance and digital collectibles.

4.4 Layer 2 Solutions and Scalability

Scalability challenges on the Ethereum network prompted the rise of Layer 2 solutions. Optimistic Ethereum, a Layer 2 scaling solution, demonstrated its efficacy in reducing transaction costs and increasing throughput. Projects like Synthetix implemented Layer 2 solutions to enhance the efficiency of their synthetic asset trading platforms. The adoption of Layer 2 solutions has become a pivotal trend in addressing the scalability limitations of Ethereum, offering users faster and more cost-effective transactions.

4.5 Cross-Chain Compatibility

The trend of cross-chain compatibility gained prominence as projects sought to address the fragmentation of the DeFi space across different blockchains. Polkadot, with its interoperable blockchain network, exemplifies this trend by enabling the seamless transfer of assets and data between different blockchains. The interoperability facilitated by Polkadot and similar projects fosters collaboration and enhances the overall efficiency of the DeFi ecosystem by allowing users and assets to move seamlessly across disparate networks.

5. FINANCIAL INCLUSION AND ACCESSIBILITY IN DEFI

Through DeFi, financial inclusion has been made possible for many more people, especially those who were previously unbanked or underserved. This section explores how this type of initiative is helping to make the financial landscape more inclusive.

Unbanked and underbanked people around the world have various difficulties when it comes to accessing credit, investing, and saving. Some barriers preventing these individuals from participating in the financial sector include high fees, geographical restrictions, and formal documentation (Patwardhan, 2018). Through DeFi platforms, anyone, anywhere, can now benefit from the financial services industry. This is excellent news for people in underdeveloped regions who need more banking infrastructure. Its 24/7 availability eliminates the need for people to bank at certain hours of the day or night, especially for those who work in different time zones (Grassi et al., 2022).

Through DeFi platforms, people can now borrow and invest their assets. They can also earn interest from their loans. These financial services are ideal for people who cannot access traditional banking (Alexandre et al., 2011). Its protocols allow users to save and borrow small amounts, benefiting those needing financial assistance. One of the most common financial services that people in developing countries rely on is cross-border remittances. Through DeFi, users can now send and receive funds more efficiently and at a lower cost (Sriman & Kumar, 2022).

According to Mason & Harrison (2001), DeFi offers accessibility; users should have the necessary digital literacy to navigate the platform. This can be done through education programs. Regulatory frameworks should also adapt to address users' needs and prevent innovation from stifling. DeFi can assist small businesses by granting them access to capital through its investment and lending platforms. Individuals can also invest their assets to enhance their financial security and wealth (Schueffel, 2021). Projects of DeFi are working on lowering transaction costs and increasing the number of users. They are also exploring collaborating with financial institutions to bridge the gap between new and old economic systems. Doing so can help disadvantaged groups get the best possible deal (Akkizidis & Khandelwal, 2007).

The ability of decentralized financial services (DeFi) to expand financial inclusion and improve accessibility is a central component of its potential. According to Ali et al. (2020), through blockchain technology, these platforms can help individuals who have previously been excluded from the mainstream financial system. However, they also expose various risks and challenges that must be resolved.

6. TOKENIZATION OF ASSETS IN DEFI: TRANSFORMING THE INVESTMENT LANDSCAPE

Asset tokenization is a fundamental feature of Decentralized Finance, changing how people view and invest in assets. This section explores the significance of this innovation and how it is transforming traditional investment approaches.

Benedetti Rodríguez-Garnica (2021) refers to asset tokenization as representing real-world assets such as stocks, properties, art, and commodities on a distributed ledger. Each token represents a fraction of an asset's market value. Through asset tokenization, people can now own high-value assets, even if they can't afford to purchase them at total market value. The accessibility and liquidity of tokenized assets are two of the main advantages of this innovation. They allow global investors to trade them on DeFi platforms easily. Moreover, they eliminate the need for high capital requirements in traditional markets. Blockchain technology provides transparency in the ownership history of an asset, reduces fraud, and enhances trust (Notheisen et al., 2017). Its smart contracts can automate various administrative tasks, such as asset management and dividend distribution (Zhitomirskiy et al., 2023).

According to Norta et al. (2018), tokenizing properties is one of the most common ways to democratize the real estate market. This allows investors to buy and sell properties without going through the laborious process of building a whole new infrastructure. Equities and stocks can be tokenized to enable for instant borderless trading. Physical storage is unnecessary for certain commodities, such as gold and silver, which can be tokenized. In addition, high-value art and collectibles can be easily fractionalized, making them accessible to a broader audience.

Stein Smith (2019), the regulatory landscape for asset tokenization is still evolving. Complying with international standards and local laws is crucial for any company to operate successfully. Security is one of the most critical factors a company must consider regarding tokenization.

The advent of asset tokenization has posed a threat to traditional financial markets. It offers various advantages, such as reduced fees and accessibility. Due to the increasing number of exchanges and financial institutions working with DeFi platforms, the gap between traditional finance and decentralized solutions is being bridged. One of the most essential advantages of asset tokenization is its ability to diversify an investor's portfolio. It allows them to manage risk by distributing their stakes across multiple assets. Moreover, DeFi platforms are creating new financial products based on tokenized indices and yield-generating tokens (Trabelsi, 2020).

7. REGULATORY CHALLENGES AND COMPLIANCE

Due to the decentralized nature of DeFi, it has a unique set of challenges when it comes to navigating the regulatory landscape. This section looks into some obstacles and how they can be overcome.

The decentralized nature of DeFi makes it challenging for conventional regulations to address its borderless and complex character. Its centralization and lack of intermediaries present significant challenges to traditional consumer protection and financial oversight models.

Zhitomirskiy et al. (2023) explored that it can be challenging to comply with anti-money laundering and Know Your Customer regulations in a decentralized environment. The same goes for securities regulations. They are determining if trading activities and token offerings qualify as securities can be complex. Taxation and Reporting for transactions involving capital gains and income can be difficult. Changes in legislation have to be considered when it comes to handling sensitive information collected in DeFi applications.

Developers of decentralized identity systems are exploring ways to ease compliance with anti-money laundering and Know Your Customer (KYC) regulations. They can also use blockchain technology to integrate real-world data to make it easier to meet external legal obligations, such as tax reporting and securities laws. Some of the projects and organizations working on blockchain technology are actively participating in developing regulations and contributing to the discussions. One of the most common practices in the DeFi ecosystem is auditing smart contracts for compliance (Frick, 2019).

The US has been taking a cautious approach when it comes to regulating the activities of financial firms and individuals using blockchain technology. Several agencies, such as the Securities and Exchange Commission (SEC), the Commodity Futures Trading Commission (CFTC), and the Financial CEN, are currently involved in developing regulations for DeFi. Europe is also working on a comprehensive framework for the regulation of cryptocurrencies. In Asia, some countries, such as South Korea and China, have strict rules regarding the operation of DeFi. On the other hand, countries such as Japan and Singapore have supportive frameworks (Massad & Jackson, 2023).

Several DeFi protocols operate through community governance, which allows users to participate in the development of the system. This type of self-regulation can address various security concerns. The evolution of the regulatory system faces multiple challenges and opportunities. Regulators must balance the protection of consumers, investors, and financial stability with the pursuit of innovation. In light of DeFi's cross-border nature, international collaboration is necessary to establish effective regulations (Massad & Jackson, 2023).

8. RISKS AND SECURITY CONCERNS IN DEFI: SAFEGUARDING A DECENTRALIZED LANDSCAPE

Although decentralized finance offers numerous advantages, it also comes with security and risk issues. This section covers the different risks associated with this type of technology.

Ellul et al. (2020) explored the vulnerability in intelligent contracts, considered the heart of DeFi. These are vulnerable to exploitation through various code flaws, which can lead to theft, manipulation, and hacks. Examples of notable incidents highlighting this vulnerability include the DAO hack 2016 and the incident on bZx in 2020.

Despite the need for more trust in the system, the developers who maintain and create smart contracts are still expected to perform their duties properly. Some of the issues that have been highlighted include the creation of scams and rug pulls. Automated market maker (AMM) participants may experience permanent losses due to token price fluctuations. Moreover, borrowers may face liquidation if their collateral's value falls below certain levels during periods of extreme volatility (Mohan, 2022).

Al-Breiki et al. (2020) explored that DeFi utilizes the Oracles to distribute real-world data, but manipulation can lead to price inaccuracies. Mishandling of these oracles can also cause losses. Front-running, conversely, can benefit malicious actors who exploit the delay in price information. Due to the nature of DeFi's business, it is subject to a constantly changing regulatory landscape. This can affect the operations of the platform and its users. Potential legal consequences for users and projects include violating existing regulations (Zhou et al., 2023).

The security and risks associated with DeFi are some of the most critical factors that need to be considered when it comes to becoming a financial ecosystem. Its decentralized nature provides tremendous opportunities but requires regular monitoring to prevent attacks and vulnerabilities. This can be done by establishing effective security measures and education (Carter & Jeng, 2021).

9. DECENTRALIZED EXCHANGES (DEXS)

The emergence of decentralized exchanges has disrupted the traditional framework of digital asset trading. This section will explore the various facets of DEXs and their growth, supported by real-world examples and proof.

Brasse & Hyun (2023). Digital asset exchange platforms (DEXs) are web-based platforms that allow users to swap digital currencies without requiring an intermediary. The number of such platforms has increased significantly over the past few years. Some of the prominent ones include PancakeSwap, UniSwap, and SushiSwap. They support various asset classes, such as non-fungible tokens and cryptocurrencies. DEXs are decentralized, eliminating the need for a central authority and fostering trustlessness. They allow users to manage their assets directly without depositing them into a major exchange. Global accessibility makes them attractive to users worldwide (Ghosh et al., 2023).

Due to the competition that DEXs bring to the table, centralized exchanges have to lower their fees and innovate to remain competitive. They also eliminate the need for banks, brokers, and other intermediaries. They operate 24/7, which enables them to provide continuous trading. DEXs may encounter issues with low liquidity for certain assets, but the growth of AMM protocols has helped increase this liquidity. In addition, they may be subject to regulatory confusion in certain jurisdictions, which can result in legal and compliance problems. The user experience of DEXs may still be inferior to that of centralized exchanges (Pandian et al., 2023).

(Ghosh et al. (2023) explored one of the most prominent DEXs, UniSwap, which processes billions of dollars in transactions. Its governance token is also a major asset. Another famous DEX is SushiSwap, which has gained popularity due to its innovative features, such as Onsen pools. DEXs are also exploring operating on multiple blockchains to improve their interoperability. One of the main advantages of this approach is that it allows them to reduce their transaction fees. In addition, they can integrate with other DeFi protocols, which makes them an integral part of the ecosystem.

The evolution and proliferation of DEXs have been regarded as a seismic shift in the exchange industry. Their growth has been attributed to the increasing demand for privacy and control. Despite their challenges, DEXs are expected to continue playing a vital role in the decentralized finance industry (Brasse & Hyun, 2023).

10. ENVIRONMENTAL CONCERNS AND SUSTAINABILITY

As the rapid emergence and growth of decentralized finance continues, it is becoming more apparent that the sector's environmental impact is becoming more concerning. This section aims to explore the various challenges faced by the industry and look into possible solutions.

Akbar et al. (2021), several decentralized applications (DeFi), such as Ethereum, utilize Proof-of-Work (PoW) consensus protocols, which consume much energy. The mining process alone can contribute to carbon emissions. The high gas fees associated with transactions on DeFi networks can cause energy consumption to increase. In addition, the constant need for new equipment in Proof-of-Work (PoW) mining can result in electronic waste.

Wang et al., 2022, the energy consumption of Ethereum, a significant platform for decentralized exchange, has been the subject of discussion. In some cases, it has been compared to the energy consumption of entire nations. Also, the carbon footprint of networks that operate on Proof-of-Work (PoW) blockchains has been criticized.

Several DeFi platforms, like Ethereum, are transitioning to a Proof-of-Stake (PoS) consensus system. Compared to a traditional Proof-of-Work (PoW) system, Proof-of-Stake (PoS) is more efficient and requires less energy-intensive mining. Another feature that DeFi networks can benefit from is Layer 2 scaling solutions. Some projects are actively exploring the utilization of green DeFi initiatives. These include offsetting their carbon footprint or adopting eco-friendly practices, which can attract users and investors with a keen environmental consciousness. As a competitive advantage, sustainability-oriented projects can also gain a significant advantage over others (Gudgeon et al., 2020).

Kshetri & Voas (2022) users and projects can contribute to carbon offset initiatives by investing in projects that minimize their carbon emissions, such as wind or solar energy generation. Some DeFi projects, such as Algorand or Tezos, are built on more eco-friendly blockchains

Through transparent reporting, DeFi projects can provide users with more information about their energy consumption and carbon emissions. This allows them to make more informed decisions regarding their platform's environmental impact. Priority should be given to projects working toward reducing their carbon footprint. Moreover, users can choose DeFi protocols and networks that offer energy-saving transactions (Kshetri & Voas, 2022).

Due to the increasing number of transactions and the complexity of the financial transactions industry, the environment and sustainability issues are becoming more critical for the sector. As a result, the various steps taken by the DeFi ecosystem, such as the transition to PoS and the development of green initiatives, are aimed at reducing its carbon footprint. By focusing on sustainability, the platform can become a more eco-friendly and sustainable financial community (Gola & Sedlmeir, 2022).

11. REAL CASES

11.1 Aavegotchi's Commitment to Sustainability

Aavegotchi, a decentralized finance (DeFi) project built on the Aave lending protocol, is an illustrative example of a platform actively addressing environmental concerns within the blockchain space. In response to the increasing environmental impact of traditional proof-of-work (PoW) consensus mechanisms, Aavegotchi made a strategic decision to build on the Polygon (formerly Matic) network, which employs a proof-of-stake (PoS) consensus mechanism. This transition significantly reduced the carbon footprint associated with the platform's operations, contributing to a more sustainable blockchain ecosystem. Aavegotchi's commitment to sustainability is a noteworthy example of how DeFi projects can make conscientious choices in their technological infrastructure to minimize environmental impact (Jiang et al., 2022).

11.2 Renewable Energy Credits on MakerDAO

MakerDAO, a prominent decentralized autonomous organization responsible for creating the DAI stablecoin, has taken innovative steps to address environmental concerns within its ecosystem. In a groundbreaking move, MakerDAO partnered with a renewable energy credit (REC) platform to purchase carbon offsets and renewable energy credits. These credits compensate for the energy consumption associated with the Ethereum blockchain, where MakerDAO primarily operates. MakerDAO is committed to mitigating its environmental impact by integrating these offsets into its operational model. This case study highlights the potential for DeFi platforms to proactively engage in environmentally conscious practices and contribute to the broader sustainability movement (Brennecke et al., 2022)

11.3 Curve Finance and Gas Efficiency

Curve Finance, a decentralized exchange (DEX) optimized for stablecoin trading, exemplifies how gas efficiency measures can enhance sustainability within the DeFi space. Recognizing the environmental concerns associated with high gas fees on the Ethereum network, Curve Finance implemented gas-efficient strategies to minimize transaction costs. The platform utilizes algorithmic techniques to optimize token swaps, reducing the computational workload and subsequently lowering the associated carbon footprint. This case study underscores the importance of user-friendly and environmentally conscious design choices in DeFi protocols, promoting sustainability without compromising functionality (Wiyono et al., 2021)(Fan et al., 2022).

12. SUCCESSFUL DEFI PROJECTS AND THEIR IMPACT IN THE FINANCE INDUSTRY

One of the pioneering success stories in the decentralized finance (DeFi) space is Compound Finance. This decentralized lending protocol revolutionized the lending landscape by introducing algorithmic interest rates that dynamically adjust based on supply and demand. Compound's impact has been profound, attracting users seeking efficient and decentralized lending solutions. Its success has been pivotal in influencing the broader DeFi lending space, catalyzing the proliferation of similar platforms that prioritize algorithmic interest rate mechanisms for decentralized lending and borrowing (Auer et al., 2023).

Uniswap, a decentralized exchange (DEX), stands out as another exemplar in the DeFi realm. Uniswap introduced the automated market maker (AMM) model, eliminating the need for traditional order books and enabling users to trade various ERC-20 tokens directly from their wallets. Its user-friendly interface and permissionless trading have significantly contributed to the rise of decentralized exchanges, impacting the way users interact with digital assets in a decentralized and efficient manner (Aigner & Dhaliwal, 2021).

Aave, a decentralized lending and borrowing protocol, has left an indelible mark with its introduction of flash loans. These innovative loans allow users to borrow without collateral, provided the borrowed funds are returned within the same transaction block. Aave's pioneering work in introducing flash loans has influenced the DeFi landscape by opening new possibilities for arbitrage and complex financial strategies, showcasing the potential for uncollateralized borrowing within decentralized financial ecosystems (Gudgeon et al., 2020).

MakerDAO, a decentralized autonomous organization (DAO) governing the stablecoin DAI, has played a transformative role in the DeFi space. It introduced decentralized stablecoins and collateralized debt positions (CDPs), providing users with an alternative stablecoin option not reliant on traditional banking. MakerDAO's impact has been instrumental in reshaping how users perceive and utilize stablecoins within decentralized financial environments (Sun et al., 2022).

Yearn Finance, a yield aggregator that automates fund movements across different lending platforms, has significantly influenced the DeFi landscape with its optimization strategies. Yearn Finance introduced the concept of yield farming, allowing users to earn governance tokens by providing liquidity. Its success has influenced how users interact with and optimize yield within the DeFi ecosystem, contributing to the trend of decentralized yield optimization (Cousaert et al., 2022).

Curve Finance, a decentralized exchange optimized for stablecoin trading, has addressed impermanent loss concerns for stablecoin liquidity providers. Its focus on low-slippage swaps has influenced subsequent decentralized exchanges, emphasizing the importance of specialized platforms within the DeFi ecosystem to cater to specific asset classes (Wiyono et al., 2021).

Synthetix, a decentralized synthetic asset issuance platform, has expanded the DeFi landscape by enabling users to mint synthetic assets representing various real-world and crypto assets. Its success has contributed to the growth of decentralized derivatives within DeFi, providing users with exposure to a wide range of assets without directly holding them (Liu et al., 2021).

These real-world case studies illustrate the diversity of impact that successful DeFi projects have had on users and the financial industry, ranging from decentralized lending and borrowing to automated market making, yield optimization, and the creation of synthetic assets. The lessons learned from these projects continue to shape the evolution of the decentralized finance ecosystem, driving innovation and transforming traditional financial paradigms (Liu et al., 2021).

13. CONCLUSION

In conclusion, the emergence and rapid evolution of decentralized finance (DeFi) have undeniably transformed the global financial architecture, ushering in a paradigm shift acknowledged by various researchers and industry experts. Hacioglu and Aksoy (2021) emphasize the transformative force of

DeFi, highlighting its significant impact on the global financial landscape. The disruptive nature of DeFi is evident in the innovations introduced, such as the establishment of a resilient financial infrastructure and the substantial increase in financial transactions, as noted by Wieandt and Heppding (2023).

Zhou et al. (2023) provide empirical evidence of the growing value within the DeFi ecosystem, with the total value locked (TVL) skyrocketing from a few hundred thousand dollars to tens of billions in just a few years. This surge in value is not only indicative of the financial potential of DeFi but also reflects its widespread applicability across various sectors. Users now have access to a spectrum of financial services, including insurance, loans, derivatives, and stablecoins, transcending geographical and socio-economic barriers (Schueffel, 2021).

The decentralized exchange platforms within the DeFi space, as highlighted by Sriman and Kumar (2022), have revolutionized the way global finance operates. As discussions around sustainability and environmental issues gain prominence, DeFi stands out as a beacon of possibility, opening up new avenues for the worldwide finance industry. The impact of DeFi is immense, and its innovative, adaptable, and decentralized principles position it as a significant player in shaping the future of finance (White, 2023).

In summary, it is crucial to recognize that the journey towards a decentralized financial future must also address sustainability concerns. As detailed earlier, the examples of Aavegotchi, MakerDAO, and Curve Finance showcase that sustainable practices can coexist with the innovative nature of DeFi. The DeFi community can contribute to a more sustainable financial ecosystem by acknowledging and actively working towards environmentally conscious choices. The transformative influence of DeFi on the global financial landscape is profound. Its ability to innovate, adapt, and provide financial services to diverse users underscores its potential to redefine traditional financial systems. As we navigate this revolutionary era in finance, we must recognize the importance of addressing sustainability concerns within the DeFi ecosystem to ensure a responsible and inclusive financial future. This chapter has explored the journey of DeFi, its impact on various sectors, and the potential it holds for shaping the future of finance on a global scale.

REFERENCES

Aigner, A., & Dhaliwal, G. (2021). Uniswap: Impermanent loss and risk profile of a liquidity provider. SSRN *Electronic Journal*. doi:10.2139/ssrn.3872531

Akbar, N. A., Muneer, A., ElHakim, N., & Fati, S. M. (2021). Distributed hybrid double-spending attack prevention mechanism for proof-of-Work and proof-of-Stake blockchain consensuses. *Future Internet*, *13*(11), 285. doi:10.3390/fi13110285

Akkizidis, I., & Khandelwal, S. (2007). *Financial risk management for Islamic banking and finance*. Springer.

Al-Breiki, H., Rehman, M. H., Salah, K., & Svetinovic, D. (2020). Trustworthy blockchain oracles: Review, comparison, and open research challenges. *IEEE Access : Practical Innovations, Open Solutions*, *8*, 85675–85685. doi:10.1109/ACCESS.2020.2992698

Alexandre, C., Mas, I., & Radcliffe, D. (2011). Regulating new banking models to bring financial services to all. *Challenge*, *54*(3), 116–134. doi:10.2753/0577-5132540306

Alexandre, C., Mas, I., & Radcliffe, D. (2011). undefined. *Challenge, 54*(3), 116–134. doi:10.2753/0577-5132540306

Ali, O., Ally, M., Clutterbuck, & Dwivedi, Y. (2020). The state of play of blockchain technology in the financial services sector: A systematic literature review. *International Journal of Information Management, 54*, 102199. doi:10.1016/j.ijinfomgt.2020.102199

Auer, R., Haslhofer, B., Kitzler, S., Saggese, P., & Victor, F. (2023). The technology of decentralized finance (DeFi). *Digital Finance*. Advance online publication. doi:10.1007/s42521-023-00088-8

Bartoletti, M., Chiang, J. H., & Lluch-Lafuente, A. (2022). A theory of automated market makers in DeFi. *Logical Methods in Computer Science, 18*(4), 8955. Advance online publication. doi:10.46298/lmcs-18(4:12)2022

Bellavitis, C., Fisch, C., & Momtaz, P. P. (2022). The rise of decentralized autonomous organizations (DAOs): A first empirical glimpse. *Venture Capital, 25*(2), 187–203. doi:10.1080/13691066.2022.2116797

Benedetti, H. E., & Rodríguez-Garnica, G. (2021). Tokenized assets and securities. SSRN *Electronic Journal*. doi:10.2139/ssrn.4069119

Brasse, A., & Hyun, S. (2023). Cryptocurrency exchanges and the future of Cryptoassets. *The Emerald Handbook on Cryptoassets: Investment Opportunities and Challenges*, 341-353. https://doi.org/doi:10.1108/978-1-80455-320-620221022

Brennecke, M., Guggenberger, T., Schellinger, B., & Urbach, N. (2022). The de-central bank in decentralized finance: A case study of MakerDAO. *Proceedings of the Annual Hawaii International Conference on System Sciences*. 10.24251/HICSS.2022.737

Cai, C. W. (2019). Triple-entry accounting with blockchain: How far have we come? *Accounting and Finance, 61*(1), 71–93. doi:10.1111/acfi.12556

Carter, N., & Jeng, L. (2021). DeFi protocol risks: The paradox of DeFi. SSRN *Electronic Journal*. doi:10.2139/ssrn.3866699

Chen, Y., & Bellavitis, C. (2020). Blockchain disruption and decentralized finance: The rise of decentralized business models. *Journal of Business Venturing Insights, 13*, e00151. doi:10.1016/j.jbvi.2019.e00151

Cheng, H. K., Hu, D., Puschmann, T., & Zhao, J. L. (2021). The landscape of blockchain research: Impacts and opportunities. *Information Systems and e-Business Management, 19*(3), 749–755. doi:10.1007/s10257-021-00544-1

Cousaert, S., Xu, J., & Matsui, T. (2022). Sok: Yield Aggregators in DeFi. *2022 IEEE International Conference on Blockchain and Cryptocurrency (ICBC)*. 10.1109/ICBC54727.2022.9805523

Darlin, M., Palaiokrassas, G., & Tassiulas, L. (2022). Debt-financed collateral and stability risks in the DeFi ecosystem. *2022 4th Conference on Blockchain Research & Applications for Innovative Networks and Services (BRAINS)*. 10.1109/BRAINS55737.2022.9909090

Dhanambhore. (2023, April 8). *Decentralized finance (DeFi) and challenges to traditional financial system*. GeeksforGeeks. https://www.geeksforgeeks.org/decentralized-finance-defi-and-challenges-to-traditional-financial-system/

Eikmanns, B. C., Mehrwald, P., Sandner, P. G., & Welpe, I. M. (2023). Decentralised finance platform ecosystems: Conceptualisation and outlook. *Technology Analysis and Strategic Management*, 1–13. doi:10.1080/09537325.2022.2163886

Ellul, J., Galea, J., Ganado, M., Mccarthy, S., & Pace, G. J. (2020). Regulating blockchain, DLT and smart contracts: A technology regulator's perspective. *ERA Forum, 21*(2), 209-220. 10.1007/s12027-020-00617-7

Fan, S., Min, T., Wu, X., & Wei, C. (2022). Towards understanding governance tokens in liquidity mining: A case study of decentralized exchanges. *World Wide Web (Bussum)*, *26*(3), 1181–1200. doi:10.1007/s11280-022-01077-4

Financial ecosystem and strategy in the Digital Era. (2021). *Contributions to Finance and Accounting*. doi:10.1007/978-3-030-72624-9

Frick, T. A. (2019). Virtual and cryptocurrencies—regulatory and anti-money laundering approaches in the European Union and in Switzerland. *ERA Forum, 20*(1), 99-112. 10.1007/s12027-019-00561-1

Ghosh, B., Kazouz, H., & Umar, Z. (2023). Do automated market makers in DeFi ecosystem exhibit time-varying connectedness during stressed events? *Journal of Risk and Financial Management*, *16*(5), 259. doi:10.3390/jrfm16050259

Gola, C., & Sedlmeir, J. (2022). Addressing the sustainability of distributed Ledger technology. SSRN *Electronic Journal*. doi:10.2139/ssrn.4032837

Grassi, L., Lanfranchi, D., Faes, A., & Renga, F. M. (2022). undefined. *Qualitative Research in Accounting & Management*, *19*(3), 323–347. doi:10.1108/QRAM-03-2021-0051

Gudgeon, L., Moreno-Sanchez, P., Roos, S., McCorry, P., & Gervais, A. (2020). Sok: Layer-two blockchain protocols. *Financial Cryptography and Data Security*, 201-226. doi:10.1007/978-3-030-51280-4_12

Hacioglu, U., & Aksoy, T. (2021). *Financial ecosystem and strategy in the Digital Era: Global approaches and new opportunities*. Springer Nature. doi:10.1007/978-3-030-72624-9

Harvey, C. R., Ramachandran, A., & Santoro, J. (2021). *DeFi and the future of finance*. John Wiley & Sons.

Herbane, B. (2010). The evolution of business continuity management: A historical review of practices and drivers. *Business History*, *52*(6), 978–1002. doi:10.1080/00076791.2010.511185

Jensen, J. R., Pourpouneh, M., Nielsen, K., & Ross, O. (2021). The homogeneous properties of automated market makers. SSRN *Electronic Journal*. doi:10.2139/ssrn.3807820

Jiang, Y., Min, T., Fan, S., Tao, R., & Cai, W. (2022). Towards understanding player behavior in blockchain games: A case study of Aavegotchi. *Proceedings of the 17th International Conference on the Foundations of Digital Games*. 10.1145/3555858.3555883

Jung, K. (2023). Cryptocurrency in practice. *The Quiet Crypto Revolution*, 65-99. doi:10.1007/978-1-4842-9627-1_4

Kitzler, S., Victor, F., Saggese, P., & Haslhofer, B. (2023). Disentangling decentralized finance (DeFi) compositions. *ACM Transactions on the Web*, *17*(2), 1–26. doi:10.1145/3532857

Koulu, R. (2016). Blockchains and online dispute resolution: Smart contracts as an alternative to enforcement. *Script-ed*, *13*(1), 40–69. doi:10.2966/script.130116.40

Kshetri, N., & Voas, J. (2022). Blockchain's carbon and environmental footprints. *Computer*, *55*(8), 89–94. doi:10.1109/MC.2022.3176989

Lee, D. K., Yan, L., & Wang, Y. (2021). A global perspective on Central Bank digital currency. *China Economic Journal*, *14*(1), 52–66. doi:10.1080/17538963.2020.1870279

Lenz, R. (2016). Peer-to-peer lending: Opportunities and risks. *European Journal of Risk Regulation*, *7*(4), 688–700. doi:10.1017/S1867299X00010126

Liu, B., Szalachowski, P., & Zhou, J. (2021). A first look into DeFi oracles. *2021 IEEE International Conference on Decentralized Applications and Infrastructures (DAPPS)*. 10.1109/DAPPS52256.2021.00010

Lohr, M., Skiba, K., Konersmann, M., Jurjens, J., & Staab, S. (2022). Formalizing cost fairness for two-party exchange protocols using game theory and applications to blockchain. *2022 IEEE International Conference on Blockchain and Cryptocurrency (ICBC)*. 10.1109/ICBC54727.2022.9805522

Mahadeva, M. (2008). Financial growth in India. *Margin - the Journal of Applied Economic Research*, *2*(2), 177–197. doi:10.1177/097380100800200202

Makarov, I., & Schoar, A. (2022). *Cryptocurrencies and decentralized finance*. DeFi., doi:10.3386/w30006

Mason, C. M., & Harrison, R. T. (2001). 'Investment readiness': A critique of government proposals to increase the demand for venture capital. *Regional Studies*, *35*(7), 663–668. doi:10.1080/00343400120075939

Massad, T., & Jackson, H. (2023, June 24). *How to improve regulation of crypto today—without congressional action—and make the industry pay for it.* Brookings. https://www.brookings.edu/articles/how-to-improve-regulation-of-crypto-today-without-congressional-action-and-make-the-industry-pay-for-it/

Mohan, V. (2022). Automated market makers and decentralized exchanges: A DeFi primer. *Financial Innovation*, *8*(1), 20. Advance online publication. doi:10.1186/s40854-021-00314-5

Momtaz, P. P. (2022). Decentralized finance (DeFi) markets for startups: Search frictions, intermediation, and efficiency. SSRN *Electronic Journal*. doi:10.2139/ssrn.4020201

Navaretti, G. B., Calzolari, G. M., Mansilla-Fernandez, J. M., & Pozzolo, A. F. (2018). Fintech and banking. Friends or foes? SSRN *Electronic Journal*. doi:10.2139/ssrn.3099337

Nenov, C. (2023, August 14). *DeFi vs. traditional banking: A comparative analysis of efficiency and transparency*. Medium. https://medium.com/@christonenov/defi-vs-traditional-banking-a-comparative-analysis-of-efficiency-and-transparency-8a529dfc4aaf

Norta, A., Fernandez, C., & Hickmott, S. (2018). Commercial property Tokenizing with smart contracts. *2018 International Joint Conference on Neural Networks (IJCNN)*. 10.1109/IJCNN.2018.8489534

Notheisen, B., Cholewa, J. B., & Shanmugam, A. P. (2017). Trading real-world assets on blockchain. *Business & Information Systems Engineering*, *59*(6), 425–440. doi:10.1007/s12599-017-0499-8

Osmani, M., El-Haddadeh, R., Hindi, N., Janssen, M., & Weerakkody, V. (2020). Blockchain for next generation services in banking and finance: Cost, benefit, risk and opportunity analysis. *Journal of Enterprise Information Management*, *34*(3), 884–899. doi:10.1108/JEIM-02-2020-0044

Ou, W., Huang, S., Zheng, J., Zhang, Q., Zeng, G., & Han, W. (2022). An overview on cross-chain: Mechanism, platforms, challenges and advances. *Computer Networks*, *218*, 109378. doi:10.1016/j.comnet.2022.109378

Ozili, P. K. (2022). Decentralized finance research and developments around the world. *Journal of Banking and Financial Technology*, *6*(2), 117–133. doi:10.1007/s42786-022-00044-x

Paliwal, A. (2022). Analysis between different decentralized lending and borrowing protocols. *Journal of Business Analytics and Data Visualization*, *3*(1), 15–23. doi:10.46610/JBADV.2022.v03i01.003

Pandian, K., Pfeiffer, D., & Qian, S. (2023). Decentralized finance. *The Emerald Handbook on Cryptoassets: Investment Opportunities and Challenges*, 141-156. doi:10.1108/978-1-80455-320-620221010

Patwardhan, A. (2018). undefined. Handbook of Blockchain, Digital Finance, and Inclusion, 1, 57-89. doi:10.1016/B978-0-12-810441-5.00004-X

Phalan, B., Green, R. E., Dicks, L. V., Dotta, G., Feniuk, C., Lamb, A., Strassburg, B. B., Williams, D. R., Ermgassen, E. K., & Balmford, A. (2016). How can higher-yield farming help to spare nature? *Science*, *351*(6272), 450–451. doi:10.1126/science.aad0055 PMID:26823413

Piñeiro-Chousa, J., Cabarcos, Á. L., & González, I. (2022). DeFi and Start-UPS: Revolution in finance. *Financing Startups*, 163-184. doi:10.1007/978-3-030-94058-4_10

Popescu, A. D. (2022). Understanding FinTech and decentralized finance (DeFi) for financial inclusion. *FinTech Development for Financial Inclusiveness*, 1-13. doi:10.4018/978-1-7998-8447-7.ch001

Qin, K., Zhou, L., Livshits, B., & Gervais, A. (2021). Attacking the DeFi ecosystem with flash loans for fun and profit. *Financial Cryptography and Data Security*, 3-32. doi:10.1007/978-3-662-64322-8_1

Saengchote, K., Putnins, T. J., & Samphantharak, K. (2022). Does DeFi remove the need for trust? Evidence from a natural experiment in Stablecoin lending. SSRN *Electronic Journal*. doi:10.2139/ssrn.4161945

Saha, P., Hossain, M. E., Prodhan, M. M., Rahman, M. T., Nielsen, M., & Khan, M. A. (2022). Profit and loss dynamics of aquaculture farming. *Aquaculture (Amsterdam, Netherlands)*, *561*, 738619. doi:10.1016/j.aquaculture.2022.738619

Schär, F. (2021). Decentralized finance: On blockchain- and smart contract-based financial markets. *RE:view*, *103*(2). Advance online publication. doi:10.20955/r.103.153-74

Schueffel, P. (2021). DeFi: Decentralized finance - An introduction and overview. *Journal of Innovation Management*, *9*(3), I–XI. doi:10.24840/2183-0606_009.003_0001

Sriman, B., & Kumar, S. G. (2022). Decentralized finance (DeFi): The future of finance and Defi application for Ethereum blockchain based finance market. *2022 International Conference on Advances in Computing, Communication and Applied Informatics (ACCAI)*. https://doi.org/10.1109/ACCAI53970.2022.9752657

Stein Smith, S. (2019). Data as an asset. *Blockchain, Artificial Intelligence and Financial Services*, 213-239. doi:10.1007/978-3-030-29761-9_17

Sun, X., Chen, X., Stasinakis, C., & Sermpinis, G. (2022). Multiparty democracy in decentralized autonomous organization (DAO): Evidence from MakerDAO. SSRN *Electronic Journal*. doi:10.2139/ssrn.4253868

Trabelsi, N. (2020). Impact of crypto-asset trade on financial stability. *Advances in Finance, Accounting, and Economics*, 210–232. doi:10.4018/978-1-7998-0039-2.ch011

Umar, Z., Polat, O., Choi, S., & Teplova, T. (2022). Dynamic connectedness between non-fungible tokens, decentralized finance, and conventional financial assets in a time-frequency framework. *Pacific-Basin Finance Journal*, *76*, 101876. doi:10.1016/j.pacfin.2022.101876

Van der Merwe, A. (2021). Cryptocurrencies and other digital asset investments. The Palgrave Handbook of FinTech and Blockchain, 445-471. doi:10.1007/978-3-030-66433-6_20

Waltl, B., Sillaber, C., Gallersdörfer, U., & Matthes, F. (2018). Blockchains and smart contracts: A threat for the legal industry? *Business Transformation through Blockchain*, 287-315. doi:10.1007/978-3-319-99058-3_11

Wang, D., Wu, S., Lin, Z., Wu, L., Yuan, X., Zhou, Y., Wang, H., & Ren, K. (2021). Towards a first step to understand flash loan and its applications in DeFi ecosystem. *Proceedings of the Ninth International Workshop on Security in Blockchain and Cloud Computing*. 10.1145/3457977.3460301

Wang, Y., Lucey, B., Vigne, S. A., & Yarovaya, L. (2022). An index of cryptocurrency environmental attention (ICEA). *China Finance Review International*, *12*(3), 378–414. doi:10.1108/CFRI-09-2021-0191

Werner, S., Perez, D., Gudgeon, L., Klages-Mundt, A., Harz, D., & Knottenbelt, W. (2022). Sok: Decentralized finance (DeFi). *Proceedings of the 4th ACM Conference on Advances in Financial Technologies*. 10.1145/3558535.3559780

White, E. (2023). What does finance democracy look like?: Thinking beyond fintech and regtech. *Transnational Legal Theory*, *14*(3), 1–25. doi:10.1080/20414005.2023.2204777

Wieandt, A., & Heppding, L. (2023). Centralized and decentralized finance: Coexistence or convergence? *The Fintech Disruption*, 11-51. doi:10.1007/978-3-031-23069-1_2

Wiyono, A., Saw, L. H., Anggrainy, R., Husen, A. S., Purnawan, Rohendi, D., Gandidi, I. M., Adanta, D., & Pambudi, N. A. (2021). Enhancement of syngas production via Co-gasification and renewable densified fuels (RDF) in an open-top downdraft gasifier: Case study of Indonesian waste. *Case Studies in Thermal Engineering*, *27*, 101205. doi:10.1016/j.csite.2021.101205

Wronka, C. (2021). Financial crime in the decentralized finance ecosystem: New challenges for compliance. *Journal of Financial Crime*, *30*(1), 97–113. doi:10.1108/JFC-09-2021-0218

Zetzsche, D. A., Arner, D. W., & Buckley, R. P. (2020). Decentralized finance. *Journal of Financial Regulation*, *6*(2), 172–203. doi:10.1093/jfr/fjaa010

Zhang, Z., Gu, Y., Jiang, L., Yu, W., & Dai, J. (2023). Internet of things and blockchain-based smart contracts: Enabling continuous risk monitoring and assessment in peer-to-peer lending. *Journal of Emerging Technologies in Accounting*, *20*(2), 181–194. doi:10.2308/JETA-2022-003

Zhitomirskiy, E., Schmid, S., & Walther, M. (2023). Tokenizing assets with dividend payouts—A legally compliant and flexible design. *Digital Finance*. Advance online publication. doi:10.1007/s42521-023-00094-w

Zhou, L., Xiong, X., Ernstberger, J., Chaliasos, S., Wang, Z., Wang, Y., Qin, K., Wattenhofer, R., Song, D., & Gervais, A. (2023). Sok: Decentralized finance (DeFi) attacks. *2023 IEEE Symposium on Security and Privacy (SP)*. 10.1109/SP46215.2023.10179435

Chapter 7
The Digital Disruption of Distribution:
An Automobile Industry Perspective

Kirtikumar Tolani
https://orcid.org/0009-0008-5546-9574
Chitkara Business School, Chitkara University, Punjab, India

Asif Saraiya
Santander UK PLC, UK

Sridhar Manohar
https://orcid.org/0000-0003-0173-3479
Chitkara Business School, Chitkara University, Punjab, India

ABSTRACT

Digital transformation across human activity has emerged as a critical evolutionary trend. It has not only led to the emergence of neo-business models but has disrupted traditional industries like automobile manufacturers and stakeholders. A ripple effect is visible in the distribution phase of the automobile value chain due to the rapidly evolving needs of automobile customers, awareness and access to digital platforms, and the need to enhance customer lifetime value by manufacturers. In this chapter, the digitalization of automobile distribution has been discussed in three aspects – customer acquisition, retail experience, and customer life-cycle management. Digitalization has not only helped automobile dealers reduce their customer acquisition costs but also driven higher customer engagement and satisfaction. As more customers become tech-savvy and digitally aware, automobile dealers and manufacturers will need to quickly adapt to the new normal of digitalization in order to remain relevant to customers, innovate sustainable business models, and unlock customer lifetime value.

DOI: 10.4018/979-8-3693-1762-4.ch007

Figure 1. Digital transformation across the world
Source: Fukuyama (2018)

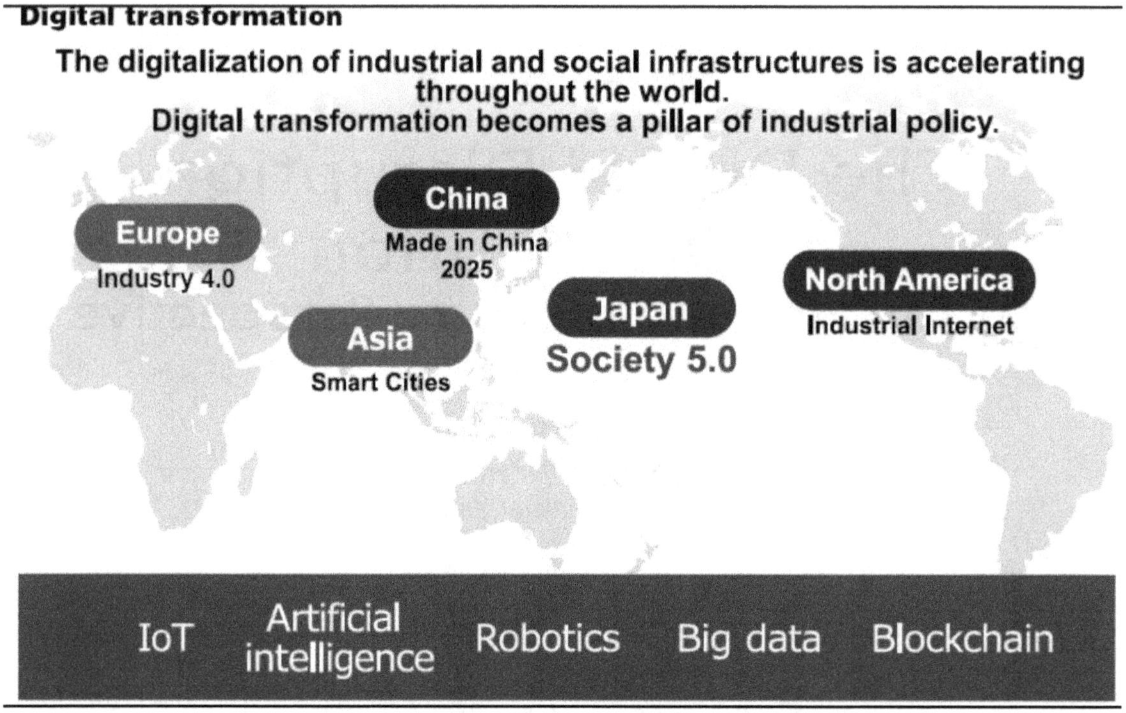

1. INTRODUCTION

The digital transformation sweeping across industries has fundamentally altered the way businesses engage with customers and market their products. In no sector is this transformation more evident than in the automobile industry, where digitalization has disrupted traditional distribution methods and consumer behaviour. The swift transition towards information-centric and technology-driven cultures has led to the democratization, decentralization, and de-globalization of data, giving rise to concepts like Society 5.0 and the Fourth Industrial Revolution (Fukuyama, 2018). This paradigm shift has not only spurred the emergence of innovative business models but also posed significant challenges to traditional players, including automobile manufacturers and dealers. The various initiatives of digital transformation reshaping the world are highlighted in Figure 1.

One of the ripple effects of this digitalization wave is profoundly evident in the distribution phase of the automobile value chain. Automobile dealers and manufacturers face the pressing need to increase customer lifetime value as consumer needs quickly change as a result of increased digital awareness and access to online platforms (Hanelt et al., 2015; Khan et al., 2021). This study embarks on an exploration of digital technology, as a pivotal force in innovation that is reshaping the industry (Shalender and Shanker, 2023).

The objectives of this paper are twofold: first, to comprehensively understand how digitalization has disrupted automobile distribution networks; and second, to shed light on how digital technology has empowered automobile dealers to build scalable and sustainable business models. The challenges and way forward for digitalization in automobile distribution have also been discussed. To achieve these

Figure 2. Dimensions of digitalization in automobile distribution

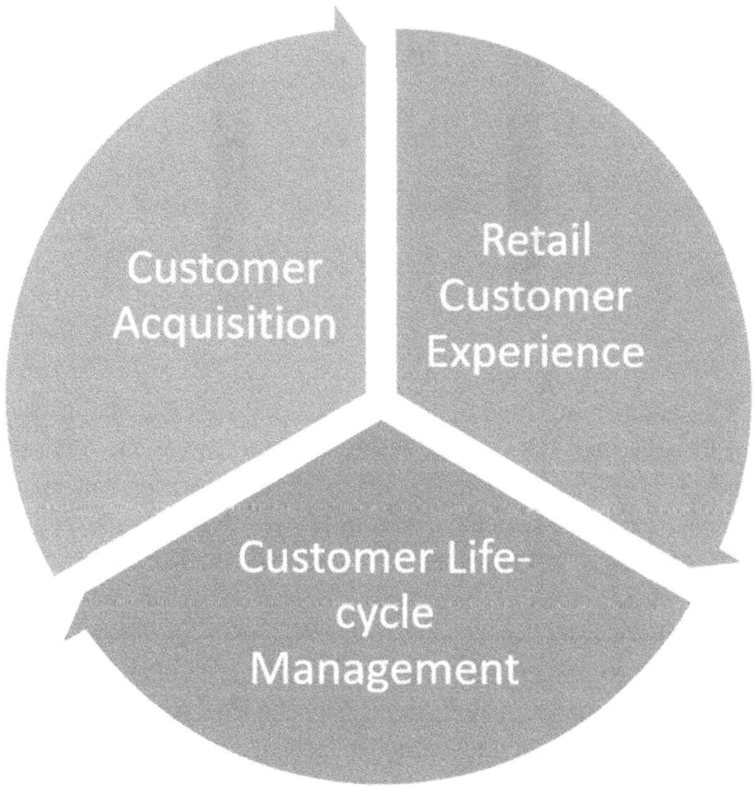

objectives, the paper examines the impact of digitalization across three crucial dimensions: customer acquisition, retail customer experience, and customer life-cycle management, as illustrated in Figure 2.

2. LITERATURE REVIEW

A systematic literature review was conducted to identify the broad impact of digitalization on the Automobile industry, in reference to the dimensions of customer acquisition, retail customer experience, and customer life-cycle management. The review results were classified broadly in the dimensions as shown in Figure 3.

The systematic literature review was conducted using PRISMA (Preferred Reporting Items for Systematic Reviews and Meta-Analyses) technique to identify and shortlist relevant literature. The search criteria were defined based on the research objectives and the above dimensions, to determine the inclusion or exclusion of articles from the analysis. The study used one search query "digitalization AND automobile AND distribution". The second search query used the keywords "digitalization AND automobile AND retail" while the last query used was "digitalization AND automobile AND CRM". The eligibility of literature was extended beyond peer-reviewed papers, to relevant reports published by government organizations, reputed consulting agencies, news and automobile websites and automobile industry associations. The timeline for the literature reviewed was publications in English from January

Figure 3. Dimensions of digitalization

2014 onwards. The scholar databases used in this present study were the Scopus-indexed journals from publishers such as IEEE, Elsevier, IGI Global, MDPI, Science Direct and Taylor-Francis, along with other sources like Google Scholar and Industry Reports from reputed consultancies and service providers like SalesForce. The outcome of the search resulted in 150 articles with titles, abstracts and keywords. The collected articles were cross-checked to eliminate duplicates and articles without full text, resulting in the exclusion of 80 articles.

The balance of 70 articles was further scanned to examine their relevance and quality. The literature that did not meet the relevant eligibility criteria was excluded and resulted in 50 articles, along with additional websites with relevant content, being selected for the final review. These articles were further clustered into relevant dimensions of digitalization for effective study.

3. CUSTOMER ACQUISITION

Customer Acquisition is the process of identifying potential customers for a product, attracting and engaging them, and finally getting the customers to buy the products. Customer acquisition is a term often associated with the Marketing Funnel, which envisages the movement of customers from awareness to purchase, further extending to loyalty and advocacy (Hoban and Bucklin, 2015) as shown in Figure 4.

In this journey through the digital landscape of the automobile industry, it becomes evident that digital marketing, specifically website marketing, is not just a choice but a necessity for dealers and manufacturers to thrive in the new digital era. It is a transformative force that has redefined the rules of engagement, driving us towards a future where the automotive sector is as much about virtual experiences as it is about physical touchpoints.

In this section, we delve into the multifaceted world of customer acquisition in the automobile industry, unravelling the lead generation process, the technology and techniques at play, website design innovations, AI/ML applications, remarketing strategies, and the enablers and challenges of implementation. Real-world industry examples are presented to illustrate the practical application of these strategies.

3.1 Lead Generation Through Website Design

In the competitive landscape of the automobile industry, website design serves as a vital tool for attracting and engaging potential leads. An effective website design not only captivates visitors but also enriches

Figure 4. The marketing funnel

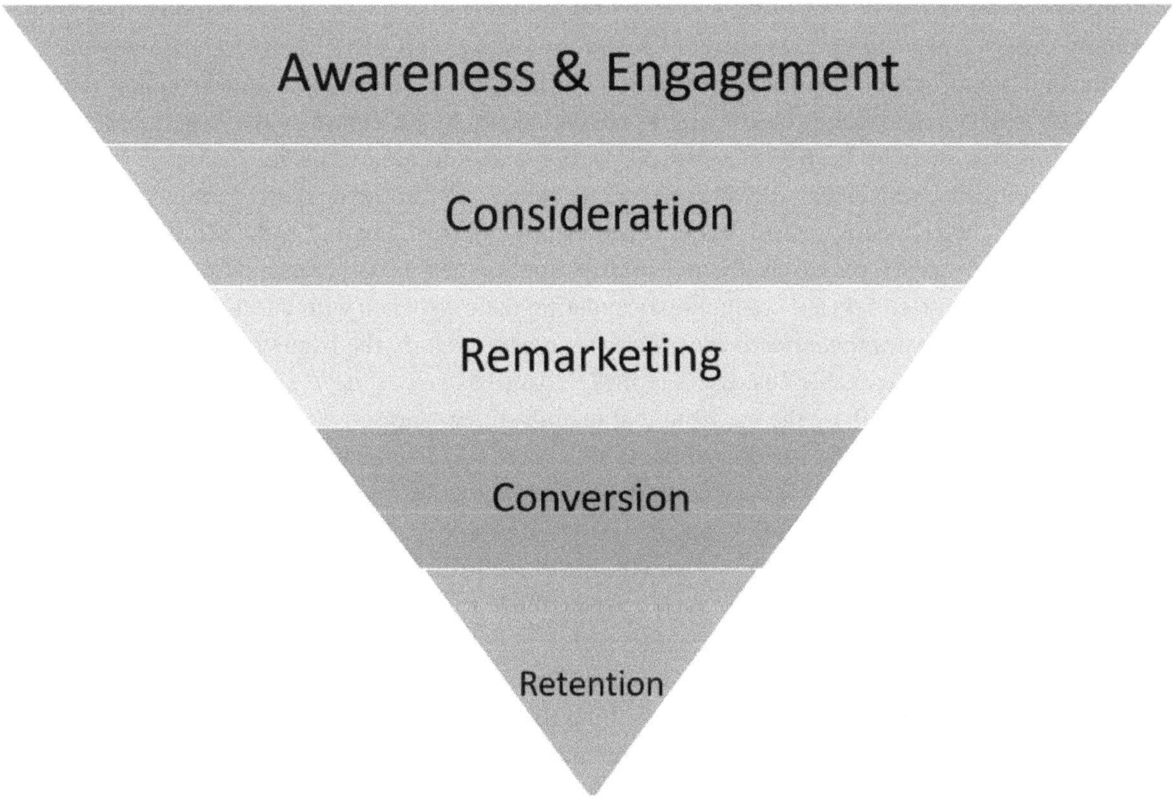

Figure 5. SSI India 2022 factors
Source: Data: JD Power, 2022, Figure: Authors' own

their experience, encouraging lead generation. As per the JD Power 2022 Sales Satisfaction Survey of car customers in India, over 73% of vehicle shoppers in India are aware of the exact model they wish to purchase, and 88% of customers research online before visiting a physical Showroom to experience the product. The brand website of any manufacturer contributes 14% to the factors determining the Sales Satisfaction Index of automobile customers in India as shown in the Figure 5 (JD Power, 2022).

According to a study by Esch & Stewart (2021), organisations are extensively using AI (hereinafter Artificial Intelligence)-enabled marketing to generate content and automate their digital marketing campaigns. This has helped them to reach out to their customer base effectively and optimise their costs of marketing. Online media tools are an effective information source for customers looking to purchase their next car, hence manufacturers are continuously updating their websites with latest product information and features. The websites that effectively enable customers to study the features, and specifications of cars and compare them to enable customers to make effective choices, have a positive influence on the customer's decision to purchase the car from that manufacturer (Samson et al., 2014).

Key elements of website design contribute to this success. Visual appeal sets the stage, with high-quality images and a clean layout creating an immediate and engaging impression. High-resolution images offering detailed views of vehicles allow potential leads to explore products in great detail. User-friendly navigation ensures visitors can effortlessly find the information they seek, with clear menus and intuitive navigation bars simplifying the browsing experience. Calls-to-action (hereinafter CTAs) are strategically placed to guide visitors toward lead capture forms and other conversion points. Well-placed CTAs strategically positioned throughout the website drive lead generation by guiding visitors toward desired actions, such as requesting a test drive or subscribing to updates. Many automobile websites also offer the services of chatbots, which are AI-enabled interactive digital tools that communicate with customers and answer their queries, take suggestions and also help customers navigate through the website toward the final purchase stage of products. These chatbots offer engaging and personalised customer experiences thereby enhancing the retention of customers on the website and pushing them for purchase (Chung et al., 2020)

Figure 6 shows a page of the Hyundai Cars website in India (Hyundai, 2023). The specific page facilitates the end-to-end purchase of a car online through the company's 'Click-to-buy' initiative for customers. The website offers an attractive image of the product, clear CTAs, and also offers the services of a Virtual Assistant to answer customer queries and further engage with customers.

Mobile responsiveness is imperative in an era when users access websites from various devices, ensuring seamless adaptation to different screen sizes. A mobile-responsive design enhances the user experience, allowing potential leads to explore and engage with content effortlessly, regardless of their device.

On Automobile websites, vehicles are prominently displayed, and tools like 360-degree views and virtual showrooms make it possible to explore vehicles in great detail. Content accessibility ensures that valuable information such as vehicle specifications, reviews, and pricing is easily accessible, facilitating efficient information retrieval. Interactive tools like configurators empower users to customize vehicles according to their preferences, promoting engagement and decision-making. Automobile manufacturers are increasingly adopting e-commerce online retailing techniques like Augmented reality (AR) and Virtual reality (VR) technologies to showcase cars to customers. These technologies are easily available and accessible to consumers via their smartphones (Sharma and Bansal, 2023). Mishra et al. (2021) conducted three experiments to study customer responses to online technology interfaces (AR/VR and mobile apps) for hedonic and utilitarian products. The results showed that AR was more user-friendly, and customers found AR more responsive when buying a hedonic (vs. utilitarian) product. Contrastingly,

Figure 6. Website page for Hyundai Cars, India
Source: www.clicktobuy.hyundai.co.in, 2023

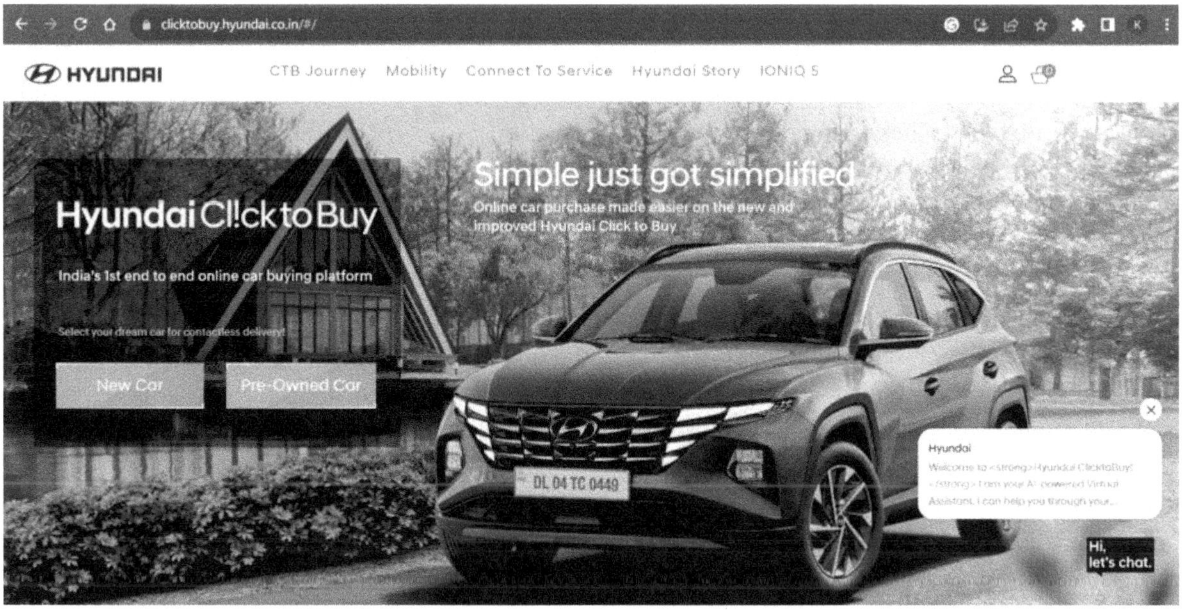

Figure 7. Virtual showroom concept of Kia Motors
Source: www.kia.com, 2023

while buying a utilitarian product (cars or durables), Touch interface users may have a better experience and greater purchase intentions, as compared with AR. Figure 7 below shows a Virtual Showroom concept hosted by Kia Motors on its website (Kia, 2023). This concept offers customers an immersive and 360-degree experience of viewing cars through the website, customizing the cars using different colours or accessories and understanding the features of the cars through videos and other visual aids. Customers are then requested to register for a call-back or a Test Drive by filling up their contact details, which are then shared with the nearest Dealer to service the request.

While designing the websites, speed and performance optimization are essential, as slow-loading pages can deter visitors. Optimizing images and utilizing content delivery networks (CDNs) improve page load times, enhancing the user experience. Building trust through customer reviews, ratings, industry awards, certifications, and affiliations instils confidence in potential leads. Security measures such as SSL certificates and secure payment gateways protect visitor data, which is crucial for establishing trust and encouraging contact sharing. According to Bulsara and Vaghela (2022), customers, specifically millennials, place a high value on trust, e-service quality and subjective norms while browsing for products online. Intuitive lead capture forms request essential information without overwhelming visitors, and progressive profiling collects additional details over time, striking a balance between data collection and a seamless experience.

In conclusion, website design serves as a critical component of successful lead generation in the automobile industry, enhancing the user experience, fostering engagement, and ultimately contributing to the growth and success of automobile dealers and manufacturers in the digital landscape.

3.2 Lead Generation Using Remarketing Techniques

Remarketing, a potent digital marketing strategy in the automotive industry, involves re-engaging potential leads who have previously visited automobile dealers' or manufacturers' websites. This strategy relies on cookies, tracking codes, and personalized ad campaigns to entice visitors to return and convert into valuable leads. The process begins with cookie-based tracking, where small data pieces are stored in visitors' browsers to monitor their behaviour, pages visited, and preferences. This data enables visitor segmentation, categorizing individuals based on interests and engagement, and facilitating highly targeted remarketing campaigns (Olga and Vlad, 2014). Personalized ads are then deployed to these segments, appearing on other websites and social media platforms, showcasing vehicles and offers aligned with prior interactions. Dynamic ads take personalization further, displaying specific vehicle models or features previously of interest to the visitor. According to Sahni et al. (2019), remarketing and retargeting lead to 14.6% of customers returning to the website within 4 weeks of the first visit. The chances of revisiting the website also decrease as the delay from the first visit increases; hence the need for automation of remarketing campaigns is felt (Esch and Stewart, 2021).

Remarketing also involves creating retargeting campaigns for specific visitor groups and tailoring ads to their particular needs or interests, such as financing options for those who left during the financing stage. To prevent overexposure, frequency capping limits ad displays. Cross-device remarketing ensures a seamless experience as users switch between devices. Crafting compelling ad creatives and messaging is crucial, as is aligning ads with prior interactions and providing incentives to revisit the website. Conversion optimization techniques, like streamlined forms, are employed to facilitate lead generation. Privacy compliance is essential, with businesses offering transparent information on cookie usage and opt-out options. Remarketing, fueled by cookies and personalized ad campaigns, reignites leads' interest, improving the likelihood that they will convert and boosting the success of the automotive sector's digital marketing.

3.3 Lead Generation Through Social Media Marketing

The extensive use of social media among digitally connected customers has opened up new avenues of marketing for the Automobile industry. Social networks enable dealers to interact with their target audience, segmenting ads based on customer profiles to streamline marketing efforts. Social media advertising also builds the brand's image, and increases the visibility of products to get closer to customers (Cordova et al., 2022). Dealers, in tandem with automobile manufacturers, are leveraging their social media presence to attract and engage with customers on social media channels like Facebook, Instagram and LinkedIn (Barlow, 2020). The use of Social Media Marketing not only enables dealers to communicate the latest updates or offers to customers in real-time, but it also helps them to maintain engagement with customers or gauge their response to new offers or products. Social media marketing also helps dealers keep marketing and customer acquisition costs low compared to traditional offline marketing (Johnson and Leonard, 2020). Automobile dealers are also using location-based mobile targeting techniques for attracting hyperlocal customers with customised offerings (Manohar et al., 2021) Figure 8 shows a social media post on the LinkedIn channel by an authorized dealer of Skoda Auto in India (LinkedIn, 2023).

Thus, the successful implementation of website marketing for lead generation in the automobile industry hinges on leveraging enablers while addressing the challenges inherent in the digital landscape. A well-rounded approach, coupled with adaptability and adherence to privacy regulations, positions businesses to thrive in the competitive online space.

3.4 Industry Practices: Digital Customer Acquisition

Several prominent automobile dealers and manufacturers have effectively harnessed website marketing for lead generation. These examples demonstrate the diverse strategies employed within the industry:

Tesla, Inc.: Tesla, known for its innovative electric vehicles, has leveraged website marketing to generate leads effectively. The website has product features, virtual test drives and enables customers to reserve vehicles online. It also features personalised vehicle suggestions based on user preferences.

Ford Motor Company: Ford runs a strong website marketing strategy that includes full vehicle specifications, interactive configurators and a 'build and price' feature. It uses personalisation driven by data to suggest vehicles and offers to visitors whilst they browse.

BMW Group: BMW's website is well-known for its 360-degree views, AR (Augmented Reality) test drives and virtual showrooms. These tools make the customer experience immersive and allows the users to explore their entire basket of their offerings.

Toyota Motor Corporation: Toyota provides in-depth information about its line-up of vehicles in great detail including hybrid and electric options. The website focuses on its sustainable mobility options and provides functionality to request a quote and schedule test drives.

Audi AG: The sophisticated design and user experience of Audi's website ensures that it stands out from the crowd. Using emerging technologies like AI Chatbots to interact with visitors, respond to enquiries and guide customer towards lead capture forms are evident on the website.

Hyundai Motor Company: Hyundai's website uses dynamic remarketing, displaying models that visitors have shown interest in. Tools for comparing vehicles are also offered on the website.

Figure 8. Social media post of an automobile dealer on LinkedIn

Volkswagen Group: Volkswagen's website focuses on user-generated content, featuring customer reviews and ratings alongside detailed vehicle information. This transparent approach builds trust and credibility among potential leads.

Mercedes-Benz USA: Mercedes-Benz employs comprehensive lead capture forms and uses visitor data to personalize future interactions. They emphasize customer-centric experiences, offering virtual showroom tours and AI-powered chat support.

These industry examples showcase the versatility of website marketing in the automobile sector. Each brand implements unique strategies to capture, engage, and convert potential leads, aligning its digital presence with the evolving expectations of consumers in the automotive marketplace.

Case Study: 3 trends driving the Auto industry's shift to dealer digitization (*Thinkwithgoogle.com*, 2020)

According to a report on the Indian Automobile Industry by Google in 2020, the trend of digitalization had picked up strongly post-2015. The study also offered the following insights into online customer behaviour:

Preference for Online: The average number of customer visits to showrooms had also fallen by 50% from 2016 to 2019. 4 out of 5 people would like to purchase a car online if the option was available. Nearly 33% of customers would pre-pone their car purchases if an online option was available. 24% of customers would prefer a Test Drive of their intended car at their home. This shows a clear shift towards digitalization.

Online product discovery: More than 90% of prospective car customers seek information online from brand websites, YouTube, professional reviews & customer testimonials. 56% of customers visit the car brand's website during their journey and over 60% search for their local car dealerships online. This necessitates that automobile dealers and brands also align themselves to the new normal of digitalization.

Industry trends: In the case of Maruti Suzuki India Ltd., a leading car manufacturer in India, the share of web-based enquiries has surged to 39% from 3% over three years. Digital enquiries contributed over 30% of the total enquiries for Hyundai Motor India, another car manufacturer. Nearly 21 out of 26 touchpoints of customers with dealerships are digital, as per the Google report.

4. RETAIL CUSTOMER EXPERIENCE

With the growing popularity of e commerce in the retail market, many traditional "brick-and-mortar" retailers are also pivoting their business models. According to Palmie et al. (2022), traditional retailers engage with specialised digital service providers to make the shift towards digital business models, to create a meta-ecosystem of retailers competing in the digital universe. A similar impact of digitalization is felt in the automobile sector as well and transcends the sphere of customer acquisition to customer interaction and engagement as well. Globally, the trend of having digital interfaces in car showrooms or retail touchpoints is catching up. The use of digital interfaces either reduces bottlenecks in product discovery or substantially enhances customer experience (Gauri et al., 2021). The techniques of virtual reality (VR), mixed reality (MR), and augmented reality (AR) are helping automobile dealers to offer immersive and personalized purchase experiences to customers (Sharma and Bansal, 2023). At the same time, using compact screens and optimized virtual stores helps automobile dealers reduce real estate costs compared to having larger showrooms with multiple car displays. The advent of digital business

Figure 9. Impact of digitalization on automobile retail formats
Source: ET Auto.com (2020)

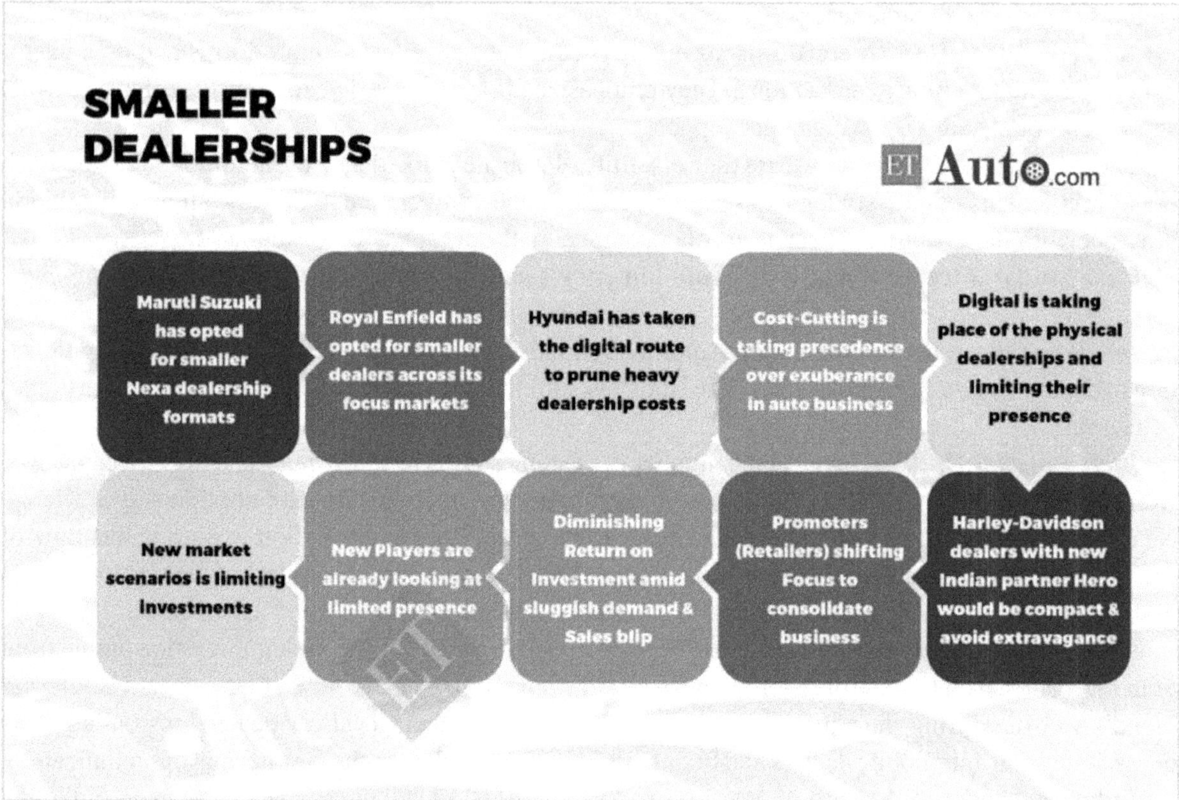

led to the erosion of the brick-and-mortar dealership models and COVID-19 expedited the transition process as shown in Figure 9 (ET Auto.com, 2020).

Unlike traditional car showrooms, digital showrooms do not have cars on display. Instead, these showrooms have large interactive screens where customers can experience the cars and all their features before taking a test drive. The screen shows a 3D (three-dimensional) view of the car, capturing its attributes from multiple angles. Buyers can see colour options, check out various security features, and learn about the car's features. Some automobile dealers use AR (augmented reality) on an iPad to project a virtual 3-D image of the car in the digital store. A trend catching up across retail interfaces globally is the use of personalized mixed reality to customize purchase experiences for customers. Automobile dealers, in line with this trend, offer customers a mixed reality experience by way of personal wearable devices like smart glasses or AR/VR headsets that enable customers to explore cars in the virtual world (Mishra et al., 2021)

Automobile manufacturers, in collaboration with their dealers, are innovating new retail formats for customers. One such format is the Experience Centre concept, which aims to engage Car customers strongly with their favourite brands and products by offering a state-of-the-art immersive digital experience (Bauer, 2018).

4.1 Advantages of Digital Automobile Showrooms

Improved Product Presentation: Digital screens can be used by car dealerships to showcase their products in a dynamic and engaging way, using visuals, videos, and interactive presentations.

Lee and Fiore (2006) in their research have suggested that customers' perceptions of product features and quality are substantially impacted by dynamic displays, improving their purchase probability.

Interactive and Immersive Experiences: Interactive and immersive digital interfaces enable customers to explore different facets of the vehicles like variants, configurations and features. This enables higher personalization of options, leading to higher engagement and satisfaction (Roy et al., 2023).

Information Access: Customers in Car showrooms can access real-time pricing information, detailed product information, and comparison tools via digital screens. This enables clients to study vehicles specifications before speaking with sales consultants and make well-informed selections on their own, free from time limitations. (Kim et al., 2018).

Virtual Test Drives: Customers can take virtual test drives without ever leaving the showroom thanks to the integration of virtual reality (VR) technology-enabled digital screens. Research has demonstrated that VR-enabled Test Drives can positively influence both purchase intention and consumer satisfaction (Fan et al., 2020).

Considering the above factors, automobile dealerships looking to boost sales and enhance the customer experience may find that digital screens are a useful tactic. Dealerships can give clients a more interesting and engaging experience by incorporating digital screens into the flow and design of the showroom and the sales process. Dealerships may also maximise sales conversion rates and customise the showroom experience by utilising client data obtained from digital screen interactions.

4.2 Industry Practices: Digital Automobile Showrooms

Kia Motors: The Beat 360 brand experience centre is a cutting-edge, digital-first showroom model created by Kia Motors shown in Figure 10 below. It has a "Surround Media" Zone with a Kia car on a turntable with an 11-meter-wide digital screen. It also has a 'Mixed Reality (MR)' zone, in which customers can explore the cars in the virtual world through mixed reality technology, wearing head-mounted display units. Customers can also visualise and create their own car using a Kia digital configurator. (AutocarIndia.com, 2019)

MG Motors: MG Motors has experimented with a car-less digital Showroom called the 'MG Digital Studio'. It has innovative digital elements such as a digital running façade, a video wall configurator, and a mega visualiser with a Virtual Reality (VR) / Augmented Reality (AR) Zone wherein customers can configure cars and visualise their choices in an immersive experience. (MG Motor, 2019).

Skoda Auto: Skoda Auto has launched an Experience Centre which has a variety of digital interfaces like touchscreen kiosks, where customers can configure their cars. It also has large digital screens inside the showroom that feature information about the brand, the company's heritage and the latest product updates and offers on cars (Skoda Auto, 2023).

Citroen: Citroen has tried to bridge the physical-digital divide of Automobile Showrooms by launching the 'Phygital' (physically digital) concept of retail. Phygital is an innovative format of omnichannel retailing that combines physical and digital elements of retail, focusing on the human touch to satisfy social and symbolic consumer needs (Pangarkar et al., 2021). These showrooms have a digital screen on

Figure 10. Kia motors digital showroom
Source: AutocarIndia.com, 2019

Figure 11. Citroen motors digital showroom
Source: Automotiveleadnews.com, 2021

the exterior, and numerous interior screens and a high-definition 3D car configurator allowing customers a 360-degree view of their choice of cars, shown in Figure 11.

5. CUSTOMER LIFE-CYCLE MANAGEMENT

The rapid pace of digitalization in the customer acquisition and retail interface stages of the customer journey generates an immense amount of data. This data, if captured and analysed properly, can lead to rich insights into customer behaviour, that can be leveraged by automobile firms to extract value. Further, as e-commerce and D2C (direct-to-consumer) sales channels reshape vehicle distribution, manufacturers are venturing into innovative models such as subscription-based services and digital sales platforms. While these changes are profound, they have ripple effects that extend beyond the supply chain. The transformation in vehicle distribution has direct implications for marketing strategies within the automotive sector.

In this context, customer relationship management (CRM) software emerges as a lynchpin in the arsenal of tools that automobile dealers and manufacturers must wield. CRM software is not only a luxury in the digital age of the automobile business, but also a need due to its ability to streamline, organise, and improve client interactions. (Jhamb et al., 2022). This section explores the value and advantages of CRM software for the automotive industry, explaining how it helps with relationship optimisation, business growth, and navigating the ever-changing landscape.

5.1 The Evolving Automotive Supply Chain: A Catalyst for CRM Adoption

It is essential to understand the significant changes taking place in the automotive supply chain before diving into the details of CRM software. Conventional supply chain models are being reassessed; they are being redefined by complex networks of manufacturers, service providers and dealers. This transition is being driven by several factors:

Changing Customer Demands: As per a report from Salesforce in 2023, 18% of car customers would prefer to buy a car online given the availability of such an option. This statistic underscores the growing importance of online tools and platforms that facilitate research, comparison, and even negotiation of prices and terms. Customers value their time and are increasingly turning to mobile-enabled digital tools for a more efficient car shopping experience. Consumer expectations have shifted significantly. Today's automotive customers are digitally savvy, demanding convenience, personalization, and seamless purchasing experiences. They expect the ability to research, configure, and even purchase vehicles online. This shift necessitates a reimagining of how vehicles are distributed and marketed.

Rise of E-Commerce: E-commerce is no longer confined to retail goods; it has permeated the automotive industry. Virtual showrooms, 360-degree views, and augmented reality experiences have made online car shopping more comfortable for customers. The future business models will include the sale and operation of vehicles. (PwC, 2018). E-commerce platforms and direct-to-consumer sales channels have emerged as viable distribution methods, bypassing traditional intermediaries (Tandon et al., 2021)

Subscription-Based Services: Innovative models, such as subscription-based services, have gained traction. Customers can now subscribe to vehicles, accessing the convenience of ownership without the long-term commitment. These services require a different approach to customer engagement, as subscribers may switch vehicles frequently based on their needs.

Digital Sales Platforms: The ordering, configuring and delivery of vehicles on a digital platform is an option that is actively explored by manufacturers around the world. The transparency and customisation on offer require a solid digital marketing and customer relationship strategy. The supply chain in the automotive industry continues to undergo fundamental changes where the evidence stacks up in favour of having more customer-focused strategies. The CRM software helps exploit this feature.

5.2 Understanding CRM Software

Customer Relationship Management (CRM) software is a solution designed to help with the managing and nurturing of relationships with customers and prospective customers. It gives you a single view or interface for capturing, organising and leveraging customer data to improve interactions and increase sales. At a broad level, the CRM software allows businesses in the following aspects:

Information Database: CRM software functions as a repository for customer data with important information like contact details, purchase history, interactions and preferences. This centralised view ensures that the sales and marketing teams are always ready with the most up-to-date view of the customer. Organisations are now using AI integrated with CRM systems to analyze the huge volumes of customer data. Automobile organizations are developing competencies and processes to harness the power of AI-based decision-making (Chatterjee et al., 2021)

Automation of Tasks: CRM software has communication capabilities, including integration with email systems and other directory or contact management systems. This allows the CRM tools to have effortless interactions with customers, guaranteeing timely responses.

Communication: CRM software offers communication tools, including email integration and contact management. These features enable seamless communication with customers, ensuring that inquiries are addressed promptly.

Understanding Consumer Behaviour: CRM systems have the ability to provide insights into customer behaviour and preferences. The data gathered can then be used for various marketing campaigns, product promotions and personalised recommendations to customers.

Slice and Dice Target Customers: CRM software provides businesses with the ability to segment customers based on multiple criteria. The segments can be used in targeted marketing and promotions and increase the relevance of the interactions.

Performance Tracking: With a suite of reporting and analytics capabilities, the CRM software helps businesses see trends and patterns in data that can be used to measure the effectiveness of marketing campaigns, sales and performance tracking. In this context, an innovative process of social CRM has emerged which enables advertisers to analyse large volumes of data to evolve an efficient, effective and cost-effective marketing strategy (Lamhrari et al., 2022).

5.3 Importance of CRM Software in the Automotive Industry

CRM software becomes crucial in light of how the automotive supply chain is evolving. A PriceWaterhouseCoopers study from 2017–2018 centres on how the market will evolve by 2030, starting with users and their patterns of mobility. It emphasises both autonomous and shared mobility, and CRM represents a firm step in that direction. CRM software is essential to the automotive sector because of the following features:

Personalization at Scale: In an era where customers expect personalized experiences, CRM software equips automobile dealers and manufacturers with the tools to deliver just that. By leveraging customer data, businesses can create tailored marketing campaigns, recommendations, and offers. For instance, a CRM system can analyze a customer's previous purchases and preferences to suggest vehicle models that align with their interests. Personalization enhances customer engagement and fosters loyalty.

Seamless Omnichannel Engagement: The automotive customer journey is no longer confined to the showroom. Customers interact with brands across various touchpoints, including websites, mobile apps, social media, and email. CRM software enables businesses to orchestrate seamless multichannel engagement. Whether a customer submits an inquiry on the website, engages with the brand on social media, or visits a physical dealership, the CRM system ensures that their interactions are tracked, integrated, and responded to cohesively.

Customer Lifecycle Management: CRM software excels in managing the entire customer lifecycle, from acquisition to retention. It aids in lead generation by capturing potential customers' information through web forms, chatbots, or other digital touchpoints. As leads progress through the sales funnel, CRM systems facilitate lead nurturing, automating email campaigns, and personalized communications. This continuous engagement is critical in an industry where the purchase cycle can be lengthy and customer relationships endure beyond a single transaction.

Enhanced Customer Support: In the automotive industry, excellent customer support is paramount. CRM software enhances customer support by centralizing customer inquiries, complaints, and service requests. Service teams can access customer histories, making it easier to diagnose issues and provide timely solutions. Moreover, CRM systems enable proactive maintenance reminders and warranty tracking, enhancing the overall customer experience.

Data-Driven Decision-Making: The automotive sector is not just about selling vehicles; it's also about optimizing operations and strategies. CRM software provides valuable data and analytics that inform decision-making. Businesses can analyze sales performance, marketing campaign effectiveness, and customer behaviour to refine strategies continuously. This data-driven approach is essential in a rapidly changing industry where agility and adaptability are paramount.

5.4 Benefits of CRM Software in the Automotive Industry

The adoption of CRM software in the automotive industry yields a plethora of benefits, including:

Improved Customer Retention: CRM software helps businesses foster long-term relationships with customers. By understanding their preferences and providing personalized experiences, businesses can enhance customer loyalty and reduce churn rates. Repeat customers are not only more profitable but also serve as brand advocates.

Streamlined Lead Management: Efficient lead management is critical in the automotive sector. CRM software streamlines lead capture, nurturing, and distribution to sales teams. It ensures that leads are not lost or overlooked, increasing the likelihood of conversion.

Enhanced Marketing Campaigns: CRM software enables data-driven marketing campaigns. Businesses can segment their customer base and target specific segments with relevant promotions. This level of targeting increases the effectiveness of marketing efforts and maximizes return on investment.

Increased Sales Efficiency: CRM systems automate routine sales tasks, such as data entry and follow-up. This automation frees up sales teams to focus on high-value activities, such as building relationships and closing deals. The result is increased sales efficiency and productivity.

Real-time Analytics: Access to real-time analytics is invaluable in an industry where market conditions can change rapidly. CRM software provides up-to-the-minute insights into customer behaviour, sales performance, and marketing ROI. This real-time data empowers businesses to make agile decisions and adapt to changing circumstances.

Enhanced Customer Service: CRM software facilitates responsive and efficient customer service. Service teams can access customer information, service histories, and warranties, enabling them to provide quick and accurate solutions to customer inquiries and issues.

5.5 Case Study

CRM Success Coupled with Data Analytics in the Automotive Industry: *American Honda Motor Co., Inc. turns service repair data into cost savings.* (SAS 2023)

Background: When a car customer brings his vehicle into an Acura or Honda dealership in the USA, a lot of data is generated. During each visit, the service technicians generate data on the repairs, including any warranty claims made previously to Honda Motor Co., Inc., that feed directly into its database. This includes the type of work performed, customer payments, service advisor comments, and many other data points.

Examining warranty data to make maintenance more efficient: Like all automobile companies, American Honda works with a set of dealerships that perform warranty repair work on its cars. This can be a significant cost for the company; hence American Honda uses analytics superimposed on their customer data to ensure that warranty claims are complete and accurate upon submission. Before the implementation of CRM, dealership personnel used to take up to 1 week every month to examine the warranty data for defects; currently, this data is available online in real-time, freeing up substantial time for other tasks.

Identifying suspicious claims accurately: The Advanced Analytics group used SAS Analytics to analyze warranty data, which allowed the Claims group and field personnel to quickly and accurately identify incomplete, inaccurate, or non-compliant claims. Earlier the dealership team needed 3 minutes to identify a suspicious warranty claim and could get it confirmed as correct with a success rate of 35%. After implementing the CRM software, they need under 1 minute to identify any suspicious claim, while the success rate of identification has more than doubled to 76%.

To decrease warranty expenses: The Advanced Analytics team used SAS Analytics to create a proprietary process to surface suspicious warranty claims for scrutiny daily to make sure they comply with existing guidelines. Proper identification of warranty claims coupled with training at dealerships, has helped to halve the warranty expenses of American Honda by 52% over the previous phase.

Using service data to forecast future needs: The Advanced Analytics team also uses service and parts data to develop strong relationships with customers by ensuring dealers have regular parts available for customer repairs.

Customer feedback that drives the business: American Honda uses analytics to rapidly assess customer survey data. The Advanced Analytics team uses SAS to mine survey data to gain insights into how vehicles are being used and identify design changes that are most likely to improve customer satisfaction.

Thus, in an automotive industry undergoing rapid transformation, CRM software and data analytics emerge as pivotal tools for dealers and manufacturers. It enables personalization at scale, facilitates multichannel engagement, supports customer lifecycle management, enhances customer support, and empowers data-driven decision-making. The benefits include improved customer retention, streamlined lead management, enhanced marketing campaigns, increased sales efficiency, and real-time analytics.

As the supply chain evolves to accommodate changing customer demands and digitalization, CRM software serves as a driving force in aligning marketing strategies with the new distribution landscape. It is not merely a tool but a strategic imperative for businesses seeking to thrive and lead in the dynamic and competitive automotive industry of the digital era. With CRM software as a trusted ally, the road ahead becomes not just one of change but also of opportunity and growth.

6. CHALLENGES IN DIGITALIZATION

While digitalization in distribution can support manufacturers and dealers to improve their marketing efforts and connect closely with customers, it faces several challenges in implementation as outlined below:

Data Privacy: Real-time data collection and analysis is the lifeblood of the digitalization process. A huge amount of user data is generated during the interaction of customers with brand websites, dealership visits and other virtual interactions. Further, digitalization has also become a part of the product (automobiles) now, leading to the emergence of 'connected' cars. These connected cars are not only interfacing with user devices like smartphones but also interacting with the manufacturer's ecosystem online, churning out a high volume of user data during their interaction with customer devices (Buck and Reith, 2020). This raises the concerns of misuse and security of sensitive and personal user data generated during such interactions.

Developed countries like Europe and the USA have strong data protection regulations that ensure that the security and consent of users are secured before and after data is collected. Developing countries like India however are still evolving data privacy laws, leaving customers vulnerable to compromise of user data. Contrastingly, while data protection is essential, according to Farrugia et al. (2022), stringent data privacy regulations like GDPR and CCPA, impact customer data collection and usage, thereby compromising the efficiency of digitalization.

Viability of digitalization: Automobile dealers may need to make a sizeable initial investment in digitalisation as well as provide training to their employees. Dealerships need to commit resources and make sure staff members are proficient to get the most out of implementing digital interfaces (Pantano et al., 2017). Integrating various technologies within the marketing stack can be complex, requiring seamless data flow. Adequate resource allocation for website marketing efforts, including budget, technology, and personnel, presents a continuous challenge.

Overexposure to digital marketing: Keeping up with evolving search engine algorithms and advertising platforms is an ongoing challenge. The widespread use of ad blockers limits the visibility of remarketing ads, and intense competition for online user attention necessitates creative and engaging content. Too many remarketing ads can exhaust or aggravate customers, so it is important to find the right balance between the frequency and relevance of remarketing.

Technology Reliability: In the case of the digitalisation of retail formats, digital screen stability and reliability must be guaranteed to prevent technical issues that could irritate customers. To manage such issues, regular upkeep, system backups, and effective technical support are necessary (Roy et al., 2023).

Organizational change: The Organizational transformation required to change over from a

traditional business model to a digital one is not easy and may face resistance from within the organization. (Esteller-Cucala et al., 2020). It requires substantial upskilling of the employees from the transition phase to keep them updated about the latest trends in digitalization, as well as an acceptance of the change that the transition brings.

User preferences: While a trend is seen in the shift of automobile customers to digitalization, some customers would still prefer offline interactions in the product experience and delivery stage. Customers would still prefer to visit car dealerships and test-drive the cars physically instead of using AR/VR interfaces. Additionally, many customers would prefer to get their cars delivered at dealer showrooms with their families (JD Power, 2022). Servicing and maintenance of cars would still require visit to authorized dealer service centres, and cannot be substituted digitally.

7. WAY FORWARD

Data governance: To protect data privacy, data governance is an essential process to be taken up by institutional regulators, dealers, service providers and manufacturers as well. It is important to ensure the right balance between government regulation of user data mining, balanced by the need for manufacturers and dealers to track customer enquiries and their driving habits to offer better services to them. Robust privacy compliance measures build trust with visitors and ensure data handling aligns with legal requirements.

Support from manufacturers: Digitalization can help automobile dealers transition to smaller showrooms and reduce their real estate costs. Manufacturers need to accept this reduction in format sizes in line with the industry trends. A skilled and trained workforce proficient in digital marketing, data analysis, and web development is essential at dealerships. Manufacturers also need to support dealers transitioning to the digital phase in terms of sharing their costs and providing support in terms of upskilling and training of manpower to extract maximum advantage from digitalization.

Seamless integration of visual interfaces: There are several ways that digital screens can be a part of the sales process and showroom design. They can be used, for instance, to show product details, offer interactive demos, or even take orders from customers. Dealerships may give consumers a more interesting and engaging experience by incorporating digital screens into the showroom setting (Lee et al., 2018).

Flexible and updated content: To avoid user fatigue from digital marketing, digital screens with content should be updated frequently to feature the newest models, features, and special deals. Customers will be more engaged and likely to return if the content is updated regularly. Customer testimonials can also be shown on digital screens, which can contribute to credibility and confidence (Roy et al., 2023).

Data analytics for personalisation: Dealerships can personalise the showroom and website experience by offering customised recommendations and targeted promotions based on customer preferences, by utilising customer data obtained from digital screen interactions. This can maximise the rate of sales conversion. For instance, the dealership might message a customer displaying a specific car model or promotional offer if they are browsing on a screen. (Kim et al., 2018). Enablers include the availability of robust data analytics tools like Google Analytics and Adobe Analytics, which provide valuable insights into visitor behaviour. A well-integrated marketing technology stack, comprising CRM systems, marketing automation platforms, and lead management tools, streamlines processes. User-friendly Content Management Systems (CMS) facilitate content updates.

Omnichannel approach: Depending on the user preferences of customers, the optimal approach would be an omnichannel approach, where the best features of conventional and digital channels are integrated to offer optimal experiences to customers. According to Bijmolt et al. (2021), firms need to co-ordinate their activities across the digital and traditional channels across different stages of the customer journey. This will lead to a cycle of continuous evolution of customer-centric retail formats through experimentation and innovation, ultimately resulting in an integrated or omnichannel format that best serves customer preferences (Gauri et al., 2022).

8. DISCUSSION

The digitalization of the automobile industry's distribution phase has brought forth a transformative shift with far-reaching implications. The observations and findings presented in this paper, focusing on digital marketing strategies within the industry, warrant a discussion of several key facets. The research underscores the pervasiveness of digitalization in the automobile sector. It is no longer a mere trend but a fundamental aspect reshaping how automobile dealers and manufacturers engage with their audience. The COVID-19 pandemic acted as a catalyst, expediting the industry's adaptation to digital technologies due to the lockdowns and interruptions in economic growth. This shift has been sustained post-pandemic, emphasizing its enduring impact.

Due to digitalization, the traditional customer journey of new car sales is in a stage of change, and the traditional distribution triangle between manufacturers, dealers and customers is changing. The impact of the digital transformation process on new car sales is noticeable (Bacher, 2020). While Digitalization offers the advantages of better tracking of customers and enhancing customer experience, the product discovery experience is still incomplete without customers physically driving the car. A comparison of both the retail formats is given in Table 1.

Table 1. Comparison of conventional and digital automobile retail channels

Stage of Retail	Conventional Channel	Digital Channel	Comparison	Advantage
Customer Awareness	ATL & BTL Campaigns (Print Ads, Brochures, Field Events)	Digital Marketing campaigns (Google Ads, Social Media)	Digital channels offer better targeting, diverse and economical media for product discovery	Digital Channel
Customer Acquisition	Walk-ins, Tele-ins, Referrals	Online Leads, Website, Social Media Leads	Better tracking and qualification of customer enquiries	Digital Channel
Product Discovery	Showroom Visits, Home visits by dealer Sales Team	Website, Online consultations, Online Brochures, Videos	Safety, Convenience, Flexibility, Speed of digital channels is higher	Digital Channel
Product Experience	Showroom visit, Test Drives	Online simulation	Conventional channels offer better 'touch-and-feel' experience through Test Drives	Conventional Channel
Finance & Documentation	Showroom Visits, Financer visits	Bank Website, Virtual documentation	Safety, Convenience, Flexibility of Digital channel is higher	Digital Channel
Pre-purchase validation	Friends, Family, Sales Consultants	Online customer Reviews, 3rd party Websites	Trustworthy validation, informed decision making with Digital channels	Digital Channel
Purchase Deal Finalisation	Showroom Visits	Website, Telephone	Customers prefer to visit Showrooms and negotiate for best offers on Cars	Conventional Channel
Product Delivery	Car Showroom	Customer choice of location	Customers prefer to receive car delivery at Showrooms to celebrate the purchase	Conventional Channel
Post-purchase Services	Car Service Centre	Not feasible	Aftersales Maintenance / Repair of Cars require physical presence through authorized Service centres	Conventional Channel

An analysis of the Table 1 shows that Digital Channels for Automobile Retail have their advantages, primarily in the product awareness and discovery stage, in terms of flexibility, convenience and better control for dealers on marketing spends. The advent of digitalization has also helped to manage aspects like documentation and neutral validation of products before purchase. However, the traditional channels are still required for customers to physically experience the product in a comfortable environment and also engage in negotiation for the best offers on the products. Further, the final two stages in the customer retail journey - product delivery and post-purchase services - validate the existence and requirement for brick-and-mortar stores in Automobile Retail. Therefore, the future of automobile distribution lies in the synergy of digital and physical touchpoints—the "phygital" approach.

Depending on their location, industry sector, and clientele, auto dealers and manufacturers adopt varying degrees of digitalization. Developed countries have seamlessly transitioned to digital dealership models, benefiting from high digital awareness among their consumers. In contrast, developing countries face challenges due to limited digital awareness and device penetration. Adopting an omnichannel approach, where digital and physical touchpoints complement each other, becomes essential in such contexts.

The research highlights the nuanced preferences of automobile dealers and manufacturers based on the segment of the industry they cater to. Passenger car dealers exhibit a higher propensity for digitalization due to the digital awareness of their customers. However, dealers dealing with vehicles further down the value chain, such as commercial vehicles or 3-wheelers, face lower digitalization preferences due to the limited choice of vehicles and lower customer digital awareness, and hence lower propensity towards this aspect.

The competitive nature of the automobile industry necessitates a continuous investment in digital tools, both hardware and software. While digitalization offers efficiency and customer engagement benefits, staying up-to-date with the latest digital consumer technologies and behaviours is resource-intensive. The need for upskilling and training dealership personnel further adds to the investment required. As the industry evolves, manufacturers and dealers have the opportunity to explore innovative retail distribution models. These models can disrupt traditional practices and create new opportunities. For instance, direct-to-consumer sales and subscription-based services are emerging trends that have the potential to redefine the industry's landscape.

9. CONCLUSION

The digital transformation sweeping through the distribution phase of the automobile value chain represents a profound and enduring shift. This article has delved into the multifaceted impact of digitalization on the industry, with a particular focus on customer acquisition, retail experience and customer relationship management strategies. Digitalization is no longer a passing trend but a cornerstone of the automobile sector. The rapid adoption of digital technologies, further accelerated by the COVID-19 pandemic, has reshaped how automobile dealers and manufacturers engage with their audiences. This transformation shows no signs of abating, highlighting its enduring significance. While digitalization has ushered in a new era, it cannot entirely supplant the physical experience. Many consumers still value hands-on interaction with vehicles, especially when contemplating a significant purchase. Also, the scale of digitalization of automobile retail varies based on the geography and the segment of vehicles being catered to. The rapid technology change necessitates that automobile dealerships continuously upgrade their digital interfaces and also upskill their manpower to adapt to new technologies. The challenge lies in striking the right balance between resource allocation and return on investment.

For Automobile distribution then, the way forward lies in the seamless integration of digital and physical touchpoints—a "phygital" approach. The key is, thus, an omnichannel strategy that accommodates both digital-savvy and digitally hesitant consumers. Dealerships that master this synergy are poised for success.

In conclusion, digitalization has indelibly transformed the distribution phase of the automobile value chain. Challenges and regional disparities persist, yet embracing digitalization while bridging the gap between digital and physical experiences is essential for the continued relevance and success of automobile dealers and manufacturers in this new digital era. The industry's future hinges on adeptly navigating this dynamic landscape, combining the strengths of digital technologies with the enduring appeal of physical interactions in the automotive marketplace. By embracing the "phygital" paradigm and addressing the challenges head-on, the automobile industry is poised to not only thrive but also lead in the evolving digital era.

REFERENCES

AutoS. (2023). https://briteskoda.com/experience-center/

Autocar. (2019, October 10). *New Kia experience centre inaugurated in New Delhi.* Autocar India. https://www.autocarindia.com/car-news/new-kia-experience-centre-inaugurated-in-new-delhi-414447

Bacher, N., & Manowicz, A. A. (2020). *Digital auto customer journey-An analysis of the impact of digitalization on the new car sales process and structure.* www.ijsrm.com

Bansal, N. (2020). *3 trends driving the auto industry's shift to dealer digitization.* https://www.thinkwithgoogle.com/intl/en-apac/consumer-insights/consumer-journey/3-trends-driving-the-auto-industrys-shift-to-dealer-digitization/

Barlow, C. (2020). *Social Media Marketing 2020: A Guide to Brand Building Using Instagram, YouTube, Facebook, Twitter, and Snapchat, Including Specific Advice on Personal Branding for Beginners.* Independently Published.

Bauer, H. (2018). *The Digital Customer Journey in the Automobile Industry - A Quick-Check for the Retail Environment.* Seinajoki University of Applied Sciences. Retrieved from https://urn.fi/URN:NBN:fi:amk-201805219194

Bijmolt, T. H., Broekhuis, M., De Leeuw, S., Hirche, C., Rooderkerk, R. P., Sousa, R., & Zhu, S. X. (2021). Challenges at the marketing–operations interface in omni-channel retail environments. *Journal of Business Research, 122*, 864–874. doi:10.1016/j.jbusres.2019.11.034

Buck, C., & Reith, R. (2020). Privacy on the road? Evaluating German consumers' intention to use connected cars. *International Journal of Automotive Technology and Management, 20*(3), 297–318. doi:10.1504/IJATM.2020.110408

Bulsara, H. P., & Vaghela, P. S. (2022). Millennials Online Purchase Intention Towards Consumer Electronics: Empirical Evidence from India. *Indian Journal of Marketing, 52*(2), 53–70. doi:10.17010/ijom/2022/v52/i2/168154

Chatterjee, S., Chaudhuri, R., Vrontis, D., Thrassou, A., & Ghosh, S. K. (2021). Adoption of artificial intelligence-integrated CRM systems in agile organizations in India. *Technological Forecasting and Social Change*, *168*, 120783. doi:10.1016/j.techfore.2021.120783

Chauhan, C. (2020, December 30). *Smaller dealerships are here. Did COVID expedite them?* ET Auto. https://auto.economictimes.indiatimes.com/news/aftermarket/smaller-dealerships-are-here-did-covid-expedite-them/80019721

Chung, M., Ko, E., Joung, H., & Kim, S. J. (2020). Chatbot e-service and customer satisfaction regarding luxury brands. *Journal of Business Research*, *117*, 587–595. doi:10.1016/j.jbusres.2018.10.004

Cordova-Buiza, F., Urteaga-Arias, P. E., & Coral-Morante, J. A. (2022). Relationship between Social Networks and Customer Acquisition in the Field of IT Solutions. *IBIMA Business Review.*, *2022*, 631332. Advance online publication. doi:10.5171/2022.631332

Esteller-Cucala, M., Fernandez, V., & Villuendas, D. (2020). Towards data-driven culture in a Spanish automobile manufacturer: A case study. *Journal of Industrial Engineering and Management*, *13*(2), 228–245. doi:10.3926/jiem.3042

Fan, X., Chai, Z., Deng, N., & Dong, X. (2020). Adoption of augmented reality in online retailing and consumers' product attitude: A cognitive perspective. *Journal of Retailing and Consumer Services*, *53*, 101986. doi:10.1016/j.jretconser.2019.101986

Farrugia, C., Grima, S., & Sood, K. (2022). The General Data Protection Regulation (GDPR) for risk mitigation in the insurance industry. In Big Data: A game changer for insurance industry (pp. 265-302). Emerald Publishing Limited.

Fukuyama, M. (2018). Society 5.0: Aiming for a New Human-Centered Society. *Japan Spotlight*. https://www.jef.or.jp/journal/

Gauri, D. K., Jindal, R. P., Ratchford, B., Fox, E., Bhatnagar, A., Pandey, A., Navallo, J. R., Fogarty, J., Carr, S., & Howerton, E. (2021). Evolution of retail formats: Past, present, and future. *Journal of Retailing*, *97*(1), 42–61. Advance online publication. doi:10.1016/j.jretai.2020.11.002

Hanelt, A., Piccinini, E., Gregory, R. W., Hildebrandt, B., & Kolbe, L. M. (2015). Digital Transformation of Primarily Physical Industries - Exploring the Impact of Digital Trends on Business Models of Automobile Manufacturers. *Wirtschaftsinformatik Proceedings, 88*. https://aisel.aisnet.org/wi2015/88

Hanson, R. (2016). The Age of. In *Work, Love, and Life when Robots Rule the Earth (Illustrated edition)*. OUP Oxford.

Hoban, P. R., & Bucklin, R. E. (2015). Effects of Internet Display Advertising in the Purchase Funnel: Model-Based Insights from a Randomized Field Experiment. *JMR, Journal of Marketing Research*, *52*(3), 375–393. doi:10.1509/jmr.13.0277

Hyundai Motor India. (2023). www.clicktobuy.hyundai.co.in

Johnson, Z., & Leonard, T. (2020). *Performance branding: Borrow from the past to win the future.* Henry Stewart Publications.

Kampani, N., & Jhamb, D. (2020). Analyzing the role of e-crm in managing customer relations: A critical review of the literature. *Journal of Critical Review*, *7*(4), 221–226.

Khan, R., Taqi, M., & Saba, A. (2021). The role of digitization in automotive industry: The Indian perspective. *International Journal of Business Ecosystem & Strategy, 3*(4), 20-29. doi:10.36096/ijbes.v3i4.277

Kim, D., Suh, Y. H., & Lee, J. (2018). The impact of digital signage on customer envy, store patronage, and purchase decision. *Information & Management*, *55*(6), 735–747.

Kuhnert, F., Stürmer, C., & Koster, A. (2017-2018). Five trends transforming the Automotive Industry. *PricewaterhouseCoopers GmbH Wirtschaftsprüfungsgesellschaft*. www.pwc.com/auto

Lamrhari, S., El Ghazi, H., Oubrich, M., & El Faker, A. (2022). A social CRM analytic framework for improving customer retention, acquisition, and conversion. *Technological Forecasting and Social Change*, *174*, 121275. doi:10.1016/j.techfore.2021.121275

Lead, A. (2021, February 28). *Citroën India launches 'La Maison Citroen' phygital showrooms in Bangalore*. Automotive Lead. https://automotiveleadnews.com/2021/02/28/citroen-india-launches-la-maison-citroen-phygital-showrooms-in-bangalore/

Lee, H. H., Fiore, A. M., & Kim, J. (2006). The role of the technology acceptance model in explaining effects of image interactivity technology on consumer responses. *International Journal of Retail & Distribution Management*, *34*(8), 621–644. doi:10.1108/09590550610675949

LinkedIn. (2023). https://www.linkedin.com/company/%C5%A1koda-sga-cars-dealer/?originalSubdomain=in

Livemint.com. (2019). *Emerging tech helps reshape how vehicles are showcased*. Retrieved from https://www.livemint.com/technology/tech-news/emerging-tech-helps-reshape-how-vehicles-are-showcased-11573143279548.html

Manohar, S., Sharma, V., & Mittal, A. (2023). Reinforcing Requirements and Stimulating the Purchase Intentions: Growing Location Based Mobile Targeting Techniques. In Enhancing Customer Engagement Through Location-Based Marketing (pp. 56-65). IGI Global.

Mishra, A., Shukla, A., Rana, N. P., & Dwivedi, Y. K. (2021). From "touch" to a "multisensory" experience: The impact of technology interface and product type on consumer responses. *Psychology and Marketing*, *38*(3), 385–396. doi:10.1002/mar.21436

Mitchell, W. J., Borroni-Bird, C. E., & Burns, L. D. (2010). *Reinventing the Automobile: Personal Urban Mobility for the 21st Century*. The MIT Press. doi:10.7551/mitpress/8490.001.0001

MotorK. (2023). https://www.kia.com/in/vr/showroom/index.html#/showroom

Motors, M. G. (2019, October 31). *MG Motor Unveils India's First Digital Car-Less Showroom: Mg Digital Studio*. MG Motor. https://www.mgmotor.co.in/media-center/newsroom/mg-motor-indias-first-digital-car-less-showroom

Olga, B., & Vlad, M. (2014). Remarking as a Tool in Online Advertising. *Ovidius University Annals, Series Economic Sciences, 14*(2).

Palmié, M., Miehé, L., Oghazi, P., Parida, V., & Wincent, J. (2022). The evolution of the digital service ecosystem and digital business model innovation in retail: The emergence of meta-ecosystems and the value of physical interactions. *Technological Forecasting and Social Change*, *177*, 121496. doi:10.1016/j.techfore.2022.121496

Pangarkar, A., Arora, V., & Shukla, Y. (2022). Exploring phygital omnichannel luxury retailing for immersive customer experience: The role of rapport and social engagement. *Journal of Retailing and Consumer Services*, *68*, 103001. doi:10.1016/j.jretconser.2022.103001

Pantano, E., Priporas, C. V., & Sorace, S. (2017). Enhancing the showrooming experience: A conceptual framework for augmenting mirrors in clothing e-stores. *Journal of Retailing and Consumer Services*, *35*, 1–10.

Power, J. D. (2022, November 9). *In Era of Digital Information Search, Vehicle Shoppers in India Still Sensitive to Product Discovery at Showrooms*. JD Power. https://www.jdpower.com/business/press-releases/2022-india-sales-satisfaction-index-study-ssi

Roy, S. K., Singh, G., Sadeque, S., Harrigan, P., & Coussement, K. (2023). Customer engagement with digitalized interactive platforms in retailing. *Journal of Business Research*, *164*, 114001. doi:10.1016/j.jbusres.2023.114001

Sahni, N. S., Narayanan, S., & Kalyanam, K. (2019). An Experimental Investigation of the Effects of Retargeted Advertising: The Role of Frequency and Timing. *Journal of Marketing Research*. doi:10.1177/0022243718813987

Salesforce. (2023). *OEM EVOLUTION: A new customer journey for new customer expectations*. https://www.salesforce.com/content/dam/web/en_us/www/documents/industries/manufacturing/manufacturing-auto-oem-evolution_v2.pdf

Samson, R., Mehta, M., & Chandani, A. (2014). Impact of online digital communication on customer buying decision. *Procedia Economics and Finance*, *11*, 872–880. doi:10.1016/S2212-5671(14)00251-2

SAS Institute Inc. (2023). *American Honda Drives Personalization at Scale with SAS*. https://www.sas.com/en_gb/customers/american-honda.html

Shalender, K., & Shanker, S. (2023). Building Innovation Culture for the Automobile Industry: Insights from the Indian Passenger Vehicle Market. *Constructive Discontent in Execution: Creative Approaches to Technology and Management*, 131.

Sharma, A., & Bansal, A. (2023). Digital Marketing in the Metaverse: Beginning of a New Era in Product Promotion. In Applications of Neuromarketing in the Metaverse (pp. 163-175). IGI Global.

Tandon, U., Mittal, A., & Manohar, S. (2021). Examining the impact of intangible product features and e-commerce institutional mechanics on consumer trust and repurchase intention. *Electronic Markets*, *31*(4), 945–964. doi:10.100712525-020-00436-1

Van Esch, P., & Stewart Black, J. (2021). Artificial intelligence (AI): Revolutionizing digital marketing. *Australasian Marketing Journal*, *29*(3), 199–203. doi:10.1177/18393349211037684

Chapter 8
Social Media Advertising: A Dimensional Change Creator in Consumer Purchase Intention

N. S. Bharathi
Christ University, India

Deep Jyoti Gurung
https://orcid.org/0000-0003-4438-3210
Christ University, India

ABSTRACT

The chapter discusses the importance of social media platforms, especially Instagram and YouTube, in advertising for influencing the consumer purchase intention of Generation Z. Various methods of business expansion through social media advertising have been explored. The authors examine in detail several key characteristics of social media advertising that affect consumer purchase intentions, including emotional appeal, interactivity, trust, creativity, and the role of e-word of mouth. The impact of Web 2.0 technologies on the development and effectiveness of social media advertising is emphasized, highlighting the close relationship between the Web 2.0 systems to enable the delivery of advertisements based on usage and preferences to accurately target, increasing the efficiency and effectiveness of advertising campaigns mobile, Web 2.0. Implicit communication, mobile-related advertising, and mobile-specific content have become important parts of social media advertising. Researchers from various fields have drawn on rapidly evolving social media platforms, each with its own unique perspective.

1. INTRODUCTION

Social media is no longer a simple platform that connects people and was created to cut down communication barriers. social media slowly grew as a fun-creating, messaging, and information-sharing platform. Social media platforms did bring up new challenges for people, with regards to sustainability it may be Facebook, Twitter, Instagram, YouTube, LinkedIn, Reddit, Snapchat, and so on. Digital marketing gradually incorporated social media as a platform for using it to personally reach people with

DOI: 10.4018/979-8-3693-1762-4.ch008

advertisements. These sites now serve as revenue-generating tools for both users and developers. The rise of a new class of powerful and sophisticated customers who are challenging to influence, persuade, and keep is proof that social media have altered the power dynamics in the marketplace. Social media is used as a platform to search and find information about a product or brand.

There are numerous methods to use social media as a marketing tool. They are inexpensive but priceless sources of "live" client feedback that help businesses hone their marketing strategies and frequently avert disasters. Social media platforms can be used as public relations and marketing tools, as well as customer influencers and tools that let customers personalize their online shopping experiences. Last but not least, social media presents a wide range of opportunities for businesses as a platform for leveraging the intelligence and creativity of the general public. Social media can be used to encourage consumer engagement with brands through participation in the creation and innovation of new products. To address the emerging realities in the customer-dominated marketplace, a strategic reorientation and frequently drastic shift of company and management attitudes is required. The impact of the Internet and, in particular, the Social Media movement on the market and consumer behaviour should be understood by marketers. It is also crucial to recognize and comprehend the function of social media as a marketing tool and a component of the overall marketing strategy.

Consumer plays the most pivotal role in the success story of any platform and business. Majority of the Consumers have received social media advertising with a positive attitude. However, there are threats of authenticity and privacy concerns. These are further taken into consideration to overcome. A recent study on social media usage found that the average global user has accounts on 8.4 different social platforms and spends 2 hours and 25 minutes per day on social media. Social media advertising is becoming a competitive strategy setter as the large number of influencers are coming up with new ideas, and innovations to attract customers. The challenge of creating successful social media advertising is becoming harder day by day.

1.1 Objectives

- The chapter discusses the importance of social media platforms, especially Instagram and YouTube, in advertising to influence the consumer purchase intention of Generation Z.
- This chapter also covers several social media advertising characteristics that have a favorable effect on customer purchase intention, including emotional appeal, interactivity, credibility, creativity, time, trust, privacy, and so forth. Electronic word-of-mouth did play a significant part in raising consumer knowledge of products, but it first raised awareness of various social media platforms.
- The chapter also covers what steps should be taken to sustain consumer trust and how to retain consumer connections.
- The elements of social media that affect consumer intention are the main topic of this chapter. Intention is measured using a variety of psychological variables; in this case, we concentrate on attitude, emotions, reviews, and content production on social media.

2. SOCIAL MEDIA PARADIGM SHIFT

With the proliferation of social media marketing strategies, we now require the addition of a fifth P: participation. It's reasonable to argue that these new channels are changing how marketers operate in

Table 1. Important of social media as a prominent marketing communication tool

Year	Study	Importance of Social Media as a Prominent Marketing Communication Tool
2011	Hanna et al.	not only a source of information but also an important source of influence
2011	Sinclair and Vogus	a platform for expressing the likes and dislikes, opinions and feelings about the products/ brands by the consumers
2012	Akrimi and Khemakhem	quick and smooth transition of information
2012	Pandya	an important medium of advertising the products as it influences the masses
2013	Gao et al	cost-effective in terms of reach
2013	Okazaki and Taylor	a capable tool for international advertising by making use of three capabilities, viz. "networking capability", "image transferability", and "personal extensibility"
2013	Saxena and Khanna	reach of these sites to people is much wider as compared to traditional forms of advertising
2014	Abzari et al	social media advertising has more significant effect than traditional advertising
2014	Fuller et al.	the customers invited in co-creation projects
2015	Vishnoi and Padhy	co-creation with the users gives the users a greater sense of involvement
2017	Kyriakopoulou and Kitsios	the virtual interaction with customers, companies have the chance to improve their products, their brand and total appearance on social media aiming to satisfy consumers' needs
2018	Ahmed and Raziq	can develop their promotional strategies in a better way

the same way that social media is affecting how consumers live their daily lives. regarding their company. Regardless of whether our goal is to uphold or enhance customer partnerships, educate customers about our advantages, and advertise a brand or associated promotion, Innovative social media platforms are useful for developing new products and influencing brand attitudes. Using social media platforms, channels, and technologies is known as social media marketing. software for producing, exchanging, delivering, and communicating offerings that are valuable for stakeholders in the organization. This definition is demonstrated by new developments in social

press. While social media marketing first had an impact on firms' marketing strategies, more

Social funding, such as Kickstarter, has been used in recent years for business purposes to finance new business endeavors) and social indexing (such as source choice data from social users such as Google)

3. WEB 2.0 AND SOCIAL MEDIA

The next evolution of the web was viewed as Web 2.0. Many-to-many content was Web 2.0's central idea. People may create their own blogs and webpages, upload videos, and contribute to the abundance of user-generated content on the internet. The Web 2.0 package included an easier-to-use platform. Using third-party software, individuals with no HTML knowledge may now create a respectable website for the first time. Since a lot of these platforms were fully web-based, practically any computer could be used. Next came social media, which is a perfect fit for the Web 2.0 evolution as a whole because it focuses on user-friendly platforms that let people create content. Among Web 2.0 breakthroughs, social media stands out due to the introduction of web-based sharing. Before Facebook and Twitter, you could always send your friends an interesting article via email. However, social media reduced this procedure

Figure 1. Web 2.0 and social media

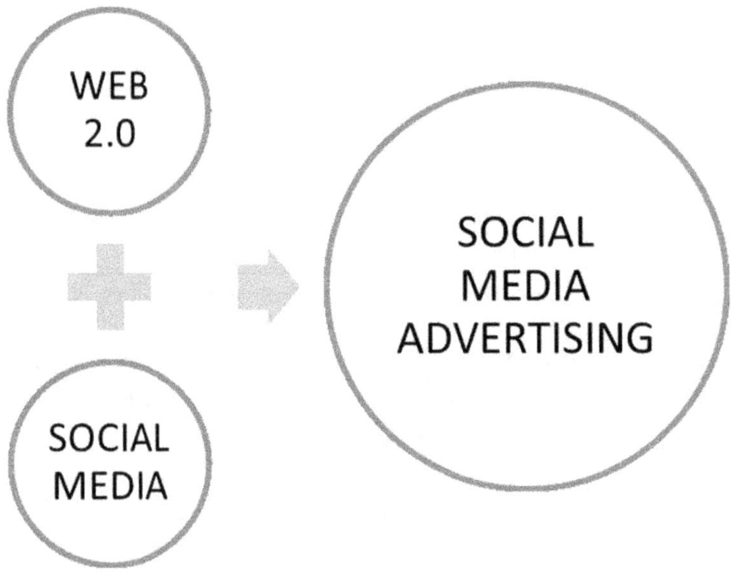

to the touch of a button. Thus, social media is a Web 2.0 innovation that focuses on the user by enabling them to select other content to share with their networks, in addition to promoting user-generated content.

Social media are the channels via which information is shared between people and organizations as well as the material itself (Kietzman, Hermkens, McCarthy, & Silvestre, 2011). Nonetheless, it is mostly a consumer product. Broadcast media monologues (one to many) become social media dialogues (many to many) thanks to Web 2.0 technologies. Web 2.0's technological underpinnings are expanded upon by Internet-based applications to create social media.

We offer five axioms that marketers should adhere to in order to assist international marketing strategists in using social media effectively and engaging creative customers in a positive manner. These axioms deal with the emergence of social media and creative consumers, the local and general components of the information shared on social media, and the historically dependent nature of the technological infrastructure supporting social media.

1. A nation's technology, culture, and political system all influence social media usage.
2. Local events rarely stay local in the era of social media.
3. In the era of social media, broad topics rarely stay broad; instead, big issues are frequently (re) interpreted locally.
4. The technology, culture, and political system of a certain nation frequently influence the decisions and works produced by creative consumers.
5. Technology tends to be historically dependent; that is, rather than evolving along unique paths because they are the best solution, technologies in different countries progress along diverse trajectories due to inertia.

4. SOCIAL MEDIA ADVERTISING AND PURCHASE INTENTION

Social media has introduced novel online marketing practices and provided new opportunities for brands to communicate with their consumers. The perception of advertisements is crucial to their success. Prior research has tried to determine the causes of consumers' perceptions of online advertising, and it has been discovered that an increase in consumer perception is related to an increase in online advertising. This revealed a strong and positive relationship between customer perception and internet advertising. Furthermore, it was found that all sub-dimensions of consumer perception positively and significantly influenced online advertising and its dimensions (Nasir et al., 2021).

Fundamentally, businesses must consider how social media influences consumer behavior, especially purchase intent (Lim et al., 2017). Customers who purchase items, products, and services from businesses expect them to deliver with integrity, honesty, and competence because trust is an important determinant of consumer buying intention, (Isip, Maria & Lacap, Jean Paolo, 2021). In particular, a recent study by Bigne et al. (2018) highlighted the importance of social media in the airline business. Factors such as entertainment and engagement on social media can influence attitudinal factors such as confidence and perceived value, which in turn influence consumers' purchase intentions.

Social media has the power to influence how customers feel about a particular service or good. In other words, marketers can facilitate contact and interaction with consumers via online platforms. Numerous studies have revealed that marketers can use social media to influence consumers' attitudes

Figure 2. Social media advertising and purchase intention

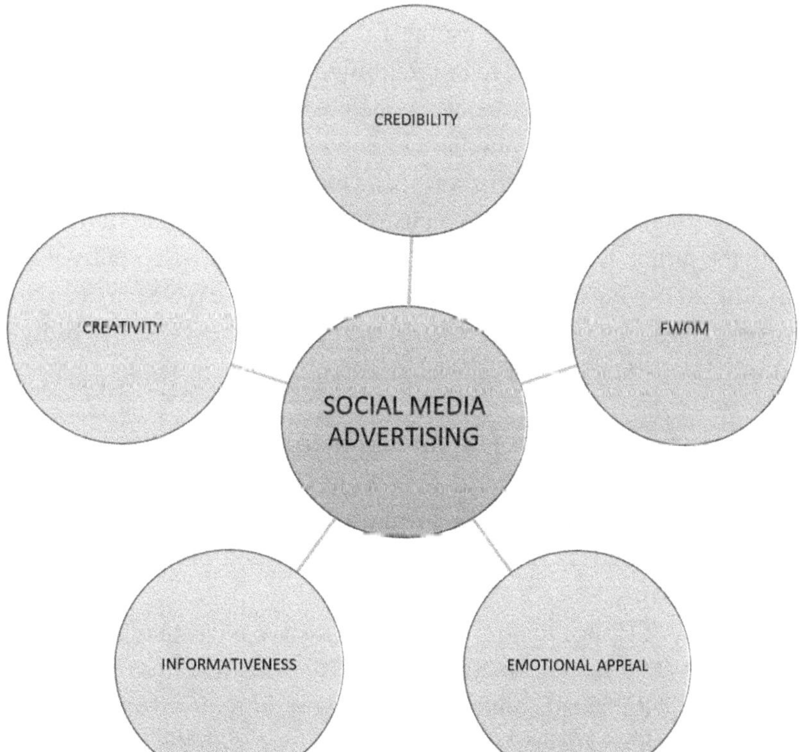

toward their goods or services (Alalwan et al., 2017; Baum et al., 2019). In particular, a recent study by Bigne et al. (2018) highlighted the importance of social media in the airline business. Factors such as entertainment and engagement on social media can influence attitudinal factors such as confidence and perceived value, which in turn influence consumers' purchase intentions.

4.1 Effect of Electronic Word of Mouth (eWOM) on Purchase Intention

Word-of-mouth (WOM) can be defined as individual-to-individual, a conversation between a receiver and a sender. It is well known for framing consumer purchase decisions (Dash, Manoj & Sahu, Rajendra & Pandey, Ashutosh, 2018).

There has been a plethora of studies on eWOM, and its effect on consumer purchase intention is well-established and acknowledged by many researchers on various social media platforms. Considering it as a feature of social media advertising strategy could give marketers a positive approach to influencing consumers. Social media users are exposed to an enormous quantity of eWOM information, either intentionally or unintentionally, and prior research has found such eWOM information is influential on consumers' purchase intentions. E-WOM consists of three dimensions namely intensity, the valence of opinion, and content. Social media has a role in empowering consumers to Share views and product knowledge and then build influence, both individually and collectively to other consumers (Tjhin, V.U. & Aini, Siti, 2019). When users believe that eWOM information is helpful, they are more likely to engage in it. The observation of a positive significant link between information usefulness and information adoption is also consistent with Erkan and Evans' findings (2016).

According to the findings of Xue et al., (2014) information quality has a favorable relationship with information usefulness. Because users can obtain information through EWOM, the quality of the material is a concern. Consumers find the knowledge useful if the information quality meets their expectations. The usefulness of knowledge is also related to its credibility. Hussain A, Ting DH, and Mazhar M (2022) have proved that positively engaged customers generate positive WOM in the online social community, which improves the chance of customers' purchase intentions and brand perceptions. As a result, brands should reconsider, redesign, and re-strategize their social media ads to favourablfavorablyheir customers. As a result, e-WOM has a significant impact on customer purchasing intention because customers are more concerned with the online feedback of other customers who have recently used similar products (Tjhin, V.U. & Aini, Siti, 2019). Further research could examine the eWOM in one social media website. Also, a comparison between social media websites in the context of eWOM can bring valuable theoretical and managerial insights (Erkan, Ismail & Evans, Chris, 2016).

4.2. Effect of Informativeness on Purchase Intention

Informativeness is described as the "ability to inform users about product alternatives that enable them to make choices yielding the highest value". Informativeness is a perceptual concept assessed using self-reported items. (lee,2016).

According to Lee, Jieun & Hong, Ilyoo. (2016), Informative advertising messages on an SNS will draw users' attention, motivate them to associate the ad with a positive image, and, in some cases, drive them to share the messages with friends on the social network of their own volition. Overall, the literature believes that informative SNS advertisements provide a user with the capacity to make an informed purchase decision in the future and that the user's perception of this favorable aspect of the ad leads to the formation of a positive attitude towards both the viral behavior and the ad.

Gao and Zang (2016) identified that interesting and informative advertisements can increase consumers' favorable impression of advertisements. According to previous researchers, informativeness has a positive impact on consumer purchase intention and also on consumer attitude. According to Lee and Hong (2016), informativeness influences consumers' plans to purchase the products presented in social media ads. Similarly, it was discovered that the informativeness of the social network influences confidence and that both trust and informativeness influence the purchase intentions of social media users. According to Alalwan, Ali. (2018) The level of informativeness existing in social media ads may facilitate customers to make better purchasing decisions and, as a result, increase their purchase intention.

4.3. Effect of Credibility on Purchase Intention

Advertising credibility refers to consumers' perceptions of the credibility and truthfulness of advertising content. Research has proven that credibility is an important factor in developing a favorable attitude toward advertising.(Chen, Wen-Kuo & Ling, Chia-Ju & Chen, Chien-Wen,2022). Many studies have found that credibility has a positive influence on consumer attitude which leads to purchase intention. credibility seems to be the most significant predictor of social media advertising value.(Manel Hamouda,2017). According to Wang, Y., & Genç, E. (2019) Credibility is not a luxury, but a necessity for businesses to thrive in today's market. Marketers can enhance advertising views by building credibility in mobile ads. For example, using a credible spokesperson in mobile video ads boosts source credibility. A targeted and personalized advertising message delivered to the customer at the right place and right moment increases credibility. Information given by highly credible sources is regarded as valuable and encourages knowledge transfer; it is also the initial factor in the individual persuasion process (Erkan and Evans 2016).

The credibility of information is the main determinant in consumer decision-making, and previous research has found a positive relationship between information credibility and consumer purchase intention, particularly when the information is useful and adaptable. (Leong, Choi Meng & Loi, Alexa & Woon, Steve, 2022).). Therefore, this research predicts that the credibility of information on social media is positively related to purchase intention.

4.4. Effect of Emotional Appeal on Purchase Intention

According to Lee and Hong (2016), Emotion is defined as a person's condition of feeling in the form of an effect. Emotions can be positive or negative (for example, love, and joy). (e.g., fear, anger, sadness). Emotion, according to research in the field of communication, is an essential factor that can significantly affect the effectiveness of a message.

Emotional appeal is a persuasion technique that uses emotional content to create an emotional response to a message. (Lee, Jieun & Hong, Ilyoo.,2016). An emotional appeal has a significant role in influencing consumer interest in the product. Emotional appeal is defined as clearly expressing specific interests or reasons, attracting consumers' attention to advertising products or services, and trying to persuade consumers to pay attention to or buy such products or services. Emotional appeal is concerned with how emotional responses are elicited as a result of the message's motivational relevance to people. The emotional appeal seeks to arouse the emotions of consumers and mainly involves persuasive methods to persuade consumers to agree with their advertisements. (Chen, Wen-Kuo & Ling, Chia-Ju & Chen, Chien-Wen,2022).and Lee and Hong (2016) suggested that effective emotional appeal has a positive impact on consumers' advertising attitudes Users on an SNS are more likely to respond positively to a persuasive message or other online content if it has a higher emotional appeal. Previous researchers have found that emotional appeal has a positive influence on consumer attitude, so including emotional content in social media advertising such as humor, sentiments, and horror has a positive impact on consumer purchase intention.

4.5. Effect of Creativity on Purchase Intention

The degree to which an advertisement is unique and surprising is referred to as its creativity. Divergence and relevance are widely recognized as the defining characteristics of advertising creativity; divergence refers to elements that are novel, different, or unusual, whereas relevance is concerned with elements that are meaningful, appropriate, helpful, or valuable to the audience. Creative messages capture more attention and lead to favorable attitudes toward the featured goods.

Creativity has always been an exploratory method of trying new things. Creativity is considered the oxygen of advertising. Creativity is required for businesses to handle the challenges of rising competition, demanding customers, and unpredictable environments. Because consumer requirements, wants, and expectations constantly change, marketers must maximize their creativity (Das. et, al,2023). And he believed that to provide excellent value to firms and consumers, marketers must incorporate creativity into every touchpoint.

Creativity, which is quite often emphasized in advertising literature, has a significant impact on user behavior in an SNS environment. SNS users are more interested in advertising with educational value and creative material. According to Lee, Jieun & Hong, Ilyoo. (2016), Advertising creativity was found to be a significant predictor of attitude that encourages purchase intention.

Creativity's impact on purchase intention is not well explored by researchers, hence there is a necessity to know the impact of creativity on consumer purchase intention with empirical evidence that can be an add-on knowledge to the existing literature.

5. SOCIAL MEDIA WORKS AT ALL STAGES OF PURCHASE DECISION

1. Raise awareness: By keeping an active presence in the social media platforms where target customers "live" and by incorporating social media into the marketing mix, brands can raise awareness through social media marketing. Unilever's Knorr's brand. The #LoveAtFirstTaste initiative paired singles based on shared cuisine tastes and then set them up on dates where they had to share food and consent to being photographed. The campaign featured a campaign landing page, an interactive flavour-profile questionnaire, and a number of social videos, one of which is a well-known YouTube video.
2. Influence desire: Social media marketing is comparable to traditional catalog and advertising marketing and promotional activities to get customers to identify a desire. Every new collection from the clothing line Lilly Pulitzer is shared on Facebook, Flickr, and Vimeo. Viewers are able to browse through images of its designs as soon as they are taken. It is similar to being on Vogue's pages.

3. Promote trial: Social media can also be utilized to boost loyalty and sample courses. Offering a free trial of a product is known as sampling; these are typically mailed supplied at stores, on the street, or delivered to customers' residences. Using social media can be employed to find potential candidates who are qualified for samples. Use of Celestial Seasonings uses this strategy to give away 25,000 samples of its newest tea flavors. Known as Share The Magic Facebook fans were encouraged to share the magical ways that tea brightens their days. Celestial Seasonings sent a coupon and a complimentary sample in exchange.
4. Encourage purchases: Social media is a platform for many sales promotion incentives, such as group offers and bargains, as well as a distribution route. In order to be eligible for exclusive offers, many consumers "like" or follow brands on social media. Currently trending on Taco Bell is this tweet: "We utilize @Snapchat. Login as tacobell. Include us. Tomorrow, we're going to send a covert announcement to all of our buddies! Shhhh. Coupons were given as gifts to friends.
5. Strengthen brand loyalty: Social media platforms provide users with interesting activities that encourage them to spend more time with the brand, which should raise brand loyalty levels. You only need to look at social games that give prizes to the users who are the most devoted. This is precisely what the UK's Lidl grocery business achieved with its brilliant Social Price Drop Twitter campaign. The goal of the campaign was to give its social media followers the ability to decide how much to pay for particular products over the Christmas season. The more Twitter users discussed a product, like the "Christmas

6. YOUTUBE AS A SOCIAL MEDIA ADVERTISING PLATFORM INFLUENCING CONSUMER PURCHASE INTENTION IN GENERATION Z

Although Generation Z frequently uses YouTube, a significant advertising channel, it is unknown how well these advertisements reach and influence them. Due to greater information transparency than earlier generations, the Z generation is known for having a higher level of skepticism They also exhibit some resistance to advertisements, choosing to ignore them or block them using third-party software. (Pagefair, 2015; eMarketer, 2015; Bolton et al., 2013).

Reaching users through YouTube is made more challenging by their relatively high resistance to traditional push advertising and their skepticism towards advertising in general. However, this is not the only barrier to effective YouTube marketing; many viewers simply ignore advertising on YouTube. Approximately 80% of all skippable YouTube commercials are done so by users. (Pagefair, 2015)

Soukup (2014) concentrated on the use of social media, particularly YouTube. He claims that YouTube is a sizable platform with a variety of uses, including advertising, archive work, education, entertainment, news reporting, political communication, the arts and cultures, religion, medicine, the military, fandom, interpersonal communication, and observation. The author contends that communication theories will need to be reconfigured to analyze this platform, which is always changing because it is so diverse.

Soukup (2014) explicitly states that a new study is required on the platform and that theory needs to be modified in light of what this medium delivers after analyzing previous research on the YouTube site. Despite being relatively young, online video advertising is a crucial component of Google's YouTube monetization strategy.

Previous studies have shown that a variety of elements might affect an advertisement's efficacy. Three of these criteria are particularly pertinent to influencer marketing on YouTube, Ad attitude, (Dehghani

Figure 3. Attitudinal stages of consumer purchase intention

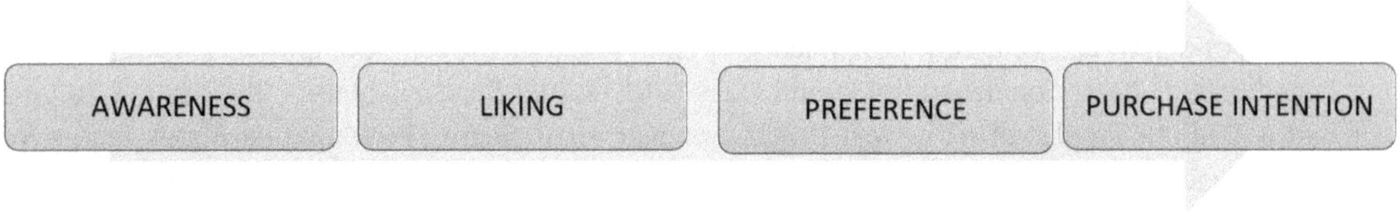

et al, 2016; MacKenzie, Lutz & Belch, 1986), purchase intention (Dehghani et al, 2016; MacKenzie & Lutz, 1989; Lee & Hong, 2016; Ott, Vafeiadis, Kumble & Wadell, 2016) and Perceived entertainment.

The widespread use of smartphones and other mobile multimedia devices, which are primarily used for viewing videos, posting pictures, instant messaging, playing online games, entertainment, social connection, seeking information, and generally perceiving the world, has helped Generation Z display the highest levels of sophistication. In emerging countries, smartphone ownership has steadily increased, particularly among younger customers.

6.1. Attitudinal Stages of Consumer Purchase Intention

6.1.1. Awareness

According to Corbett and Durfee (2004), the media can represent, convey, and help the community understand important educational information. Social media platforms like Facebook, Youtube, Instagram, Twitter, and blogs are another way to get news, and distributing information to a larger group is the most popular and quick communication medium (Hamid et al., 2017; Irwin et al., 2012; Kimmons, 2014).

Youtube advertising tries to create awareness among viewers about new products or existing products. Two views are connected to behavioural change while raising awareness. According to the first viewpoint, increasing people's knowledge of a subject and encouraging acceptable attitudes leads to a shift in behaviour. According to the second point of view, people should evaluate their options methodically before acting in a way that will help the economy (McKenzie-Mohr, 2000). By regularly associating a stimulus with other stimuli of either positive or negative valence, awareness can influence an attitude through evaluative conditioning (Sweldens, Corneille, and Yzerbyt, 2014).

6.1.2. Liking

To follow consumers' progress through the purchase decision process, several advertising models were created. However, traditional advertising was utilized to determine the relationship between like and preference in the construction of these response hierarchy models. Since social media and other online ICT platforms have grown rapidly over the past ten years, academia and advertising practitioners are increasingly interested in how traditional advertising models relate to marketing on social media.

According to Nalewajek and Macik, (2013), social media-induced behavioural changes can alter attitudes towards responsible consumption. However, they noted that there was a lack of research that quantitatively analyzed online video advertising among young consumers in both developing and developed economies. Arajo et al.(2017) discovered that Brazilians liked YouTube videos with advertising more than Americans (US) and Britons (UK) did.

According to Duffet, et al, (2019), liking has a significantly favourable effect on preference owing to YouTube advertising among Millennials and liking has a significantly larger positive effect on preference by Millennial respondents who watched 1–5 YouTube advertisements. The studies revealed that there is a positive association between liking and preference.

6.1.3. Preference

According to Bolton et al. and Zambodla, younger cohorts are heterogeneous because older cohort members are more likely to have different psychographic characteristics, lifestyles, attitudes, values, needs, interests, preferences, and desires from younger cohort members. They are also more likely to consume media differently than younger cohort members. Sharma emphasizes that additional research should examine young cohorts Ad in terms of various social media platform formats across various age groups. AAPIS model and Aspinwall's consumer acceptance theory have also emphasized the importance of preference as a mediator between cognitive and behavioural attitudinal responses.

Around the world, Generation Z has many similar traits and characteristics, yet developing countries typically have less developed digital ICT infrastructure than industrialized countries, which could lead to different attitudes because of worsening social and economic conditions. There is additionally a significant difference in access to online platforms and wealth in many developing nations.

According to Duffet, et al, (2019), Preference can create an impact on the attitude of young consumers toward a product or service. Awareness and liking are interconnected in influencing preference creation and creating a positive attitude on the advertisement on Youtube.

6.1.4. Purchase Intention

A number of organizations also use YouTubers, testimonies, influencers, and celebrity endorsers (who frequently have millions of impressionable young followers) to spread good news and support the organizations via their YT channels. According to several studies, YouTubers strongly impact their younger followers' purchasing decisions and have a significant favourable association with them.

Mir and Rehman recommend that businesses use YouTubers to advertise their products by inserting commercials into YT since younger generations tend to find UGC more credible. Rasmussen discovered that American college students were inclined to purchase items recommended by YouTubers. According to Sokolova and Kefi, among French Generation Z and Y respondents, influencers on YouTube and Instagram are associated favourably with buy intentions due to their trustworthiness and parasocial activity. According to Hwang et al., Chinese consumers' purchasing intentions are positively impacted by parasocial celebrity contacts on social media. Therefore, it is advised that businesses utilize YouTubers more frequently to promote their products in order to have a favourable impact on consumers' behavioural and attitude responses. YouTubers are not only more affordable than traditional celebrities, but they also have a larger following.

According to Duffet, et al, (2019), youtube advertising has a positive impact on awareness, liking and preferences have an impact on consumer purchase intention.

The Generation Z attitude is unpredictable because of competitive technological advancements and emerging trends, it becomes important to study the attitudinal changes of Generation Z to predict and make digital advertising an effective means of creating an impact on consumer purchase intention.

7. INSTAGRAM AS A SOCIAL MEDIA ADVERTISING PLATFORM INFLUENCING CONSUMER PURCHASE INTENTION

Instagram has grown in popularity as a social media tool for companies to advertise their goods and services. According to a study by Neil Kenneth Pais and Nirmal G, Instagram positively affects consumer trust, peer social influence, security, and electronic word of mouth, all of which are important factors in determining the rate at which Instagram users make purchases Nearly 75% of Instagram users asked in a different study by Dana Rebecca Designs stated that Instagram influenced their decisions to buy. According to previous research, Instagram significantly affects consumers' intentions to make purchases. It is crucial to remember, nevertheless, that the influence of Instagram on consumers' intentions to make purchases might differ based on several variables, including product categories, prices, and cultural norms.

Instagram, a social media network, is vital to the social milieu of its diverse user base, who regularly engage with the platform (Instagram Revenue and Usage Statistics (2021), 2021). For starters, it amuses various people who enjoy perusing other users' stuff on the platform. In addition to offering amusement and relaxation, this platform facilitates simultaneous real-time conversation between many users. Information is shared between people thanks to this communication (Mattern, 2017).

As Instagram grew, so did the influence exerted by individuals or groups known as Instagram influencers (de Veirman et al., 2017). By endorsing or promoting certain items alongside their artwork and other topics covered on social media, these people or organizations hope to influence consumers toward the use or consumption of certain things. Because of their enormous following, which defines their popularity on the platform, these Instagram influencers are likable (de Veirman et al., 2017).

Various organizations can interact with their followers on Instagram through the help of influencers (Tafesse & Wood, 2021). Additionally, there is a substantial correlation between these organizations' follower counts and their level of interaction with both their target markets and followers (Tafesse & Wood, 2021).

According to many studies, Instagram influencers are individuals with a strong online reputation on Instagram who possess a special ability to persuade their followers to buy particular goods or services (Kolarova, 2018). 9 out of 10 consumers think influencer marketing is very effective, and 93% of marketers also employ it, according to the research (Kolarova, 2018). Because consumers trust social media influencers (SMI) more than brands, they are also very important to the marketing sector (Kolarova, 2018; Sokolova & Kefi, 2020). They frequently utilize Instagram. Instagram influencers typically have a sizable following and a high interaction rate (Kolarova, 2018).

The majority of participants have been using Instagram for a considerable amount of time. It is important to keep in mind, too, that each person's time on the platform differs based on their other lives. obligations, including jobs and interests, as reported by the participants. Although a few individuals were while some could not precisely recall the moment they joined Instagram, others could recall using some of the most notable things that happened to them at almost the same period.

Time is a key factor in social media use, and one of the key components of perceiving a social media platform advertisement is trust.

All kinds of Instagram influencers have a significant impact on the various online consumers' purchase intentions on the platform; the extent of this impact is dependent on several criteria. Trust is regarded as one of the most important. Based on the research, it is evident that users' emotional feelings towards an influencer, the influencer's social obligation to society, and the influencer's fan base are the three factors that impact users' trust in Instagram influencers. When evaluating Instagram Influencers' recommendations and their marketing efforts regarding a product or service on the platform, consumers consider all of these factors to be crucial. Because it influences them throughout the pre-purchasing phase, the majority of users believe eWOM to be highly effective. Two unique forms of electronic word-of-mouth (eWOM) were identified by the research: positive and negative. From customers' perspectives, the dissemination of positive eWOM via reliable Instagram influencers is highly significant and affects their perceptions of a product or service on the platform. But if an influencer misuses the concept of eWOM by endorsing a good or service that can, inadvertently, inflict financial or emotional harm, they will become less credible, which will increase the likelihood that customers would purchase them.

Subsequent research endeavors may employ perspectives from a broader range of vocations about the same field of study, in which revenue level may surpass what pupils can allocate. As a result, it might influence their priorities and mode of thought. Other variables, such as the degree of interaction between Instagram users and the influencers they follow and the use of a bigger sample size to examine the perspectives of influencers, may also provide fresh insights into the same research question. Lastly, trust and electronic word-of-mouth were factors considered in various studies.

8. FUTURE GROWTH OF SOCIAL MEDIA

A noticeable trend in social media marketing lately has been the domination of videos, with short-form videos, live streaming, and interactive features becoming increasingly popular on sites like YouTube, Instagram Reels, and TikTok. Influencer marketing is becoming more and more popular as companies work with influencers—particularly micro-influencers—to establish genuine connections with niche audiences. Stories on Facebook, Instagram, and Snapchat are examples of temporary content that is gaining popularity because they create a sense of excitement and promote regular user interaction. The distinction between social media and online shopping experiences is becoming hazier due to the incorporation of e-commerce elements into social media platforms. Personalized businesses are using chatbots and messaging apps for instantaneous and personalized interactions, while augmented reality (AR) and virtual reality (VR) technology are improving customer experiences. Promoting user-generated content (UGC) is still a highly effective way to establish credibility and trust. Platforms have tightened their standards in response to data privacy concerns, which has affected data-driven marketing and targeted advertising. Marketers are expanding their reach across various social media channels by tailoring their content to target audiences' demographics. TikTok is becoming a significant player, especially with younger viewers, which is encouraging businesses to investigate innovative engagement tactics on the site.

9. CONCLUSION

The use of social media is substantially and favorably correlated with the intention to buy. Previous studies have looked into and demonstrated the strong and positive association between purchase intention and social media advertising. (Lacap, Jean Paolo & Isip, Maria, 2021). Advertising attitude is more influenced by advertising credibility and emotional appeal than by informativeness and irritation, according to the original web-based advertising value model's findings. Executives in marketing and advertising should take particular note of the findings. The credibility of consumer advertising is essential in the context of social media marketing to boost consumer inclination to purchase. On social media, advertisers need to establish a positive reputation for themselves among consumers. (Ling, Chia-Ju & Chen, Chien-Wen & Chen, Wen-Kuo, 2022). The research indicates that the opinions of one customer have an impact on the purchase intentions of other consumers. As a result, businesses need to be aware of these possible channels of communication and work to capitalize on them by examining and participating in eWOM communication. Consistent with the results of the prior poll, millennials' attitudes about social media advertising favorably influence their intentions to make purchases. Finding out what people like and dislike about advertisements that are shown on social media platforms, can help marketers draw in customers and make the required adjustments to produce a social media ad that works. Results show that millennial buying intentions are significantly impacted by peer communication. (Ashutosh Pandey, Rajendra Sahu, and Manoj Dash (2018). "The study demonstrates that there is a significant relationship between informativeness, entertainment, credibility, and social media advertising value," claims Manel Hamouda (2017). Customers' opinions and actions about social media advertising will be influenced by its positive value.

Future Scope

There is a gap in the research on how social media advertising affects consumers' intentions to make purchases. because a variety of elements, including the psychological, physiological, and social behaviors of consumers, vary with location, age, and gender.

Artificial intelligence is operating as a source of opinion generation, which creates a gap in the study of consumer behavior and purchase intention. Better business and consumer experiences could be achieved by the use of artificial intelligence monitoring and social media literacy research.

REFERENCES

Adiratna, H., & Wulansari, A. (2021). Factors Influencing Purchase Intention of Elancing Using UTAUT Model: A Case Study of Mahajasa. *Malaysian Journal of Social Sciences and Humanities*, *6*(9), 563–564. doi:10.47405/mjssh.v6i9.1056

Alalwan, A. (2018). Investigating the impact of social media advertising features on customer purchase intention. *International Journal of Information Management*, *42*, 65–77. doi:10.1016/j.ijinfomgt.2018.06.001

Alalwan, A. A., Rana, N. P., Dwivedi, Y. K., & Algharabat, R. (2017). Social media in marketing: A review and analysis of the existing literature. *Telematics and Informatics*, *34*(7), 1177–1190. doi:10.1016/j.tele.2017.05.008

Antonova, N. (2014). Psychological Effectiveness of Interactive Advertising in Russia. *Journal of Creative Communications*, *10*(3), 303–311. Advance online publication. doi:10.1177/0973258615614426

Araújo, C. S., Magno, G., Meira, W., Almeida, V., Hartung, P., & Doneda, D. (2017). Characterizing videos, audience and advertising in YouTube channels for kids. In *Lecture Notes in Computer Science (Including Subseries Lecture Notes in Artificial Intelligence and Lecture Notes in Bioinformatics)*. Elsevier. doi:10.1007/978-3-319-67217-5_21

Balakrishnan, J., & Manickavasagam, J. (2016). User Disposition and Attitude towards Advertisements Placed in Facebook, LinkedIn, Twitter and YouTube. *J. Electron. Commer. Organ.* . doi:10.4018/JECO.2016070102

Bandara, D. M. D. (2020). Impact of Social Media Advertising on Consumer Buying Behavior: With Special Reference to Fast Fashion Industry. In *The Conference Proceedings of 11th International Conference on Business & Information ICBI.* University of Kelaniya.

Baramidze, T. (2018). *The Effect of Influencer Marketing on Customer Behaviour. The Case of YouTube Influencers in Makeup Industry*. Vytautas Magnus University.

Barry, T. M. (1987). The development of the hierarchy of effects: An historical perspective. *Curr. Issues Res. Advert.*, *10*, 251–295. doi:10.1080/01633392.1987.10504921

Baum, D., Spann, M., Füller, J., & Thürridl, C. (2019). The impact of social media campaigns on the success of new product introductions. *Journal of Retailing and Consumer Services*, *50*, 289–297. doi:10.1016/j.jretconser.2018.07.003

Bigne, E., Andreu, L., Hernandez, B., & Ruiz, C. (2018). The impact of social media and offline influences on consumer behaviour, An analysis of the low-cost airline industry. *Current Issues in Tourism*, *21*(9), 1014–1032. doi:10.1080/13683500.2015.1126236

Bleize, D. N. M., & Antheunis, M. L. (2019). Factors influencing purchase intent in virtual worlds: A review of the literature. *Journal of Marketing Communications*, *25*(4), 403–420. doi:10.1080/13527266.2016.1278028

Boateng, H., & Okoe, A. F. (2015). Consumers' attitude towards social media advertising and their behavioral response: The moderating role of corporate reputation. *Journal of Research in Interactive Marketing*, *9*(4), 299–312. doi:10.1108/JRIM-01-2015-0012

Bolton, R. N., Parasuraman, A., Hoefnagels, A., Migchels, N., Kabadayi, S., Gruber, T., Loureiro, Y. K., & Solnet, D. (2013). Understanding Generation Y and their use of social media: A review and research agenda. *Journal of Service Management*, *24*(3), 245–267. doi:10.1108/09564231311326987

Chen, W.-K., Ling, C.-J., & Chen, C.-W. (2022). What affects users to click social media ads and purchase intention? The roles of advertising value, emotional appeal and credibility. *Asia Pacific Journal of Marketing and Logistics*. Advance online publication. doi:10.1108/APJML-01-2022-0084

Chong, A. Y.-L. (2013). Predicting m-commerce adoption determinants: A neural network approach. *Expert Systems with Applications*, *40*(2), 523–530. doi:10.1016/j.eswa.2012.07.068

Chungviwatanant, T., Prasongsukam, K., & Chungviwatanant, S. (2016). A study of factors that affect consumer's attitude toward a "skippable in-stream ad" on YouTube. *Au Gsb E J.*, *9*, 83–96.

Das, K., Patel, J., Sharma, A., & Shukla, Y. (2023). Creativity in marketing: Examining the intellectual structure using scientometric analysis and topic modeling. *Journal of Business Research*, *154*, 113384. doi:10.1016/j.jbusres.2022.113384

Dash, M., Sahu, R., & Pandey, A. (2018). Social media marketing impact on the purchase intention of millennials. *International Journal of Business Information Systems*, *28*(2), 147. doi:10.1504/IJBIS.2018.10012924

Dehghani, M., Niaki, M. K., Ramezani, I., & Sali, R. (2016). Evaluating the influence of YouTube advertising for attraction of young customers. *Computers in Human Behavior*, *59*, 165–172. doi:10.1016/j.chb.2016.01.037

Duffett, R. G. (2015). Effect of Gen Y's affective attitudes towards facebook marketing communications in South Africa. *The Electronic Journal on Information Systems in Developing Countries*, *68*(1), 1–27. doi:10.1002/j.1681-4835.2015.tb00488.x

Duffett, R. G. (2015). Facebook advertising's influence on intention-to-purchase and purchase amongst Millennials. *Internet Research*, *25*(4), 498–526. doi:10.1108/IntR-01-2014-0020

Duffett, R. G., Edu, T., & Negricea, I. C. (2019). YouTube marketing communication demographic and usage variables influence on Gen Y's cognitive attitudes in South Africa and Romania. *The Electronic Journal on Information Systems in Developing Countries*, *85*(5), 1–13. doi:10.1002/isd2.12094

Duffett, R. G., Edu, T., Negricea, I. C., & Zaharia, R. M. (2020). Modelling the effect of YouTube as an advertising medium on converting intention-to-purchase into purchase. *Transformations in Business & Economics*, *19*, 112–132.

Duffett, R. G., Petroşanu, D. M., Negricea, I. C., & Edu, T. (2019). Effect of YouTube marketing communication on converting brand liking into preference among Millennials regarding brands in general and sustainable offers in particular: Evidence from South Africa and Romania. *Sustainability (Basel)*, *11*(3), 1–24. doi:10.3390u11030604

Erkan, I., & Evans, C. (2016). The influence of eWOM in social media on consumers' purchase intentions: An extended approach to information adoption. *Computers in Human Behavior*, *61*, 47–55. doi:10.1016/j.chb.2016.03.003

Gao, S., & Zang, Z. (2016). An empirical examination of users' adoption of mobile advertising in China. *Information Development*, *32*(2), 203–215. doi:10.1177/0266666914550113

Garg, P., Raj, R., Kumar, V., Singh, S., Pahuja, S., & Sehrawat, N. (2023). Elucidating the role of consumer decision making style on consumers' purchase intention: The mediating role of emotional advertising using PLS-SEM. *Journal of Economy and Technology, 1*, 108-118. doi:10.1016/j.ject.2023.10.001

Gautam, V., & Sharma, V. (2017). The Mediating Role of Customer Relationship on the Social Media Marketing and Purchase Intention Relationship with Special Reference to Luxury Fashion Brands. *Journal of Promotion Management*, *23*(6), 872–888. Advance online publication. doi:10.1080/10496491.2017.1323262

Hamouda, M. (2018, April 09). Understanding social media advertising effect on consumers' responses: An empirical investigation of tourism advertising on Facebook. *Journal of Enterprise Information Management, 31*(3), 426–445. Advance online publication. doi:10.1108/JEIM-07-2017-0101

Harshini, C. S. (2015). *Influence of social media ads on consumer's purchase intention.* Academic Press.

Hor'akov'a, Z. (2018). *The Channel of Influence? YouTube Advertising and the Hipster Phenomenon.* Charles University.

Hussain, A., Ting, D. H., & Mazhar, M. (2022). Driving Consumer Value Co-creation and Purchase Intention by Social Media Advertising Value. *Frontiers in Psychology, 13*, 800206. doi:10.3389/fpsyg.2022.800206 PMID:35282229

IAB & PwC Outlook. (2021). *Report Urges Digital Ecosystem to Reset Consumer Value Exchange.* Author.

Isip, M., & Lacap, J. P. (2021). *Social Media Use and Purchase Intention: The Mediating Roles of Perceived Risk and Trust.* Academic Press.

Karunarathne, E. A. C. P., & Thilini, W. A. (2022). Advertising Value Constructs' Implication on Purchase Intention: Social Media Advertising. *Management Dynamics in the Knowledge Economy, 10*(3), 287-303. DOI doi:10.2478/mdke-2022-0019

Kathiravan, C. (2017). Effectiveness of advertisements in social media. *Asian Academic Research Journal of Multidisciplinary., 4*, 179–190.

Keller, K. L. (2009). Building strong brands in a modern marketing communication environment. *Journal of Marketing Communications, 15*(2/3), 139–155. doi:10.1080/13527260902757530

Lamé, G. (2019). Systematic Literature Reviews: An Introduction. *Proceedings of the Design Society: International Conference on Engineering Design., 1*(1), 1633–1642. doi:10.1017/dsi.2019.169

Lee, J., & Hong, I. (2016). Predicting positive user responses to social media advertising: The roles of emotional appeal, informativeness, and creativity. *International Journal of Information Management, 36*(3), 360–373. doi:10.1016/j.ijinfomgt.2016.01.001

Lee, J. E., & Watkins, B. (2016). YouTube vloggers' influence on consumer luxury brand perceptions and intentions. *Journal of Business Research, 69*(12), 5753–5760. doi:10.1016/j.jbusres.2016.04.171

Leong, C. M., Loi, A., & Woon, S. (2022). The influence of social media eWOM information on purchase intention. *Journal of Marketing Analytics., 10*(2), 1–13. doi:10.105741270-021-00132-9

Li, H., & Lo, H. Y. (2015). Do You Recognize Its Brand? The Effectiveness of Online In-Stream Video Advertisements. *Journal of Advertising, 44*(3), 208–218. doi:10.1080/00913367.2014.956376

Lim, X. J., Radzol, A. M., Cheah, J., & Wong, M. W. (2017). The impact of social media influencers on purchase intention and the mediation effect of customer attitude. *Asian Journal of Business Research, 7*(2), 19–36. doi:10.14707/ajbr.170035

Mazouzi, D., & Alit, N. (2023). Factors Influencing Consumers' Attitudes and Intentions Towards Online Shopping - A Survey of a Sample of Consumers in Algeria. *Malaysian Journal of Consumer and Family Economics., 31*(1), 788–814. doi:10.60016/majcafe.v31.29

Moslehpour, M., Dadvari, A., Nugroho, W., & Do, B.-R. (2021). The dynamic stimulus of social media marketing on purchase intention of Indonesian airline products and services. *Asia Pacific Journal of Marketing and Logistics*, *33*(2), 561–583. doi:10.1108/APJML-07-2019-0442

Moslehpour, M., Ismail, T., Purba, B., & Lin, P-K. (2020). *The Effects of Social Media Marketing, Trust, and Brand Image on Consumers' Purchase Intention of GO-JEK in Indonesia*. . doi:10.1145/3387263.3387282

Mukherjee, K., & Banerjee, N. (2017). Effect of Social Networking Advertisements on Shaping Consumers' Attitude. *Global Business Review*, *18*(5), 1291–1306. doi:10.1177/0972150917710153

Nasir, V. A., Keserel, A. C., Surgit, O. E., & Nalbant, M. (2021). Segmenting consumers based on social media advertising perceptions: How does purchase intention differ across segments? *Telematics and Informatics*, *64*, 101687. doi:10.1016/j.tele.2021.101687

Natarajan, T., Balakrishnan, J., Balasubramanian, S., & Manickavasagam, J. (2015). Examining beliefs, values, and attitudes towards social media advertisements: Results from India. *International Journal of Business Information Systems*, *20*(4), 427. doi:10.1504/IJBIS.2015.072738

Pais, N., & Ganapathy, N. (2021). *The Influence of Instagram on Consumer Purchase Intention*. Academic Press.

Pierre, Pitt, Plangger, & Shapiro. (2012). *Marketing meets Web 2.0, social media, and creative consumers: Implications for international marketing strategy*. doi:10.1016/j.bushor.2012.01.007

Rasmussen, L. (2018). Parasocial interaction in the digital age: An examination of relationship building and the effectiveness of YouTube Celebrities. *J. Soc. Media Soc.*, *7*, 280–294.

Rodriguez, P. R. (2017). *Effectiveness of YouTube Advertising: A Study of Audience Analysis*. Rochester Institute of Technology.

Salim. (2023). The Effect of Informativeness, Photo Colour, Visual Aesthetic, and Social Presence towards Customer Purchase Intention on Glory of fats Instagram. *International Journal of Science and Business*, *18*(1), 96–107. doi:10.5281/zenodo.7569471

Silvia, S. (2019). The Importance of Social Media and Digital Marketing to Attract Millennials' Behavior as a Consumer. *Journal of International Business Research and Marketing*, *4*(2), 7–10. doi:10.18775/jibrm.1849-8558.2015.42.3001

Siriwardana, A. (2020). Social Media Marketing: A Literature Review on Consumer Products. *Proceedings of the International Conference on Business & Information (ICBI)*. doi:10.2139/ssrn.3862924

Smith, K. T. (2019). Mobile advertising to Digital Natives: Preferences on content, style, personalization, and functionality. *Journal of Strategic Marketing*, *27*(1), 67–80. doi:10.1080/0965254X.2017.1384043

Sokolova, K., & Kefi, H. (2020). Instagram and YouTube bloggers promote it, why should I buy? How credibility and parasocial interaction influence purchase intentions. *Journal of Retailing and Consumer Services*, *53*, 1–9. doi:10.1016/j.jretconser.2019.01.011

Stein, A., & Ramaseshan, B. (2016). Towards the identification of customer experience touch point elements. *Journal of Retailing and Consumer Services*, *30*, 8–19. doi:10.1016/j.jretconser.2015.12.001

Sun, S., & Wang, Y. (2010). Examining the role of beliefs and attitudes in online advertising: A comparison between the USA and Romania. *International Marketing Review*, *27*(1), 87–107. doi:10.1108/02651331011020410

Tjhin, V.U. & Aini, S. (2019). *Effect of E-WOM and Social Media Usage on Purchase Decision in Clothing Industry*. . doi:10.1145/3332324.3332333

Viertola, W. (2018). *To What Extent Does YouTube Marketing Influence the Consumer Behaviour of a Young Target Group*. Metropolia University of Applied Sciences.

Vingilisa, E., Yildirim-Yeniera, Z., Vingilis-Jaremkob, L., Seeleya, J., Wickensc, C. M., Grushkaa, D. H., & Fleiterd, J. (2018). Young male drivers' perceptions of and experiences with YouTube videos of risky driving behaviors. *Accident; Analysis and Prevention*, *120*, 46–54. doi:10.1016/j.aap.2018.07.035 PMID:30086437

Vrontis, D., Makrides, A., Christofi, M., & Thrassou, A. (2021). Social media influencer marketing: A systematic review, integrative framework and future research agenda. *International Journal of Consumer Studies*, *45*(4), 617–644. Advance online publication. doi:10.1111/ijcs.12647

Wang, Y., & Genç, F. (2019). Path to effective mobile advertising in Asian markets. *Asia Pacific Journal of Marketing and Logistics*, *31*(1), 55–80. Advance online publication. doi:10.1108/APJML-06-2017-0112

Xu, Q. (2014). Should I trust him? the effects of reviewer profile characteristics on eWOM credibility. *Computers in Human Behavior, 33*, 136-144. . doi:10.1016/j.chb.2014.01.027

YouTube. (n.d.). *For Press*. Available online: https://www.youtube.com/intl/en-GB/about/press/

Zambodla, N. (n.d.). *Millennials Are Not a Homogenous Group*. Available online: http://www.bizcommunity.com/Article/196/424/172113.html#more

Zeng, F., Tao, R., Yang, Y., & Xie, T. (2017). How Social Communications Influence Advertising Perception and Response in Online Communities? *Frontiers in Psychology*, *8*, 1349. doi:10.3389/fpsyg.2017.01349 PMID:28855879

Zhang, T. C., Omran, B. A., & Cobanoglu, C. (2017). Generation Y's positive and negative eWOM: Use of social media and mobile technology. *International Journal of Contemporary Hospitality Management*, *29*(2), 732–761. doi:10.1108/IJCHM-10-2015-0611

KEY TERMS AND DEFINITIONS

Consumer Purchase Intention: The possibility or willingness of a customer to acquire a good or service in the future is referred to as their consumer purchase intention. Given that it sheds light on the thinking and level of purchase readiness of potential customers, it is a crucial idea in marketing and consumer behaviour. Purchase intention among consumers can be influenced by a number of factors, thus firms and marketers should take these into account when creating their marketing plans.

Generation Z: Generation Z are cohorts aged between 18-24 years and there is considerable diversity within the generation, and not all individuals share the same characteristics.

Instagram: Instagram is a well-known photo and video-sharing social networking site and smartphone application. Since its October 2010 introduction, it has grown to rank among the most popular social networking sites worldwide. Instagram users can interact with other users by leaving comments, liking posts, and sending private messages in addition to uploading, editing, and sharing images and videos.

Social Media: Digital platforms and online services that let users create, share, and engage with content as well as establish online connections with other people are referred to as social media. These platforms are intended to promote social interactions as well as the sharing of knowledge, concepts, and media content. Social media, which consists of a variety of websites and applications with different functions, has emerged as a significant component of contemporary communication.

Social Media Advertising: The use of social media platforms to market goods, services, brands, or messages to a specific audience is known as social media advertising. In order to connect and interact with potential consumers, it incorporates paid promotional content that is posted on social media networks. Depending on the platform and the particular marketing goals, social media advertising can take many different shapes and employ different tactics.

Web 2.0: Web 2.0 is the term for the second generation of the World Wide Web, which is distinguished by changes in the architecture and functionality of websites and web apps. Compared to the static and one-way communication of the early online (also known as online 1.0), it reflects a more interactive, collaborative, and user-centered approach to the internet. The introduction of various important concepts and features by Web 2.0 helped to shape the current state of the internet.

YouTube: YouTube means Users can publish, view, and share videos on the well-known social networking website and video-sharing platform. Having been founded in February of 2005, it is currently among the most popular websites on the internet.

Chapter 9
Bridging the Digital Divide:
Navigating the Landscape of Digital Equity

Priya Gupta
https://orcid.org/0009-0001-4887-7042
Indira Gandhi University, Meerpur, India

Anjali Verma
https://orcid.org/0009-0000-7911-8893
Indira Gandhi University, Meerpur, India

ABSTRACT

The issue of digital divide in India continues to be one of the critical issues in the 21st century, impacting the social, economic, and the various educational opportunities available for the citizens. This study seeks to examine the current state of digital divide in India by focusing on disparity in internet access, mobile ownership, and digital literacy. This study also aims to identify the numerous factors responsible for digital divide and to assess the key initiatives undertaken by the Government of India to mitigate the disparities. The study employs secondary data gathered from National Family Health Survey, GSMA, TRAI, and NSO. The findings of this study reveal the disparity in internet access, digital literacy, and mobile ownership rooted due to socio-economic factors, education, and geographic locations. This study underscores the need of addressing the digital gap in India and provides valuable insights for the policymakers, stakeholders, and the practitioners to take initiatives towards digital inclusion and equity in the country.

1. INTRODUCTION

In the rapidly evolving digital landscape, the concepts of digital equity and the digital divide are critical issues that significantly impact our society (Kraus et al., 2022). Digital equity is the concept that provides that digital technologies including access to the internet and device ownership should be equally accessible to all individuals. However, the reality is that there is a wide gap between those who have access to digital infrastructure and those who don't have the resources (Imran, 2022). This gap is known as the

DOI: 10.4018/979-8-3693-1762-4.ch009

digital divide, it affects people's education, health, and economic opportunities, and the overall aspects of their life. As per the Census Bureau, in 2020 about 17% of U.S. households did not have a broadband internet connection. The disparity in digital access is particularly pronounced in India, with women being 33% less likely to use mobile internet services as compared to the male segment, and rural areas having only 15% internet penetration compared to 42% in urban areas. The World Development Report 2021 emphasizes that it is very crucial to narrow down the gap in digital access to enhance quality of life. This data-driven approach can help improve public services, enabling citizen empowerment and fostering innovational practices (The World Bank, 2021). The term "digital accessibility" refers to the design and development of digital content, services, and technologies in a way that can ensure their usability by all individuals. This encompasses a wide range of considerations, including web accessibility, assistive technologies, and inclusive design principles. The digital divide encompasses disparities in access to digital resources, digital services, and digital literacy (Botelho, 2021). This divide is particularly concerning the impact on urban areas, a significant portion of the world's population. The worldwide COVID-19 pandemic has inevitably led to a surge in the use of digital technologies which drives the need to adapt novel approaches for individuals as well as organizations (De et al., 2020). The pandemic has also deepened the issues of education inequality and the digital divide without access to ICT is even more severe (Derek Chun et al., 2022). Therefore, digital equity is not only a matter of fairness but also emerges comprehensive development and progress, accessible and beneficial to all. For closing the digital divide there is a need for public-private partnerships, community-based initiatives, and policy interventions (Diebold & Castro, 2023). This study is not merely about the access to the technology, although it is all about shaping. The need for a digital agency is felt to address major issues for attaining equity in education in the future (Passey et al., 2018).

The organization of the study will be made in a total of 9 sections. The first, section will provide an introduction to the study, second section represent the rationale of the study. The third section highlights the literature review of the study, the fourth section provides the research gap, the fifth section represents the research objectives of the study, the sixth section highlights the research methodology and the seventh section provides the results and the discussion of the study. The eighth section is the conclusion and the implications of the study, Lastly, the study indicates the limitations and the future research directions.

2. RATIONALE OF THE STUDY

Studying the digital divide is vital to throw light on the disparities that exist in accessing digital resources, information, literacy, and opportunities. This study can help in identifying the barriers that hinder individuals from accessing online job opportunities, e-commerce activities, and entrepreneurship. Understanding the current status of the digital divide in India is pivotal in identifying the disparity that exists in internet access and device ownership, it can exhibit socio-economic inequalities. Moreover, the paradigm of digital access plays a vital role in the 21st century with implications for education, healthcare, and employment opportunities. Additionally, the study acknowledges the gender disparity that prevails within the digital divide, aligning with the principle of gender equality and social justice. Digital equity is directly linked to gender equality. Women face multiple barriers to digital access to education and resources. By addressing these issues this study can contribute towards addressing the gender disparities. Bridging the digital gap and addressing the disparities is not only vital to promote social inclusion and equitable access to education, resources, and opportunities but also significant for fostering innovations, more economic growth and development, and global competitiveness, and will contribute to the overall creation of a more equitable, fairer and prosperous society. This study is not only about individual access to technology, but it is about shaping the future where everyone has the opportunity to contribute to the upliftment of society.

3. LITERATURE REVIEW

3.1 Global Digital Divide and Access to Technology

In the 21st century, the digital divide has evolved to encompass the disparities among individuals, businesses, households, and geographical regions in various socio-economic groups. These disparities pertain to significant differences in their access to information and communication technology and utilization of these resources (OECD, 2013). The concept of the digital divide is multi-dimensional across countries and various studies made a significant attempt to quantify the digital divide as the composite indices which have been a very useful technique for understanding this concept (Korovkin et al., 2022). The digital divide on a large scale and individual level, arises from a variety of factors. At the macro level, these factors include inadequacy of infrastructure, the nation's economic prosperity, high costs associated with the ownership of computers and having internet access, the complexities between the internet and politics, quality of education, and digital literacy(Okunola et al., 2017). The dimensions of digital equity have been identified such as access to quality, meaningful, and culture-relevant local language content, sharing, creating, and exchanging the content, hardware access, connectivity and software effectiveness, quality access to research and digital technologies application for enhancing learning environment (Resta et al., 2018; Sweidan & Areiqat, 2020). Digital divide by identifying the disparities using advanced technology, access to the internet, gender-based differences in device ownership, variations in access to the internet in different age groups, and disparities in rural and urban areas, also highlights the socio-economic variability in the digital divide in every field particularly in education sector representing a significant challenge (Yuan Sun & Metros, 2011). Low-income regions face difficulties in accessing digital resources and implementing digital equity all over the world (Williams, 2022). There is an important need for establishing public and private partnerships and technology access for all to bridge this technological gap in society, especially in the field of education (Afzal et al., 2023). In the online higher education system, the concept of trust plays a very crucial role in the student-instructor relationship and it is influenced by several factors like casual behavior of teaching staff and non-traditional student identities, performativity, neoliberalism, and the digital divide (Gudmundsdottir, 2010). The challenges associated with achieving digital accessibility on an urban scale are multidimensional (Moore et al., 2018). They not only involve the development of inclusive digital services and infrastructure but also public awareness campaigns, policy and regulatory frameworks, and partnerships with the relevant stakeholders. Online discussions and performance metrics in digital environments especially regarding their capacity to enable cognitive and emotional trust among students (Payne et al., 2023). Another critical dimension of the digital divide in an urban environment is digital accessibility for people with disabilities who represent a significant and diverse demographic within cities and their ability to access essential urban services, experiences, and information. By addressing this issue, cities can work towards a more inclusive and equitable urban environment (Kolotouchkina et al., 2022).

3.2 Digital Divide and Access to Technology in India

India represents the large socio-economic disparities across different caste groups basis and the research studies are still not more. The study highlighted the first level digital divide having ownership of computers and accessibility to the internet and the second level of using computers as well as having individual skills in different vulnerable groups. The findings of the study indicated that the variations in attainment of education and income between these groups were more than half by using the non-linear decomposition method (Rajam et al., 2021). In India, there is a lack of infrastructure access and learning skills which leads to the failure of implementation of changes in comparison with the developed countries with better access to infrastructure and fast learning. The study sheds light on the progress of India in transformation into a digital society (Gudmundsdottir, 2010). The findings of the study represent the significant correlation between socio-economic disparities which serves as the primary factor responsible for the extensive digital gap within the country. Moreover, the digital divide is not only evident between urban localities and prosperous residential zones, but also within this area (Hasan Laskar, 2023). The rural and Urban digital divide in high-density puts the country at a disadvantageous position and endangers economic growth and social inclusion. To address the gap of the digital divide the barriers of illiteracy, rural areas investment, and lack of skills should be addressed (Singh, 2010). In India, the digital divide is a significant concern, with only 34% of the population having access to the internet and using the efficacy of digital resources (Vassilakopoulou & Hustad, 2023). The concept of digital equity includes affordability of digital literacy, the ability to use technology effectively, and equal access to digital resources (Aissaoui, 2021). Despite the encouragement of digital equity, the nation is still devoid of the term "network" and has issues with using banking services. Lack of integrity has been seen in nations with and without the use of information and communication technology skills (Tripathy & Raha, 2019). To address the digital equity gap it is important to attend to human abilities that foster individuals for generating valuable outcomes (Aguilar, 2020).

4. RESEARCH GAP

The research gap within the context of the digital divide highlights the need for in-depth research work to analyze the current status of the digital divide in India to analyze regional and demographic variations in internet access and device ownership more comprehensively (Prathapagiri, 2019). The deeper analysis of multiple and complex factors responsible for the digital divide and a comprehensive evaluation of government initiatives aimed at addressing the gender disparity and to develop of tailored and evidence-based recommendations that account for the development of tailored and evidence-based interventions that can account the regional, technological and behavioral differences to bridge the digital divide in India effectively (Deepa et al., 2023).

5. RESEARCH OBJECTIVES

Based on the literature review, the research objective of this study is:

1. To assess the current status of the digital divide in India by identifying the disparities in internet access, device ownership, and availability of digital resources among different demographic groups and regions.
2. To conduct a comprehensive and integrated review of existing literature on digital equity and the digital divide to gain insights into the factors that are primarily responsible for the digital divide and digital inequity.
3. To examine the numerous initiatives taken by the government of India to Bridge the digital divide and to ensure digital equity in India.
4. To Propose some recommendations or strategies to address the identified disparity and to promote digital equity.

6. RESEARCH METHODOLOGY

Secondary data will be collected from the various reports including National Family Health Survey 2019-21, Reports TRAI[1], Report by NSO, and GSMA's reports to gather the data related to internet access, digital literacy, and device ownership to examine the current status of the digital divide in India and to identify the factors that are responsible for the digital divide in India. Various kinds of charts, graphs, tables, and pictures will be used to convey the outcomes of the study in a precise and accessible manner.

7. RESULTS AND DISCUSSION

7.1 Current Status of Digital Divide in India

7.1.1 Exploring Internet Usage in India

According to the database of the International Telecom Union, having ICT-based indicators provides that internet adoption and usage in India currently stands at 43% with around 58% of males and 42% of females using the internet facility unveiling a persistence of notable gender gap.

Gender disparity in accessing the Internet facility

57.1% of male has ever used the internet facility in contrast to only 33.3% of females highlighting the gender disparity. This gender gap persists across all the states as depicted in Figure 1.

Regional disparity in accessing the Internet facility

The data of the National Family Health Survey offers insights into the context of the regional digital divide. In urban areas, around 72.5% of males and 51.8% of females have access to and use the internet facility at the same point in time. In Contrast, in rural areas, only 48.7% of males and a mere 24.6% of females fulfill the same criteria. This analysis examines the trends across all states and territories, in which urban males consistently exhibited a higher usage rate, while rural females consistently exhibited a lower usage rate as highlighted in Figure 2 and Figure 3. This analysis underscores the importance of recognizing the rural-urban digital disparity and more need to tailor such kind of initiatives to address the current variations and to ensure more equitable access to internet facilities across all the geographical regions of the nation.

Figure 1. Indicating gender disparity in access to internet facility
Source: NFHS (2021)

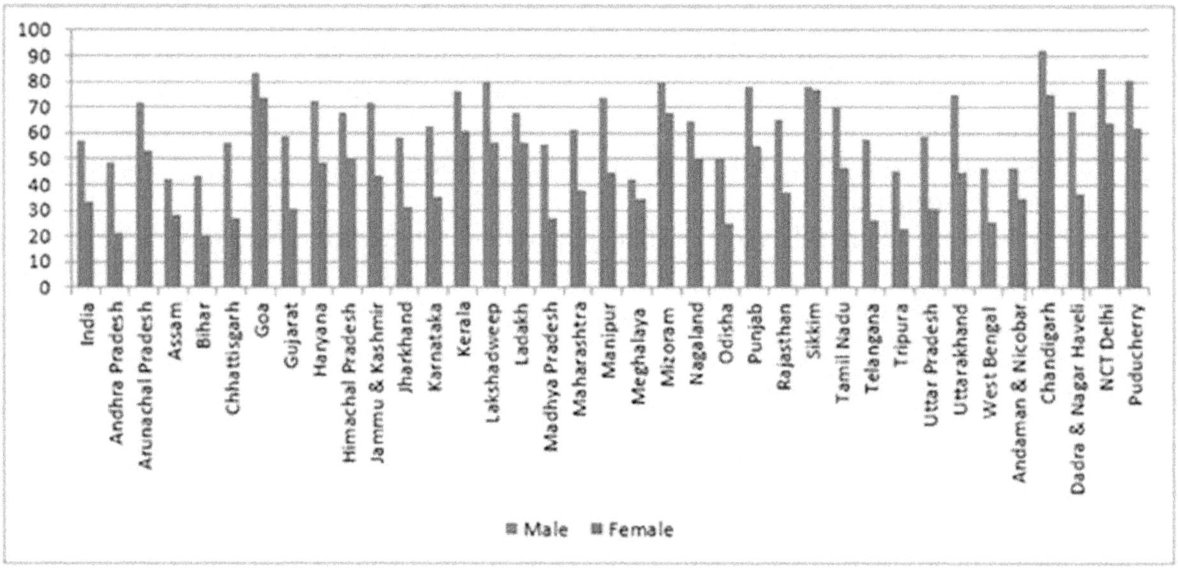

Figure 2. Indicating regional disparity in accessing Internet facility
Source: NFHS (2021)

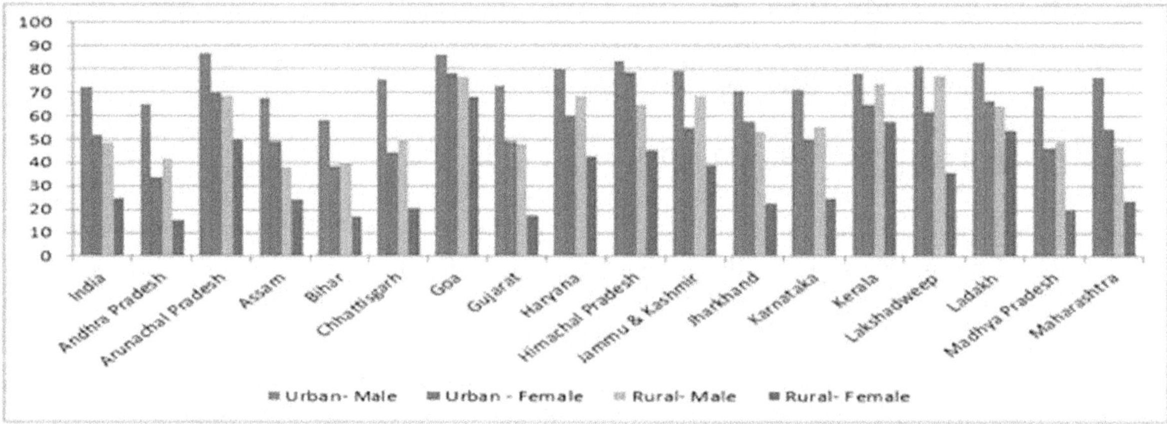

7.1.2 Mobile Ownership

Mobile phone ownership in India highlights a significant divide, with 79% of males and 67% of females owning mobile phones. Data also demonstrate a notable growth in mobile phone ownership from 2015-16 to 2019-21 as highlighted by Figure 4.

The data taken from the National Family Health Survey reveals a breakdown based on the rural-urban divide. It is profound that a significant gap prevails in mobile ownership with the prevailing pattern where a higher percentage of urban women have mobile phones as compared to their rural counterparts. However, there are some exceptions in this trend in some states and UTs such as Kerala, Ladakh, Delhi,

Figure 3. Indicating gender and regional disparity in accessing internet facility
Source: NFHS (2021)

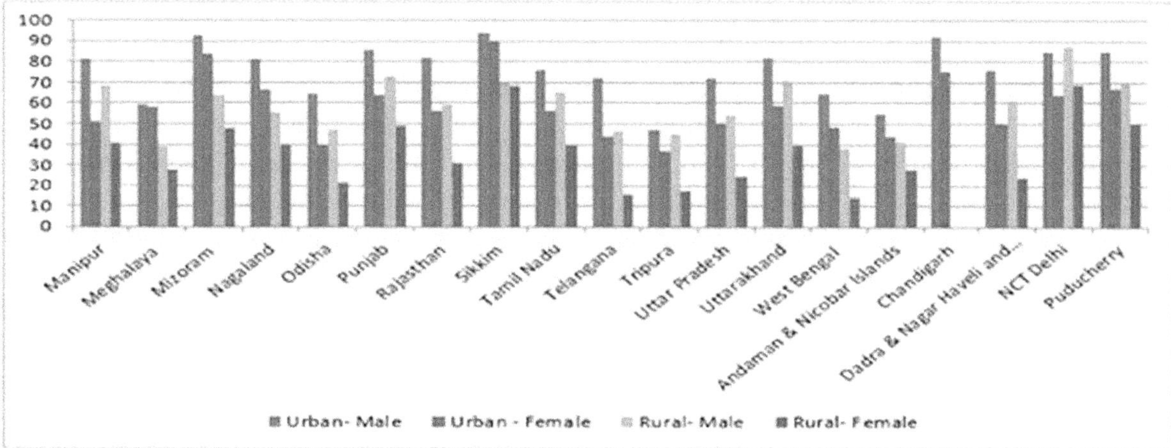

Figure 4. Growth in mobile phone ownership from 2015-16 to 2019-21
Source: NFHS (2021)

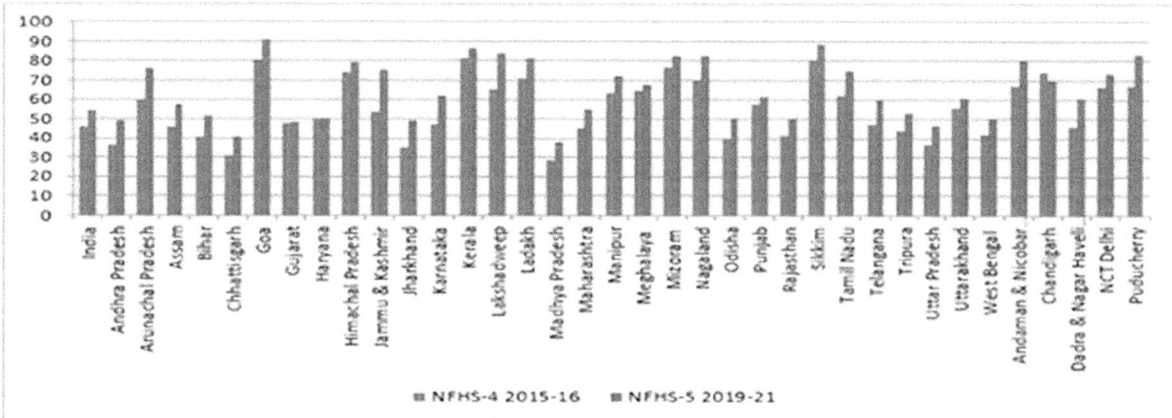

Figure 5. Regional disparity in mobile phone ownership
Source: NFHS (2021)

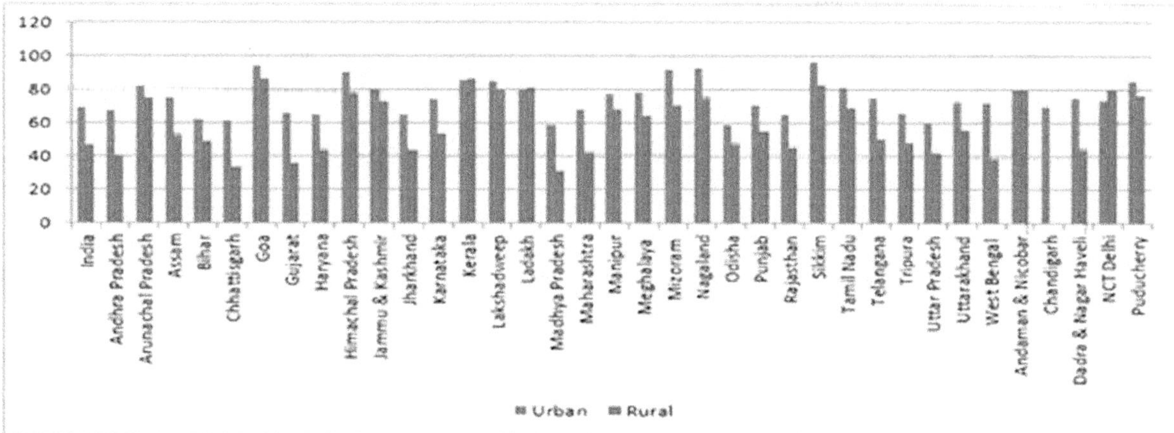

Andaman, and Nicobar exhibited that rural women have greater mobile ownership as compared to urban women. Figure 5 indicates the rural-urban disparity in females' mobile phone ownership based on state-wise data taken from NFHS-5.

7.2 Factors Responsible for Digital Divide and Digital Equity in India

The digital divide itself is a complex issue influenced by multiple factors. Based on the literature reviewed, the following factors are responsible for the issue of digital divide in India.

The first and foremost factor is the economic disparity, which serves as a significant driver of the digital divide in India, as most of the Indian population lives in poverty and is unable to afford technology and access to the internet. Regional disparity is another prominent issue, as the people residing in urban regions have better access to infrastructure as compared to those residing in rural regions, where internet connectivity is a major concern (Hasan Laskar, 2023). Apart from this, social factors like gender discrimination contribute to the digital divide, with women having very little access to technology due to numerous social and cultural norms(Abu-Shanab & Al-Jamal, 2015; Antonio & Tuffley, 2014). Moreover, a significant gap exists in digital literacy and education, with many individuals lacking the appropriate knowledge & skills to increase their ability to use digital tools and the Internet. Language barriers are another major factor that is responsible for the digital divide in India, as a large number of the Indian population is more comfortable with regional language than the English language (King & Gonzales, 2023).

Furthermore, the higher cost of digital financial services, including instruments and data plans, deter many people from investing much in their digital literacy. Numerous other infrastructural challenges prevail such as power supply and inadequate technology support.

These factors together create a more challenging landscape, where millions of Indian populations are excluded from the digital era and missing the opportunities for social, economic, and educational change that digital connectivity can offer to them. Addressing these challenges requires comprehensive efforts from the government side and can narrow the digital divide gap and ensure digital access to all citizens (Rinuan & Bohlin, 2011).

7.3 Initiative Taken by the Government of India to Bridge the Digital Divide in India

Digital India Mission

Digital India Mission is a flagship initiative of the Government of India launched in July 2015. It plays a pivotal role in addressing the digital divide by expanding the digital infrastructure including broadband connectivity and mobile networks and ensuring that more people have access to the internet.

Pradhan Mantri Jan Dhan Yojana

The main focus of this initiative is to ensure financial inclusivity by ensuring financial access to all the citizens of India. Recent data related to PMJDY indicates that around 55.59% of accounts opened under this program are owned and managed by women. This trend indicates the crucial role of this initiative in promoting gender inequality in India. Another estimate highlights that women customers of JDY (Jan Dhan Yojana) are highly profitable as compared to their male counterparts.

Digital Saksharta Abhiyan (DISHA)

National Digital Literacy Mission aimed to increase digital financial literacy, especially among rural areas. This program has been launched under the Government's digital India mission to transform India's economy. This program particularly targeted rural communities to bridge the digital divide and enhance their employability.

Bharat Net

Bharat Net is one of the transformative initiatives undertaken by the government of India to connect about 25000 rural gram panchayats through a vast fiber optic network. It aims to bridge the urban and rural divide by offering a high-speed internet connection to remote areas.

National Digital Literacy Mission

This mission plays a crucial role in addressing the digital divide in India by offering digital knowledge and skills to individuals, particularly those belonging to rural and marginalized communities. By equipping individuals with all the required digital infrastructure including computers and internet connectivity, NDLM empowers individuals to access information, government services, and online employment opportunities, thus bridging the digital divide.

Aadhar and Direct Benefit Transfer

The Government's direct benefit transfer program leverages numerous digital payment platforms to transfer subsidies, pensions, and other benefits to direct beneficiaries' accounts. It ensures that the intended recipients receive their entitlements eliminating the need for intermediaries. This streamlined approach reduces intermediaries, minimizes corruption, and promotes financial inclusion by encouraging individuals to open accounts and embrace digital payments. This initiative not only fostered financial literacy and reduced the economic disparity in the country.

7.4 Strategies and Recommendations to Address the Digital Divide

Addressing the digital divide is a multifaceted challenge that requires a vast combination of efforts and strategies to ensure equitable digital inclusion for all. Here are some strategies and recommendations to address the situation of the digital divide. More investment should be made in the expansion of high-speed internet infrastructure especially in remote and rural areas. Wi-Fi hotspot facilities should be provided in rural and urban areas to provide free and lower-cost internet access to the users. Launch various digital literacy programs to educate people on the path of using the internet, mobile phones, and various equipment like computers and empower women through the medium of digital skills to reduce the gender disparity in accessing and utilizing digital services. Implement more and more programs that specially targeted the women in empowering them through digital inclusion. Take steps to improve the mobile network connectivity, particularly in remote areas. Promote the creation of more content in regional languages that could be easily understandable by individuals. Development of digital platforms and services that can be accessible in multiple languages. Enhance the effectiveness of various initiatives taken by the government of India such as Skill India, Bharat Net, and digital India platforms. More collaborations should be undertaken private sector to extend digital access and services to underserved regions. Collaboration with international organizations to provide access to resources, knowledge, and expertise to address the digital divide in India.

Bridging the digital divide is a long-term endeavor that requires a combined commitment from the government, individuals, the public sector, the private sector, civil society, and international practitioners. A combination of these strategies and efforts adapted to local context and evolving technology trends can contribute towards a more inclusive and digitally empowered India.

8. CONCLUSION

The digital divide in India reflects the disparity in internet access and mobile ownership across various regions and demographic groups as supported by (Asrani, 2020). It is crucial to bridge the divide as online access becomes very crucial for education, healthcare, work, and more. Gender disparity is a major concern, women in India face more significant challenges in accessing and utilizing internet services, due to social and cultural norms (Oxfam India, 2022). India has made some significant progress in reducing the digital divide, but some challenges remain the same (Mukherjee et al., 2016). Rural-urban and socio-economic disparity persists, while the gender disparity in accessing digital technology and literacy requires more targeted efforts (Deepa et al., 2023; OECD, 2020). To bridge the digital divide in India, it is imperative to expand the infrastructure facilities, creation of more public Wi-Fi hotspots, and ensure more affordable access to the internet facilities. Digital literacy and skill development are crucial for implementing programs that can equip an individual with the proper knowledge and skills to navigate the digital landscape efficiently and effectively (Chandra, 2022). The creation of more and more digital content in the local language and the development of more comprehensive platforms are essential to cater to the diversified needs of India's population. International collaboration can provide individuals more access to the resources and expertise to address the digital divide effectively.

In the rapidly evolving digital environment, bridging the digital divide is not only a matter of socioeconomic development but also essential for ensuring fair and equitable opportunities to ensure the socio-economic well-being of all citizens. Implementing the strategies and recommendations outlined in this study can help in achieving a more inclusive and digitally empowered future for them.

9. LIMITATIONS OF THE STUDY AND FUTURE RESEARCH GAP

It is imperative to acknowledge the limitations of this study. The primary constraints lie in the method of data collection, which relies solely on secondary sources taken from multiple sources. The complete reliance on secondary data provides that the study's insights are based on historical data. In the rapidly evolving environment, it is difficult to capture the most recent developments and emerging trends within the field of the digital divide. As time, technology rapidly evolves, and future research work can investigate how emerging technologies like artificial intelligence, 5G, and IoT affect the digital divide. Future work can be done to understand how these advancements in the field of technology can alleviate the disparities in accessing and using digital technology. More research work can be conducted to gain deeper insights into the digital inclusion of marginalized groups. It can help in understanding the challenges faced by these groups in developing inclusive strategies. furthermore, with the growing digital footprint landscape, future research can investigate the issues related to data security issues mainly in the context of rising online activities.

8. IMPLICATIONS OF THE STUDY

The study on digital equity and the digital divide has significant implications for individuals, communities as well and societies. It can help the policymakers in designing policies and targeted interventions aimed at reducing the disparity and promoting inclusion. Firstly, it possesses the capacity to transform the landscape of educational equity by providing equal access to essential resources and opportunities. Addressing digital equity and digital divide is not just a matter of technological access but it is a key driver for social progress, economic development, and the well-being of individuals and communities in our digital world. Furthermore, it has several managerial implications that are relevant for both private and public sector organizations as well as non-profit entities. As digital access increases, it becomes imperative for managers to prioritize data security and privacy. This involves the establishment of robust cybersecurity measures for protecting both user data and organizational data, especially in those regions where digital literacy may be lower.

REFERENCES

Abu-Shanab, E., & Al-Jamal, N. (2015). Exploring the Gender Digital Divide in Jordan. *Gender, Technology and Development*, *19*(1), 91–113. doi:10.1177/0971852414563201

Afzal, A., Khan, S., Daud, S., Ahmad, Z., & Butt, A. (2023). Addressing the Digital Divide: Access and Use of Technology in Education. *Journal of Social Sciences Review*, *3*(2), 883–895. doi:10.54183/jssr.v3i2.326

Aguilar, S. J. (2020). Guidelines and tools for promoting digital equity. *Information and Learning Science*, *121*(5/6), 285–299. doi:10.1108/ILS-04-2020-0084

Aissaoui, N. (2021). The digital divide: a literature review and some directions for future research in light of COVID-19. *Global Knowledge, Memory and Communication, 71*(8/9), 686–708. https://doi.org/https://doi.org/10.1108/GKMC-06-2020-0075

Antonio, A., & Tuffley, D. (2014). The Gender Digital Divide in Developing Countries. *Future Internet*, *6*(4), 673–687. doi:10.3390/fi6040673

Asrani, C. (2020). *Bridging the Digital Divide in India : Barriers to Adoption and Usage*. Issue June.

Botelho, F. H. F. (2021). Accessibility to digital technology: Virtual barriers, real opportunities. *The Official Journal of RESNA, 33*, 27–34. https://doi.org/https://doi.org/10.1080/10400435.2021.1945705

Chandra, A. (2022). *Bridging the Digital Gender Divide*. doi:10.4018/978-1-7998-8594-8.ch002

De, R., Pandey, N., & Pal, A. (2020). Impact of digital surge during Covid-19 pandemic: A viewpoint on research and practice. *International Journal of Information Management*, *55*, 102171. Advance online publication. doi:10.1016/j.ijinfomgt.2020.102171 PMID:32836633

Deepa, M. M., Sajan, S., & Gupta, T. (2023). The intersection of technology and society: An analysis of the digital divide. *European Chemical Bulletin*, *12*(12), 1106–1116. doi:10.48047/ecb/2023.12.si12.09

Derek Chun, W. S., Siu, H. Y., Wai, M. C., & Chi, Y. L. (2022). Transformation Conference paper Bridging the Gap Between Digital Divide and Educational Equity by Engaging Parental Digital Citizenship and Literacy at Post-Covid-19 Age in the Hong Kong Context. *The Post-Pandemic Landscape of Education and Beyond: Innovation and Transformation*, 165–182.

Diebold, G., & Castro, D. (2023). Digital Equity 2.0: How to Close the Data Divide. Center for Data Collection.

Gudmundsdottir, G. (2010). From digital divide to digital equity: Learners' ICT competence in four primary schools in Cape Town, South Africa. *International Journal of Education and Development Using ICT*, *6*(2), 84–105.

Hasan Laskar, M. (2023). Examining the emergence of digital society and the digital divide in India: A comparative evaluation between urban and rural areas. *Frontiers in Sociology*, *8*, 1145221. Advance online publication. doi:10.3389/fsoc.2023.1145221

Imran, A. (2022). Why addressing digital inequality should be a priority. *The Electronic Journal on Information Systems in Developing Countries*, *89*(3), e12255. doi:10.1002/isd2.12255

King, J., & Gonzales, A. L. (2023). The influence of digital divide frames on legislative passage and partisan sponsorship: A content analysis of digital equity legislation in the U.S. from 1990 to 2020. *Telecommunications Policy*, *47*(7), 102573. doi:10.1016/j.telpol.2023.102573

Kolotouchkina, O., Barroso, C., & Sanchez, J. L. (2022). Smart cities, the digital divide, and people with disabilities. *Cities (London, England)*, *123*, 103613. doi:10.1016/j.cities.2022.103613

Korovkin, V., Park, A., & Kaganer, E. A. (2022). Towards conceptualization and quantification of the digital divide. *Information Communication and Society*, *26*(1), 2268–2303. Advance online publication. doi:10.1080/1369118X.2022.2085612

Kraus, S., Durst, S., Ferreira, J., Veiga, P., Kailer, N., & Weinmann, A. (2022). Digital transformation in business and management research: An overview of the current status quo. *International Journal of Information Management*, *63*, 102466. doi:10.1016/j.ijinfomgt.2021.102466

Moore, R., Vitale, D., & Stawinoga, N. (2018). *The Digital Divide and Educational Equity*. ACT Research & Center for Equity in Learning.

Mukherjee, B. E., Mazar, O., Aggarwal, R., & Kumar, R. (2016). *Exclusion from Digital Infrastructure and Access*. www.defindia.org/publication-2

NFHS. (2021). *Compendium of Fact Sheets India and 14 States/UTs (Phase-11)*. https://main.mohfw.gov.in/sites/default/files/NFHS-5_Phase-II_0.pdf

OECD. (2013). *Organization for Economic Co-operation Development*. OECD.

OECD. (2020). *Rural Well-being Geography of Opportunities*. https://doi.org/https://doi.org/10.1787/d25cef80-en

Okunola, O. M., Rowley, J., & Johnson, F. (2017). The multi-dimensional digital divide: Perspectives from an e-government portal in Nigeria. *Government Information Quarterly*, *34*(2), 329–339. doi:10.1016/j.giq.2017.02.002

Oxfam India. (2022). *Digital Divide India Inequality Report 2022*. Author.

Passey, D., Shonfeld, M., Appleby, L., Judge, M., Saito, T., & Smits, A. (2018). Digital Agency: Empowering Equity in and through Education. *Technology. Knowledge and Learning*, *23*(3), 425–439. doi:10.100710758-018-9384-x

Payne, A. L., Stone, C., & Bennett, R. (2023). Conceptualising and Building Trust to Enhance the Engagement and Achievement of Under-Served Students. *The Journal of Continuing Higher Education*, *71*(2), 134–151. doi:10.1080/07377363.2021.2005759

Prathapagiri, V. G. (2019). Digital Divide and Its Dimensions. *Advances in Electronic Government, Digital Divide, and Regional Development*, (April), 79–100. doi:10.4018/978-1-5225-5412-7.ch004

Rajam, V., Reddy, A. B., & Banerjee, S. (2021). Explaining caste-based digital divide in India. *Telematics and Informatics*, *65*, 101719. doi:10.1016/j.tele.2021.101719

Resta, P., Laferriere, T., McLaughlin, R., & Kouraogo, A. (2018). Issues and Challenges Related to Digital Equity: An Overview. In Second Handbook of Information Technology in Primary and Secondary Education (pp. 987–1004). Academic Press.

Rinuan, C., & Bohlin, E. (2011). *Understanding the digital divide: A literature survey and ways forward.* Innovative ICT Applications- Emerging Regulatory, Economic and Policy Issues.

Singh, S. (2010). Digital Divide in India: Measurement, Determinants and Policy for Addressing the Challenges in Bridging the Digital Divide. *International Journal of Innovation in the Digital Economy*, *1*(2), 1–24. doi:10.4018/jide.2010040101

Sweidan, N. S., & Areiqat, A. (2020). The Digital Divide and its Impact on Quality of Education at Jordanian Private Universities Case Study: Al-Ahliyya Amman University. *International Journal of Higher Education*, *10*(3), 1. doi:10.5430/ijhe.v10n3p1

The World Bank. (2021). *World Development Report 2021*. Author.

Tripathy, B., & Raha, S. (2019). Digital Divide in India. *International Research Journal of Engineering and Management Studies, 3*(5).

Vassilakopoulou, P., & Hustad, E. (2023). Bridging Digital Divides: A Literature Review and Research Agenda for Information Systems Research. *Information Systems Frontiers*, *25*(3), 955–969. doi:10.100710796-020-10096-3 PMID:33424421

Williams, D. D. (2022). *Digital equity: Difficulties of implementing the 1:1 computing initiative in low-income areas*. Academic Press.

Yuan Sun, J. C., & Metros, S. E. (2011). The Digital Divide and Its Impact on Academic Performance. *US-China Education Review*, *2*, 153–161.

ENDNOTE

1 TRAI: Telecom Regulatory Authority of India.

Chapter 10
Digital Identity and Data Sovereignty:
Redefining Global Information Flows

Kaushikkumar Patel
https://orcid.org/0009-0005-9197-2765
TransUnion LLC, USA

ABSTRACT

This chapter explores the intersection of digital identity and global data sovereignty, focusing on blockchain's impact. The authors analyze its transformative potential, security, and challenges concerning data regulations. This chapter advocates for a user-centric, ethical, and secure approach to digital identity, addressing societal implications and emphasizing user sovereignty. The chapter calls for interdisciplinary collaboration and outlines future directions for a secure digital identity framework.

1. INTRODUCTION

In the digital era, the concepts of identity and data sovereignty have become increasingly significant, influencing global information flows and raising critical questions about privacy, security, and individual rights. The surge in online activities, accelerated by global events such as the COVID-19 pandemic, has underscored the urgency of establishing robust digital identities while ensuring data protection and compliance with various regional regulations. This chapter delves into the complexities of digital identity management, emphasizing the transformative role of blockchain technology in redefining data sovereignty and the implications of these advancements on global information flows.

The discourse around digital identity has evolved beyond mere online authentication, touching on profound themes of trust, human rights, and the equitable access to digital services. Traditional models of digital identity management have often placed users at the periphery, giving institutions or centralized authorities disproportionate control over personal data. This dynamic has sparked debates and legislative actions worldwide, culminating in regulations like the General Data Protection Regulation (GDPR) in the European Union, which aims to give individuals greater control over their personal data.

DOI: 10.4018/979-8-3693-1762-4.ch010

However, the landscape of digital identity is undergoing a seismic shift with the advent of blockchain technology, promising a paradigm where individuals have true ownership—self-sovereign identity—over their digital personas. Blockchain's decentralized nature fundamentally challenges traditional power asymmetries, offering a transparent, immutable ledger that operates beyond the control of any single entity. This innovation holds the potential to revolutionize how personal data is stored, shared, and managed, consequently impacting global information flows.

Yet, the intersection of blockchain technology with existing legal frameworks presents a complex array of challenges. The immutable characteristic of blockchain appears to conflict with data protection mandates stipulated by laws like the GDPR, particularly the 'right to be forgotten.' Furthermore, the global nature of blockchain networks complicates jurisdictional and regulatory enforcement, leading to a clash between the borderless technology and geographically confined legal systems.

This chapter seeks to unravel these complexities, providing comprehensive insights into the current state of digital identity and data sovereignty. It examines the transformative potential and limitations of blockchain technology in establishing secure, user-centric digital identity systems. Additionally, it explores the intricate web of legal, technical, and ethical considerations that arise from this technological disruption, highlighting the need for a harmonized global approach. The ensuing sections will critically analyze literature and case studies, discuss prevailing challenges and propose solutions, and explore future research directions in this dynamic field.

2. LITERATURE REVIEW

The advent of the digital age has significantly transformed identity and data management, extending these concepts beyond tangible documentation into a realm characterized by digital footprints that represent both invaluable assets and potential vulnerabilities. This literature review delves deeply into the multifaceted landscape of digital identity, with a keen focus on the revolutionary role of blockchain technology and the pivotal notion of data sovereignty. The analysis is underpinned by an extensive array of scholarly articles and research papers, providing a rich, multi-dimensional understanding.

a. **Blockchain Technology: Revolutionizing Digital Identity Management:**
    i. Blockchain technology, renowned for its decentralization, immutability, and transparency, introduces revolutionary prospects for managing digital identities. References (Berman, P., 2018), (Haddouti, S. E., 2019), (Aydın, A., 2019), and (Sinha, A., 2020) provide an exhaustive exploration of blockchain's fundamental principles. They advocate that the technology's intrinsic features, such as its ability to secure transactions and its resistance to unauthorized changes, mark a significant departure from traditional centralized digital identity systems. These studies elaborate on how blockchain fosters a user-centric model, wherein individuals enjoy unprecedented control over their personal information, fundamentally challenging conventional norms of data ownership and governance.
    ii. Moreover, these references dissect the technical workings of blockchain, from its cryptographic underpinnings to its consensus algorithms, illustrating how these elements contribute to a secure, trustworthy system for digital interactions. They also debate the potential of blockchain to serve as a universal, interoperable platform that could simplify identity verification processes, thereby reducing fraud and enhancing user experience across various domains.

b. **Practical Implementations: Blockchain and Digital Identity Convergence:**
   i. Transitioning from theory to practice, references (Pöhn, D., 2021), (Digital Identity, 2023), (Dabrock, P., 2021), and (Ghaffari, F., 2021) present an array of case studies and real-world applications of blockchain within digital identity frameworks. These documents detail the implementation of blockchain-based systems in sectors such as finance, healthcare, and governmental services, highlighting improvements in efficiency, security, and user autonomy. They discuss the concept of "verifiable credentials" and how blockchain can facilitate seamless, secure sharing of certified digital attributes.
   ii. These sources also tackle the challenges encountered during these implementations, from technical obstacles to regulatory hurdles, and the solutions adopted to address them. They provide empirical evidence supporting the feasibility and benefits of blockchain applications in real-world scenarios, contributing valuable insights into the practical considerations and strategies for successful integration.

c. **Data Sovereignty: Navigating the Global Landscape:**
   i. Data sovereignty, with its implications for cross-border data flows and regulatory compliance, emerges as a critical concern in the discourse on digital identity. References (Domeyer, A., 2020), (Digital Identity, 2018), (Tan, K. L., 2022), and (Herian, R., 2020) delve into the intricate dynamics of data sovereignty, discussing the legal and ethical challenges posed by the global nature of digital data and the internet. These works examine the tension between safeguarding individual privacy and enabling state security operations, highlighting the necessity for robust legal frameworks that reconcile individual rights with national security imperatives.
   ii. These studies also explore the concept of "digital nationalism" and the trend towards data localization, critiquing the potential impact on global innovation and digital rights. They call for a balanced approach that respects sovereignty while promoting open, secure international data ecosystems.

d. **Regulatory Challenges: Compliance in a Digital World:**
   i. The literature underscores the complexities of navigating the regulatory landscape, particularly for emerging technologies. References (Zwitter, A. J., 2020), (The World Bank., 2023), (Archana V., 2022), and (Wylde, V., 2022) scrutinize the challenges of adhering to diverse, often conflicting, regional and international regulations. These sources provide an in-depth analysis of laws like the General Data Protection Regulation (GDPR), highlighting the discord between blockchain's permanent ledger system and legal mandates for data erasure and correction.
   ii. They also discuss the broader implications of these regulations for digital innovation, arguing for more harmonized, forward-looking regulatory frameworks that support technological advancement while protecting user rights. The need for multi-stakeholder dialogue and cooperation emerges as a recurring theme, emphasizing the role of various actors, from governments to tech companies, in shaping a conducive regulatory environment.

e. **Ethical Considerations: Centering the User in Digital Identity Systems:**
   i. The ethical dimensions of digital identity management receive considerable attention, with references (Shirer, M., 2023), (Hussien, Q. M., 2021), (Devi, S., 2022), and (Gilani, K., 2020) advocating for approaches that prioritize user consent, data minimization, and societal welfare. These sources critique the current data-driven economic models, highlighting the risks of

exploitation and discrimination. They propose more equitable, inclusive models that respect user agency and promote social justice.

ii. These studies also explore the psychological and cultural aspects of digital identity, discussing the concept of "digital dignity" and the right to digital self-determination. They challenge the tech industry to move beyond profit-driven incentives and embrace their role in safeguarding digital human rights.

f. **Facing the Future: Challenges and Prospects for Blockchain in Digital Identity:**

i. Despite its transformative potential, the path to integrating blockchain into digital identity systems is strewn with challenges, ranging from technical constraints to societal apprehensions. References (Quach, S., 2022), (Benkoël, D., 2020), (User Control, 2023), and (Mühle, A., 2018) outline these multifaceted challenges, offering speculative yet insightful projections on future research directions. These sources call for innovations to enhance blockchain's scalability, privacy features, and user-friendliness, urging for continued exploration into interoperable standards and frameworks.

ii. They also touch upon the broader societal implications of widespread blockchain adoption, questioning its impact on social structures, governance models, and economic systems. The discussion extends to the ethical dilemmas posed by blockchain and the need for a holistic, multidisciplinary approach to its integration into societal frameworks.

This literature review offers a comprehensive synthesis of varied scholarly perspectives on the intersection of blockchain technology, digital identity, and data sovereignty. It illuminates the transformative essence of blockchain, the global urgency surrounding data sovereignty, and the complex challenges at this crossroads. The analysis reaffirms the need for a delicate equilibrium among technological innovation, regulatory governance, ethical norms, and user-oriented design. This balance is pivotal for steering future research and policy directions, ensuring that the digital identity ecosystems of tomorrow are resilient, inclusive, and humane.

3. BLOCKCHAIN'S ROLE IN REINVENTING DIGITAL IDENTITY

The advent of blockchain technology heralds a paradigm shift in digital identity management, presenting a move towards decentralization, enhanced security, and user empowerment. This in-depth exploration, informed by an array of scholarly articles, dissects the multifaceted role of blockchain in reshaping the contours of digital identity, addressing the technical, ethical, and practical layers of this revolutionary technology. This section delves into the key aspects of blockchain technology in digital identity management, as illustrated in Table 1, exploring how blockchain is reshaping the landscape of identity verification and data control.

a. **Decentralization: The Foundation of Trust:** Blockchain's essence lies in its decentralized architecture, a groundbreaking departure from traditional centralized systems that have long governed digital identities. References (Haddouti, S. E., 2019), (Digital Identity, 2023), (Dabrock, P., 2021), and (Gilani, K., 2020) provide a deep dive into the mechanics of trustless transactions enabled by blockchain, where cryptographic functions replace institutional guarantors. This decentralization not only mitigates risks associated with data breaches and fraud by eliminating single points of failure but also introduces a new level of data privacy and security. The concept of "zero-knowledge proof," detailed in these studies, exemplifies how blockchain facilitates identity verification processes without the necessity to reveal sensitive personal information, thereby preserving individual privacy.

b. **Self-Sovereign Identity: Empowerment and Control:** At the heart of blockchain's transformative impact is the concept of self-sovereign identity (SSI), a model advocating for individual autonomy over identity credentials. References (Berman, P., 2018), (Aydın, A., 2019), (Sinha, A., 2020), and (Ghaffari, F., 2021) delve into the intricacies of SSI, highlighting the philosophical shift towards user-centric identity management systems. These sources dissect the technical frameworks that underpin SSI, emphasizing the importance of user control, data minimization, and consent. They also confront the inherent challenges in actualizing self-sovereignty, discussing the digital divide, the necessity for user-friendly interfaces, and the ongoing struggle between convenience and security.

c. **Interoperability and Standardization: Building Universal Identity Platforms:** One of blockchain's most significant promises is its potential to underpin interoperable digital identities, crucial in an increasingly globalized world. References (Pöhn, D., 2021), (The World Bank., 2023), (Best Practices, 2022), and (Archana V., 2022) scrutinize the global efforts toward standardization, which is pivotal for the seamless interaction between diverse identity systems across borders. These studies examine the concerted initiatives like the Decentralized Identity Foundation and the push for universal standards such as Decentralized Identifiers (DIDs), aiming to forge a coherent, global framework for blockchain-based digital identities.

d. **Privacy-Preserving Techniques: Balancing Verification and Anonymity:** Blockchain addresses the critical conundrum of ensuring robust identity verification while safeguarding user privacy. References (Domeyer, A., 2020), (Zwitter, A. J., 2020), (Devi, S., 2022), and (Mühle, A., 2018) explore advanced cryptographic methods, such as zero-knowledge proofs, ring signatures, and homomorphic encryption, integral to blockchain platforms. These techniques, as discussed, are instrumental in maintaining user anonymity, providing only the requisite amount of data for verification. This balance is particularly vital in sensitive sectors like healthcare and finance, where personal data security is paramount.

e. **Real-World Applications and Case Studies:** The theoretical underpinnings of blockchain in digital identity management transcend into tangible applications with far-reaching impacts. References (Shirer, M., 2023), (Benkoël, D., 2020), (Tan, K. L., 2022), and (User Control, 2023) document diverse case studies ranging from humanitarian initiatives using blockchain for refugee aid, financial inclusivity efforts for the unbanked populations, to the deployment of secure e-voting systems and healthcare data management. These applications underscore blockchain's versatility and its potential to solve real-world problems, marking a significant leap from theoretical propositions to practical solutions.

f. **Technical Challenges and Innovations:** Despite its revolutionary aspects, blockchain is not devoid of technical constraints. References (Quach, S., 2022), (Hussien, Q. M., 2021), (Herian, R., 2020), and (Wylde, V., 2022) shed light on these limitations, including scalability issues, substantial energy consumption, and the complexities involved in integrating blockchain with existing digital infrastructures. However, these studies also chart the trajectory of innovative solutions like layer 2 protocols, advancements in consensus algorithms, and the emergence of hybrid blockchain models, each designed to circumvent these technical roadblocks.

Table 1. Key aspects of blockchain technology in digital identity management

Aspect	Description
Decentralization	Eliminates single points of failure, enhancing security and privacy through a trustless system.
Self-Sovereign Identity (SSI)	Empowers individuals with control over their digital identities, ensuring autonomy and privacy.
Interoperability	Facilitates global standardization, enabling seamless interaction across various identity systems.
Privacy-Preserving Techniques	Employs advanced cryptography to balance identity verification with user anonymity.
Real-World Applications	Demonstrates versatility through use cases in humanitarian aid, financial services, healthcare, and e-voting.
Technical Challenges	Addresses scalability, energy consumption, and integration issues with innovative solutions.
Regulatory Compliance	Navigates complex legal landscapes, advocating harmonized regulations that foster innovation while protecting data rights.
Ethical Considerations	Emphasizes equitable access, prevention of surveillance, and considers societal impacts of technology adoption.

g. **Regulatory Landscape and Compliance:** Blockchain's intersection with the legal realm is a landscape of intricate compliance requirements and regulatory considerations. References (Digital Identity, 2018), (Dabrock, P., 2021), (Gilani, K., 2020), and (Ghaffari, F., 2021) navigate the legal nuances, including data ownership rights, the contentious right to erasure, and the challenges posed by cross-border data transfers. These analyses advocate for a harmonized regulatory approach that fosters innovation while safeguarding user privacy and data rights, highlighting the delicate balance regulators must maintain.
h. **Ethical Considerations and Societal Impact:** The societal repercussions of blockchain technology, especially in digital identity management, are profound and multifaceted. References (Haddouti, S. E., 2019), (Sinha, A., 2020), (Best Practices, 2022), and (Archana V., 2022) delve into the ethical quagmire, tackling issues of equitable access to digital identity solutions, the dangers of societal surveillance, and the ethical obligation to prevent technology-facilitated marginalization. These discourses emphasize a holistic adoption strategy for blockchain, one that considers its wider societal ramifications beyond technical implementation.

This extensive analysis elucidates blockchain technology's transformative role in digital identity management, drawing on a rich tapestry of scholarly work. The discussion underscores blockchain's promise in enhancing data security, user privacy, and control, while also recognizing its limitations and the complexities involved in its widespread adoption. The insights gathered here signify the importance of continued interdisciplinary research, collaborative problem-solving, and global dialogue in harnessing blockchain's full potential responsibly. This foundational understanding is critical as we venture into subsequent discussions that further unravel the complexities of digital identity in the modern world.

4. REGULATORY IMPACTS AND DATA SOVEREIGNTY IN DIGITAL IDENTITY

The intricate interplay between regulatory frameworks and technological advancements is nowhere more evident than in the realm of digital identity. As digital footprints expand, so does the need for regulations like the General Data Protection Regulation (GDPR) and others globally, designed to safeguard user privacy and data. However, these protective measures introduce several complexities, especially when considering the decentralized nature of blockchain technology and the concept of data sovereignty. This section delves into the complex world of regulatory challenges and strategic adaptations in digital identity, as presented in Table 2. It examines the profound impact of regulations on data sovereignty and identity management.

a. **The Advent of GDPR and its Global Implications:** The introduction of GDPR marked a significant shift in data privacy, emphasizing individual rights to data access, correction, and erasure. These rights, particularly the "right to be forgotten," challenge blockchain technology's foundational principle of immutability, creating a paradox where compliance seems technically unfeasible. This conflict underscores a broader global issue where technology often outpaces legal frameworks, leading to a lag in regulatory responses and adaptations (Herian, R., 2020), (Archana V., 2022).
b. **Blockchain's Decentralization Versus Data Sovereignty:** Blockchain technology, especially public blockchains, operates globally, often without physical boundaries. This characteristic raises critical questions about jurisdiction, as the data on the blockchain, though decentralized, must still comply with the laws of the country it resides in. The issue becomes more complex with data localization laws, where nations demand data storage within their territories to maintain control and enforce data sovereignty. These requirements present logistical and ethical challenges, balancing national interests with the ethos of decentralization (Archana V., 2022).
c. **Strategies for Navigating Regulatory Ambiguities:** Amid these regulatory uncertainties, various stakeholders, from blockchain developers to organizations, are pioneering methods to bridge the gap between compliance and technological constraints. Techniques include storing sensitive data off-chain, using cryptographic solutions for data masking, and explicit user consent for data processing. These strategies, while innovative, are interim solutions in the face of evolving legal landscapes and potential future litigations that could redefine compliance parameters (Herian, R., 2020), (Archana V., 2022).
d. **Self-Sovereign Identity (SSI) and Regulatory Compliance:** The concept of SSI on the blockchain promotes user autonomy over personal data, a principle that aligns with GDPR's objectives. However, the implementation of SSI poses its own dilemmas. While it circumvents reliance on centralized entities, it also emphasizes a free-market approach, often at odds with regulatory frameworks designed to protect users from the market forces SSI enables. This dichotomy highlights the tension between individual empowerment and the broader economic implications of unregulated data control (Gilani, K., 2020).
e. **Data Localization: A Double-Edged Sword:** The push for data sovereignty through localization potentially stifles global collaboration and innovation. While it strengthens national control over data, it also fragments the digital landscape, complicating operations for multinational companies and innovators. These entities face the daunting task of navigating a maze of local laws, which, while protecting data sovereignty, can also impede the seamless flow of information necessary for global operations and technological advancements (Archana V., 2022).
f. **The Dynamic Nature of Compliance and Expectations:** Regulatory compliance is not a static achievement but a dynamic requirement. With the continuous evolution of data protection laws and expectations, organizations and technologies, including blockchain, must be agile. This agility is not just about adhering to current standards but also about anticipating and adapting to future

regulatory shifts. Such foresight is essential, especially in blockchain deployment, where data once recorded is permanent and unalterable. It necessitates a balance between innovation and compliance, requiring ongoing dialogue among regulators, technologists, and users (Archana V., 2022), (Ghaffari, F., 2021).

g. **Balancing Public Interest with Technological Innovation:** The inception of regulations like GDPR stems from a need to curb unregulated data practices and protect public interest. This protective stance sometimes clashes with the blockchain community's push for unrestricted innovation. The latter's emphasis on a decentralized, free-market approach often overlooks the necessity for safeguards against potential abuses of data freedom. This scenario calls for a nuanced approach that can foster technological innovation without compromising data rights and public welfare (Herian, R., 2020).

h. **Redefining Sovereignty in the Digital Age:** Data sovereignty transcends legal and territorial boundaries, encompassing issues of user autonomy, ethical data practices, and the power dynamics inherent in data control. Achieving genuine data sovereignty challenges not only the legal and technological status quo but also the underlying economic and political systems that dictate data practices. Both GDPR and blockchain offer pathways toward greater control and privacy. However, they are part of broader structures that often compromise these very principles for broader economic or strategic interests (Herian, R., 2020).

In conclusion, the convergence of data sovereignty, regulatory frameworks, and digital identity management is a complex, multifaceted arena that continues to evolve. As nations and organizations strive to uphold data rights while fostering innovation, they must navigate the nuanced challenges posed by technologies like blockchain. The future lies in creating synergistic solutions that harmonize regulatory requirements with the transformative potentials of blockchain, ensuring that advancements in digital

Table 2. Regulatory challenges and strategic adaptations in digital identity

Challenge	Strategic Adaptation
GDPR's "right to be forgotten" vs. blockchain's immutability	Employing off-chain data storage solutions and cryptographic erasure techniques.
Global operation of blockchains vs. national data sovereignty laws	Implementing geo-specific blockchain operations and respecting data localization mandates.
Navigating regulatory ambiguities with evolving technology	Continuous stakeholder dialogue, anticipatory compliance strategies, and flexible blockchain design.
Aligning Self-Sovereign Identity (SSI) with regulatory compliance	Enhancing user consent protocols and integrating regulatory standards into SSI platforms.
Data localization hindering global innovation	Strategic data management planning, localized innovation approaches, and international regulatory cooperation.
Dynamic compliance requirements	Agile compliance frameworks, ongoing legal-technological education, and proactive policy engagement.
Balancing free-market innovation with public interest protection	Establishing ethical guidelines, fostering transparent stakeholder communication, and promoting user-centric innovation.
Redefining data sovereignty beyond territorial constraints	Advocating global data ethics standards, empowering user data autonomy, and challenging systemic data control practices.

identity equitably serve both individual rights and collective progress (Herian, R., 2020), (Archana V., 2022), (Gilani, K., 2020), (Ghaffari, F., 2021).

5. USER-CENTRIC PERSPECTIVES AND ETHICAL CONSIDERATIONS IN DIGITAL IDENTITY

The digital identity landscape is transforming with the advent of user-centric technologies, particularly Self-Sovereign Identity (SSI), which promises unprecedented control and privacy for individuals regarding their personal data. This evolution, deeply explored in (Haddouti, S. E., 2019), (Digital Identity, 2018), (Gilani, K., 2020), and (Ghaffari, F., 2021), is not without its ethical quandaries, especially as it intersects with blockchain technology's immutable characteristics. This section navigates through the ethical considerations surrounding digital identity, as outlined in Table 3. It explores the critical shift towards user-centric perspectives and the moral dilemmas they entail

SSI represents a paradigm shift in digital interactions, placing users at the helm of their identity credentials. The decentralized frameworks, as detailed in (Gilani, K., 2020) and (Ghaffari, F., 2021), eliminate the need for intermediaries, allowing individuals to manage their data transactions directly. This model's cornerstone is the empowerment it offers individuals, contrasting starkly with traditional models where corporations or entities hold authority.

However, the promise of empowerment is contingent on users' understanding and consent, raising ethical concerns about data literacy. As (Sinha, A., 2020) and (Dabrock, P., 2021) articulate, the responsibility and technical acumen required to navigate these systems are substantial, often beyond the average user's capability. This complexity introduces ethical dilemmas: it risks creating a new digital divide where only those with sufficient expertise can exercise true control, potentially marginalizing vast user segments.

The ethical discourse extends to the conflict between blockchain's immutability and the legal and moral right to data deletion, a foundational aspect of data protection regulations like GDPR (Tan, K. L., 2022). References (Herian, R., 2020) and (Archana V., 2022) delve into this tension, highlighting the contradiction between the permanent nature of blockchain records and the ethical necessity of the "right to be forgotten." This clash underscores the need for nuanced solutions that respect individuals' autonomy over their digital footprints while maintaining the security and trust that blockchain brings to digital identity systems.

Another pivotal concern is the commodification of personal data. While blockchain allows individuals to monetize their data, as discussed in (User Control, 2023) and (Wylde, V., 2022), it simultaneously encourages a marketplace mentality towards personal information. This commercial approach raises significant ethical questions: What is the intrinsic value of personal data? Are we undermining human dignity by treating personal data as a commodity? These considerations challenge the narrative of economic empowerment through data sharing, advocating a return to principles of dignity and privacy in the digital space.

The societal implications of blockchain and digital identity technologies are profound. Reference (The World Bank., 2023) warns of the potential for these structures to enhance surveillance mechanisms, compromising privacy under the pretext of security. This possibility necessitates a critical ethical examination to ensure that pursuing technological advancements does not erode fundamental human

Table 3. Ethical considerations in digital identity

Ethical Concern	Description
Data Literacy	Users' understanding and capacity to navigate user-centric digital identity systems
Right to Be Forgotten	The conflict between blockchain's immutability and the "right to be forgotten"
Data Commodification	The ethical implications of treating personal data as a commodity
Privacy vs. Surveillance	Balancing security measures with the potential for enhanced surveillance
Inclusion and Fairness	Ensuring digital identity systems do not perpetuate discrimination or exclusion

rights. The technology must not become an instrument for pervasive monitoring systems that could be exploited for control rather than serving public interest.

Furthermore, the ethical mandate for inclusion and fairness within these systems is paramount. As (Devi, S., 2022) emphasizes, there is an imperative to ensure that digital identity technologies, powered by blockchain or otherwise, do not entrench discrimination, bias, or exclusion. This goal calls for transparent, accountable algorithms and design methodologies that consciously avoid perpetuating societal inequities.

In conclusion, the journey toward a user-centric digital identity ecosystem is fraught with complex ethical considerations. The literature insists on a comprehensive approach harmonizing technological innovation with robust ethical standards. These standards must prioritize individual rights, data dignity, and societal welfare, ensuring that the digital identity mechanisms of the future are grounded in principles that transcend technological capability or economic opportunity. The path forward must be conscientiously charted, keeping human dignity, privacy, and fundamental rights at the forefront of this digital evolution.

6. USER-CENTRIC DESIGN PRINCIPLES

The foundation of digital identity systems lies in the user-centric design principles, which constitute a fundamental cornerstone for the development of ethical and effective systems. These principles revolve around the core belief that individuals should have absolute control and autonomy over their digital identities. They serve as the ethical compass and practical roadmap for designing and implementing these systems, ensuring that they prioritize the rights, interests, and experiences of users above all else. By emphasizing user-centric design, organizations can create digital identity ecosystems that empower individuals, respect their privacy, foster trust, and elevate the digital experience to new heights. These principles guide every facet of the system, from privacy and consent to transparency, security, usability, and beyond, ultimately establishing a human-centered approach that places users at the heart of the digital identity landscape.

- **Privacy:** Privacy stands as a paramount pillar within the realm of digital identity, symbolizing the inviolable right of individuals to safeguard their personal information and control its dissemination. In the context of digital identity systems, privacy demands meticulous attention, manifesting as a commitment to collect only the necessary data, utilize it exclusively for legitimate purposes, and shield it from unauthorized access or misuse. This entails robust encryption, stringent access controls, and proactive measures to deter breaches or cyber threats. Moreover, privacy champions the concept of data minimization, advocating for the retention of the least amount of information required for a specific function. It places individuals as the sovereign masters of their data, granting them the prerogative to manage and share their information on their terms. Thus, privacy acts as both a shield and a beacon, fortifying the sanctity of personal information while illuminating the path toward ethical and user-centric digital identity solutions.

- **Consent:** In the landscape of user-centric digital identity, consent emerges as an indispensable tenet, embodying the fundamental principle that individuals should have the autonomy to grant or withhold permission for the collection, processing, and sharing of their personal data. It is the cornerstone of ethical data practices, emphasizing that individuals must be informed comprehensively and transparently about how their data will be utilized, to what extent, and by whom. Consent is not merely a checkbox to be ticked; it represents a dynamic, ongoing dialogue between individuals and service providers. It necessitates clear, plain-language explanations of data practices, ensuring that users can make informed choices. Importantly, consent should be granular, allowing individuals to specify the precise purposes for which their data may be used. It embodies the concept of revocability, wherein users retain the right to withdraw their consent at any time, with their data promptly deleted or rendered inaccessible. The essence of consent lies in its empowerment of individuals, granting them control over their digital footprint and fostering trust in digital interactions. It is the bedrock upon which user-centric digital identity is built, ensuring that individuals remain the masters of their data destiny.
- **Transparency:** Transparency is a foundational pillar of user-centric design principles in the realm of digital identity and data sovereignty. It embodies the principle that individuals should have full visibility into how their personal data is collected, processed, and utilized by organizations and service providers. It is an essential aspect of fostering trust in digital interactions. Transparency necessitates clear and comprehensible communication regarding data practices, privacy policies, and terms of service. Service providers must be explicit about their data collection methods, the purposes for which data is used, and the entities with whom data is shared. Moreover, transparency entails making information accessible and understandable to individuals of diverse backgrounds and levels of digital literacy. It requires organizations to avoid obscure or convoluted language, opting for plain and straightforward explanations. In the spirit of transparency, organizations should also disclose any data breaches promptly and take appropriate measures to mitigate harm. Transparency builds a bridge between individuals and the organizations they interact with, enabling informed decision-making and engendering a sense of control over personal data. It is a cornerstone of ethical data practices and a vital element in the journey toward user-centric digital identity.
- **User Control:** User control is a fundamental principle in the realm of user-centric design for digital identity and data sovereignty. At its core, it empowers individuals to have ultimate authority over their personal data and how it is utilized by organizations and digital platforms. User control means that individuals have the right to make informed choices about the collection, storage, sharing, and deletion of their data. It emphasizes that individuals should have easy-to-use tools and interfaces that enable them to manage their privacy preferences and consent settings. This principle advocates for granular control, allowing users to specify what information they are comfortable sharing and with whom. It also includes the right to revoke consent at any time and have their data deleted or anonymized when they choose to do so. User control extends beyond mere consent; it implies that individuals can dictate the terms under which they engage with digital services. This

principle not only enhances privacy but also empowers individuals to tailor their digital experiences to align with their values and preferences. User control fosters trust by putting individuals in the driver's seat, ultimately resulting in a more equitable and respectful digital ecosystem. It underscores the importance of user-friendly interfaces that facilitate easy and intuitive control over personal data, making it a cornerstone of user-centric design in the digital identity landscape.

- **Security:** Security is an indispensable pillar of user-centric design principles in the context of digital identity and data sovereignty. It encompasses a multifaceted approach to safeguarding individuals' personal information from unauthorized access, breaches, and cyber threats. Robust security measures are paramount to ensure the confidentiality, integrity, and availability of sensitive data. This principle underscores the adoption of state-of-the-art encryption techniques to protect data both in transit and at rest. It emphasizes the importance of robust authentication mechanisms, including multi-factor authentication, to verify the identity of users and prevent unauthorized access. Security involves continuous monitoring and threat detection to promptly identify and mitigate potential vulnerabilities. Moreover, it advocates for secure storage and handling of cryptographic keys, ensuring that individuals' control over their digital identity remains uncompromised. The principle of security promotes adherence to industry best practices and compliance with relevant data protection regulations. It highlights the need for organizations to invest in cybersecurity resources, conduct regular security audits, and keep pace with evolving threats. By prioritizing security, user-centric design not only safeguards individuals' personal data but also fosters trust and confidence in digital identity systems, paving the way for a more secure and privacy-respecting online environment.
- **Interoperability:** Interoperability stands as a cornerstone of user-centric design principles in the realm of digital identity and data sovereignty. It underscores the imperative need for digital systems and technologies to seamlessly collaborate and exchange information, irrespective of their origin or platform. In the context of digital identity, interoperability is essential to ensure that individuals can assert their identity credentials across various services, applications, and platforms effortlessly. It advocates for the use of open standards and protocols that enable data portability and cross-system compatibility. This principle encourages the development of standardized data formats and communication protocols that transcend organizational boundaries, enabling secure and efficient data sharing while respecting privacy and consent. Interoperability is fundamental in preventing vendor lock-in and fostering healthy competition, as it allows individuals to choose from a diverse ecosystem of digital identity providers. It promotes the creation of ecosystems where users can control their identity data and share it securely with trusted parties, regardless of the underlying technologies. Ultimately, interoperability empowers individuals with the flexibility to manage their digital identities across a wide array of services, enhancing user convenience and reinforcing user-centricity as a guiding principle in digital identity and data sovereignty design.
- **Usability:** Usability is a fundamental user-centric design principle that plays a pivotal role in ensuring that digital identity and data sovereignty systems are accessible, efficient, and intuitive for all individuals. It emphasizes the creation of interfaces, processes, and interactions that are user-friendly and easy to navigate, regardless of a user's technical expertise or abilities. In the context of digital identity, usability dictates that identity management should be straightforward and comprehensible, even for those with limited technical knowledge. This principle encourages the development of clear and intuitive user interfaces for identity management tasks, such as cre-

ating, updating, and revoking identity credentials. It emphasizes the importance of user-centered design practices, including usability testing and feedback collection, to refine and optimize digital identity systems continuously. Usability extends beyond interface design to encompass the entire user experience, focusing on minimizing friction, streamlining processes, and reducing cognitive load. It advocates for the elimination of unnecessary complexities, jargon, and hurdles that may impede individuals from effectively managing their digital identities. By prioritizing usability, digital identity and data sovereignty systems become more inclusive and user-friendly, fostering wider adoption and ensuring that individuals can exercise control over their identity data with ease and confidence.

- **Accessibility:** Accessibility is a pivotal user-centric design principle that underscores the importance of ensuring that digital identity and data sovereignty systems are inclusive and usable by individuals of all abilities. It focuses on designing interfaces, processes, and technologies that can be accessed and understood by users with diverse needs and disabilities, including visual, auditory, motor, and cognitive impairments. In the context of digital identity, accessibility ensures that identity management tools and platforms provide equivalent access and functionality to everyone, regardless of their physical or cognitive limitations. This principle emphasizes adherence to web accessibility standards, such as the Web Content Accessibility Guidelines (WCAG), to make identity-related services and information perceivable, operable, and understandable for all users. Accessibility features may include alternative text for images, keyboard navigation, screen reader compatibility, and adjustable text size and contrast. By embracing accessibility, digital identity systems become more equitable, fostering inclusivity and empowering individuals with disabilities to exercise control over their identity data with the same level of autonomy and privacy as any other user. Accessibility considerations extend to both the design of user interfaces and the underlying technologies, ensuring that identity solutions are truly accessible to everyone, regardless of their abilities or impairments.
- **Compliance:** Compliance is a fundamental user-centric design principle that revolves around aligning digital identity and data sovereignty systems with relevant laws, regulations, and industry standards. It underscores the imperative for identity solutions to adhere to legal frameworks such as the General Data Protection Regulation (GDPR), Consumer Privacy Acts, and other data protection laws. Compliance also extends to industry-specific regulations, such as those governing healthcare (HIPAA) or financial services (PCI DSS). Designing digital identity systems with compliance in mind ensures that user data is collected, processed, and stored in a manner that respects individual rights and privacy. This principle necessitates robust data governance practices, data protection impact assessments, and mechanisms for obtaining informed consent from users. Furthermore, it involves facilitating user access to their data, allowing for data portability, and providing clear avenues for users to exercise their rights under applicable regulations, such as the right to be forgotten. Compliance-focused design mitigates legal risks, safeguards user privacy, and enhances trust in identity solutions. It involves collaboration with legal experts and ongoing monitoring of evolving regulatory landscapes to ensure that identity systems remain compliant throughout their lifecycle. By prioritizing compliance, digital identity and data sovereignty systems can navigate the complex regulatory environment while offering users the confidence that their personal data is handled in accordance with established legal standards.

In conclusion, user-centric design principles are paramount in shaping the future of digital identity and data sovereignty. By prioritizing privacy, consent, transparency, user control, security, interoperability, usability, accessibility, compliance, and cultural considerations, we can create systems that empower individuals while safeguarding their data. This holistic approach fosters trust, encourages innovation, and ensures that digital identity solutions serve the diverse needs of users worldwide. As technology evolves, an unwavering commitment to these principles is vital for ethical, inclusive, and user-friendly identity ecosystems.

7. CHALLENGES AND SOLUTIONS

Digital identity and data sovereignty presents intricate challenges that demand innovative and comprehensive solutions to ensure effective operation. This section is dedicated to dissecting the multifaceted challenges and innovative solutions in digital identity and data sovereignty, as summarized in Table 4. This section delves into these challenges in detail and offers viable and in-depth solutions:

a. Interoperability Challenges: The lack of standardized communication protocols and data formats among diverse digital identity systems and blockchain platforms poses significant hurdles. Achieving seamless data sharing and verification across these heterogeneous ecosystems is hindered by these interoperability bottlenecks. For example, a digital identity issued on one blockchain may not be easily verifiable on another, hampering cross-platform identity use cases.
    - Solutions:
        - Standardization Initiatives: Establish cross-industry consortiums and standardization bodies focused on developing and promoting open standards for digital identity and blockchain interoperability. These initiatives should encompass standardized communication protocols, data formats, and application programming interfaces (APIs) to ensure compatibility between different platforms.
        - Interoperability Middleware: Develop middleware solutions that act as bridges between disparate digital identity systems and blockchains. These middleware components would facilitate data translation and communication between systems, enabling seamless cross-platform identity verification.

b. Scalability Issues: The rapid adoption of digital identity systems raises scalability concerns, especially when integrated with blockchain technology. Traditional blockchain networks, such as Bitcoin or Ethereum, face performance limitations when processing a large volume of identity-related transactions. Ensuring that these systems can scale to accommodate a growing user base while maintaining efficiency is a critical challenge.
    - Solutions:
        - Layer 2 Solutions: Implement Layer 2 scaling solutions, such as state channels and sidechains, to alleviate the load on the main blockchain network. These solutions enable off-chain processing of identity transactions while ensuring security and finality through interactions with the main blockchain.
        - Blockchain Sharding: Explore blockchain sharding, where the network is divided into smaller partitions or shards, each capable of processing its transactions and smart con-

tracts. This approach distributes the computational load, enhancing scalability without sacrificing security.

c. Privacy Concerns: Balancing privacy and transparency within digital identity systems is a multifaceted challenge. Current digital identity systems may not adequately protect users' sensitive data, posing significant privacy risks, especially when combined with the transparency inherent in blockchain.
    - Solutions:
        - Zero-Knowledge Proofs: Incorporate advanced cryptographic techniques like zero-knowledge proofs to enable privacy-preserving identity verification. Zero-knowledge proofs allow one party to prove the validity of a statement without revealing the underlying data, ensuring user privacy.
        - Decentralized Identifiers (DIDs): Embrace decentralized identifiers, which enable users to have control over their identity information. DIDs can be paired with verifiable credentials to provide selective disclosure, allowing users to share only necessary identity attributes while keeping sensitive data private.

d. Regulatory Compliance: Ensuring compliance with data protection regulations, such as GDPR, is a complex challenge for blockchain-based digital identity systems. These systems often struggle to align with these regulatory frameworks, leading to design conflicts.
    - Solutions:
        - Smart Contracts for Compliance: Develop smart contracts that are capable of automatically enforcing compliance with data protection regulations. These contracts can manage consent, data minimization, and deletion requests while maintaining an auditable record of data processing activities.
        - Privacy by Design: Adopt a "privacy by design" approach when architecting digital identity solutions. Ensure that privacy-enhancing features are integrated from the ground up, allowing systems to align with regulatory requirements inherently.

e. User Adoption and Usability: User adoption of digital identity solutions is contingent on usability and user-friendliness. Complex identity verification processes, convoluted interfaces, or lengthy onboarding procedures can discourage individuals from embracing these solutions.
    - Solutions:
        - User-centric design: Prioritize user-centric design principles to enhance the overall usability of digital identity solutions. This entails creating intuitive and user-friendly interfaces that streamline identity verification processes.
        - Mobile-Friendly Solutions: Develop mobile-friendly digital identity applications and platforms, recognizing that mobile devices are the primary means of online interaction for many individuals.

Addressing these multifaceted challenges through innovative, practical, and nuanced solutions is imperative for harnessing the full potential of digital identity and data sovereignty. These solutions pave the way for the establishment of secure, user-friendly, and regulatory-compliant digital identity systems in our increasingly digital-centric world.

Table 4. Challenges and solutions in digital identity and data sovereignty

Challenge	Solution
Interoperability Challenges	- Standardization Initiatives: Develop open standards for interoperability - Interoperability Middleware: Create middleware solutions for cross-platform data sharing.
Scalability Issues	- Layer 2 Solutions: Implement off-chain scaling mechanisms. - Blockchain Sharding: Divide the network into shards for scalability.
Privacy Concerns	- Zero-Knowledge Proofs: Use cryptographic techniques for privacy. - Decentralized Identifiers (DIDs): Enable user-controlled identity data sharing.
Regulatory Compliance	- Smart Contracts for Compliance: Automate regulatory compliance. - Privacy by Design: Embed privacy features into the system.
User Adoption and Usability	- User-Centric Design: Prioritize intuitive user interfaces. - Mobile-Friendly Solutions: Optimize for mobile user experience.

8. FUTURE RESEARCH DIRECTIONS

In this section, we delve into potential avenues for future research in the field of digital identity and data sovereignty. This section sheds light on the future of Ethereum and blockchain technology in the realm of digital identity, elucidating the intriguing prospects outlined in Table 5. As the landscape of technology and regulation continues to evolve, these research directions offer valuable insights and areas for exploration:

a. Advanced Authentication Methods: Future research in this area should focus on advancing authentication methods beyond traditional username-password combinations. Emerging technologies such as biometrics (e.g., fingerprint, facial recognition), behavioral analysis (e.g., keystroke dynamics), and continuous authentication (monitoring user behavior in real-time) hold promise. Investigating their effectiveness, security, and user acceptance is crucial for enhancing the authentication process (Digital Identity, 2018).
b. Decentralized Identity Ecosystems: The concept of decentralized identity aims to empower individuals with control over their digital identities. Research should concentrate on developing robust decentralized identity ecosystems that facilitate secure and user-friendly management of digital credentials. This includes exploring interoperability between different identity solutions, identity recovery mechanisms, and usability enhancements (Tan, K. L., 2022).
c. Privacy-Preserving Technologies: As data privacy gains prominence, future research should explore state-of-the-art privacy-preserving technologies. Topics of interest include homomorphic encryption (allowing computations on encrypted data), secure multi-party computation (enabling joint data analysis without sharing raw data), and differential privacy (protecting individual data while allowing statistical analysis). Investigating the practicality, efficiency, and applicability of these technologies is vital (Hussien, Q. M., 2021).
d. Blockchain Scaling Solutions: Blockchain technology plays a pivotal role in digital identity, but it faces scalability challenges. Future research should delve into scaling solutions such as layer 2 protocols (e.g., Lightning Network) and sidechains. Understanding their impact on digital identity management, transaction throughput, and security is essential (Zwitter, A. J., 2020).

e. Cross-Border Data Governance: Given the global nature of data sovereignty, future research should explore cross-border data governance models and frameworks. This research should consider the diverse legal and cultural contexts within which data is governed. Topics of interest include harmonizing international data protection regulations, creating cross-border data sharing agreements, and addressing conflicts of jurisdiction (The World Bank., 2023).
f. Ethical Considerations: The ethical dimension of digital identity and data sovereignty demands attention. Research should investigate ethical considerations surrounding consent mechanisms, transparency in data usage, and algorithmic fairness. It should explore the ethical implications of identity verification methods, data sharing practices, and decision-making algorithms (User Control, 2023).
g. User-Centric Approaches: User-centric design principles should continue to guide research efforts. Future studies should focus on creating digital identity systems that empower individuals to control their personal information. Topics include user-friendly identity management interfaces, consent management tools, and personalized data-sharing preferences (Sinha, A., 2020).
h. Industry Collaborations: Collaboration between academia, industry, and regulatory bodies can drive innovation in digital identity. Future research should explore effective models of collaboration, including public-private partnerships and industry consortia. Investigating how these collaborations can contribute to the development of secure, interoperable, and user-friendly digital identity solutions is crucial (Archana V., 2022).
i. Evaluation Metrics: Establishing robust evaluation metrics for digital identity and data sovereignty systems is essential. Future research can focus on defining key performance indicators (KPIs) and metrics to assess the effectiveness, security, and trustworthiness of these systems. Developing standardized assessment frameworks will aid in comparing different solutions and ensuring their reliability (Wylde, V., 2022).
j. Interdisciplinary Research: The multidisciplinary nature of digital identity and data sovereignty calls for increased interdisciplinary research collaboration. Future studies should encourage experts from diverse fields such as technology, law, ethics, and user experience to work together. Interdisciplinary research can lead to holistic solutions that address the complex challenges in this domain (Mühle, A., 2018).

Future research in these areas holds the potential to shape the future of digital identity and data sovereignty. By addressing existing challenges and exploring new possibilities, researchers can contribute to the development of secure, privacy-respecting, and user-centric digital identity solutions.

9. CONCLUSION

The contemporary landscape of digital identity and data sovereignty is characterized by a dynamic interplay of challenges and opportunities in the ever-expanding global information ecosystem. Through an extensive exploration of the literature, this chapter has provided valuable insights into the intricate relationship between these two fundamental concepts.

The literature review undertaken in this chapter has illuminated the multifaceted nature of digital identity, revealing its pivotal role in the digital age. It has become increasingly evident that digital identity is not merely a technical construct but a cornerstone of the modern digital landscape. The review further

Table 5. Future research directions

Research Direction	Description
Advanced Authentication Methods	Explore biometrics, behavioral analysis, and continuous authentication to enhance security and user experience.
Decentralized Identity Ecosystems	Develop robust decentralized identity ecosystems focusing on interoperability, identity recovery, and usability enhancements.
Privacy-Preserving Technologies	Investigate privacy-preserving technologies such as homomorphic encryption, secure multi-party computation, and differential privacy.
Blockchain Scaling Solutions	Study scaling solutions like layer 2 protocols and sidechains to address scalability challenges in digital identity.
Cross-Border Data Governance	Research cross-border data governance models, harmonizing international data protection regulations, and addressing conflicts of jurisdiction.
Ethical Considerations	Examine ethical considerations related to consent mechanisms, transparency, algorithmic fairness, and identity verification methods.
User-Centric Approaches	Focus on user-centric design principles, creating user-friendly identity management interfaces and personalized data-sharing preferences.
Industry Collaborations	Explore effective models of collaboration between academia, industry, and regulatory bodies to drive innovation in digital identity.
Evaluation Metrics	Define key performance indicators (KPIs) and assessment frameworks to evaluate the effectiveness and security of digital identity systems.
Interdisciplinary Research	Encourage interdisciplinary collaboration among experts from various fields to address the complex challenges in digital identity.

emphasized the transformative potential of blockchain technology in redefining how digital identities are managed and controlled.

Blockchain's role in digital identity management has been scrutinized with meticulous attention to detail. This technology offers a paradigm shift by decentralizing identity verification, authentication, and access control. The emergence of self-sovereign identity has redefined the power dynamics of digital identity, placing individuals at the helm of their digital personas.

The regulatory landscape surrounding digital identity and data sovereignty has been examined closely, emphasizing the critical need for robust legal frameworks. Regulations such as GDPR have introduced both challenges and opportunities for businesses and organizations. Achieving compliance while fostering innovation remains a delicate balancing act.

Delving into user-centric perspectives and ethical considerations in digital identity, this chapter has underscored the importance of designing systems that prioritize user consent, transparency, and fairness. Ethical quandaries related to identity verification methods and algorithmic biases have necessitated careful consideration and ethical vigilance.

While presenting insights into future research directions, this chapter has identified a wide array of areas that warrant exploration and innovation. These encompass advanced authentication methods, the development of decentralized identity ecosystems, the integration of privacy-preserving technologies, solutions for scaling blockchain technology, cross-border data governance frameworks, the ethical dimensions of identity management, user-centric design principles, industry collaborations, robust evaluation metrics, and the promotion of interdisciplinary research initiatives.

In summation, the confluence of digital identity and data sovereignty encapsulates the intricate dynamics of the digital era. While technological advancements offer promising solutions for secure, user-centric identity management, the persistence of regulatory challenges and ethical considerations reminds us of the complexities inherent in this domain. As we embark on this transformative journey, interdisciplinary collaboration, ethical mindfulness, and a commitment to empowering individuals in the digital realm will be instrumental in reshaping global information flows.

REFERENCES

Aydın, A., & Bensghir, T. K. (2019). Digital Data Sovereignty: Towards a Conceptual Framework. In *2019 1st International Informatics and Software Engineering Conference (UBMYK)* (pp. 1-6). IEEE. 10.1109/UBMYK48245.2019.8965469

Benkoël, D. (2020). *What consumers really think about Trusted Digital IDs*. Retrieved from https://dis-blog.thalesgroup.com/mobile/2020/02/11/qa-what-consumers-really-think-about-trusted-digital-ids/

Berman, P. (2018). *Digital Identity As a Basic Human Right*. Retrieved from https://impakter.com/digital-identity-basic-human-right/

Best Practices for Digital Identity Verification. (2022). Retrieved from https://integrity.aristotle.com/2022/05/best-practices-for-digital-identity-verification/

Blockchain in Digital Identity. (2023). Retrieved from https://consensys.net/blockchain-use-cases/digital-identity/

Dabrock, P., Tretter, M., Braun, M., & Hummel, P. (2021). Data sovereignty: A review. *Big Data & Society*, *8*(1). Advance online publication. doi:10.1177/2053951720982012

DeviS.KotianS.KumavatM.PatelD. (2022). Digital Identity Management System Using Blockchain. SSRN. doi:10.2139/ssrn.4127356

Digital Identity Roadmap Guide. (2018). Retrieved from https://tinyurl.com/4fe4kwb9

Domeyer, A., McCarthy, M., Pfeiffer, S., & Scherf, G. (2020). *How governments can deliver on the promise of digital ID*. Retrieved from https://www.mckinsey.com/industries/public-sector/our-insights/how-governments-can-deliver-on-the-promise-of-digital-id

Ghaffari, F., Gilani, K., Bertin, E., & Crespi, N. (2021). Identity and access management using distributed ledger technology: A survey. *International Journal of Network Management*, *32*(2), e2180. doi:10.1002/nem.2180

Gilani, K., Bertin, E., Hatin, J., & Crespi, N. (2020). A survey on blockchain-based identity management and decentralized privacy for personal data. In *2020 2nd Conference on Blockchain Research & Applications for Innovative Networks and Services (BRAINS)* (pp. 97-101). IEEE. 10.1109/BRAINS49436.2020.9223312

Haddouti, S. E., & Kettani, M. D. E. C. E. (2019). *Towards an interoperable identity management framework: a comparative study*. https://doi.org//arXiv.1902.11184 doi:10.48550

Herian, R. (2020). Blockchain, GDPR, and fantasies of data sovereignty. *Law, Innovation and Technology*, *12*(1), 156–174. doi:10.1080/17579961.2020.1727094

Hussien, Q. M., & Habeeba, F. A. (2021). Survey on data security techniques in internet of things. *Al-Kunooze Scientific Journal, 2*(2). Retrieved from https://www.iasj.net/iasj/article/221409

Mühle, A., Grüner, A., Gayvoronskaya, T., & Meinel, C. (2018). A survey on essential components of a self-sovereign identity. *Computer Science Review*, *30*, 80–86. doi:10.1016/j.cosrev.2018.10.002

Overcoming the Tension Between Data Sovereignty and Accelerated Digital Transformation. (2022). Retrieved from https://tinyurl.com/mr2a4ymt

Pöhn, D., Grabatin, M., & Hommel, W. (2021). eID and self-sovereign identity usage: An overview. *Electronics (Basel)*, *10*(22), 2811. doi:10.3390/electronics10222811

Quach, S., Thaichon, P., Martin, K. D., Weaven, S., & Palmatier, R. W. (2022). Digital technologies: Tensions in privacy and data. *Journal of the Academy of Marketing Science*, *50*(6), 1299–1323. doi:10.100711747-022-00845-y PMID:35281634

Shirer, M. (2023). *IDC Survey Finds Data Sovereignty and Compliance Issues Shaping IT Decisions*. Retrieved from https://www.idc.com/getdoc.jsp?containerId=prUS50134623

Sinha, A., & Katira, D. (2020). *Digital Identity A Survey of Technologies*. Retrieved from https://digitalid.design/tech-survey-2020.html

Tan, K. L., Chi, C. H., & Lam, K. Y. (2022). *Analysis of digital sovereignty and identity: From digitization to digitalization.* https://doi.org//arXiv.2202.10069 doi:10.48550

The World Bank. (2023). *Technical Standards for Digital Identity*. Retrieved from https://tinyurl.com/4eet9zt4

User Control in Digital Identity. A Guide to Design Principles. (2023). Retrieved from https://digitalprinciples.org/principle/design-with-the-user/

Wylde, V., Rawindaran, N., Lawrence, J., Balasubramanian, R., Prakash, E., Jayal, A., Khan, I., Hewage, C., & Platts, J. (2022). Cybersecurity, Data Privacy and Blockchain: A Review. *SN Computer Science*, *3*(2), 127. doi:10.100742979-022-01020-4 PMID:35036930

Zwitter, A. J., Gstrein, O. J., & Yap, E. (2020). Digital identity and the blockchain: Universal identity management and the concept of the "Self-Sovereign" individual. *Frontiers in Blockchain*, *3*, 26. doi:10.3389/fbloc.2020.00026

KEY TERMS AND DEFINITIONS

Biometric Authentication: Biometric authentication uses physical or behavioral characteristics, such as fingerprints or facial recognition, to verify a user's identity.

Biometric Data: Biometric data includes physiological or behavioral characteristics used for biometric authentication, such as fingerprints, retina scans, or voiceprints.

Blockchain Technology: Blockchain technology is a distributed ledger system that records transactions across multiple computers, ensuring transparency, security, and immutability of data.

Compliance Frameworks: Compliance frameworks provide guidelines and requirements for organizations to adhere to relevant regulations and standards.

Consent Management: Consent management involves obtaining and managing user consent for data processing activities, ensuring compliance with privacy regulations.

Cybersecurity: Cybersecurity is the practice of protecting digital systems, networks, and data from security threats, including cyberattacks and data breaches.

Data Sovereignty: Data sovereignty is the concept that data is subject to the laws and governance of the country or region where it is stored, emphasizing data control and protection.

Decentralization: Decentralization refers to the distribution of authority and control across a network, reducing reliance on single points of control.

Digital Identity: Digital identity refers to the online representation of an individual or entity, consisting of digital attributes, credentials, and personal information.

Digital Wallet: A digital wallet is a secure application or device used to store and manage digital identity credentials and keys.

GDPR (General Data Protection Regulation): GDPR is a European Union regulation that governs data protection and privacy, imposing strict requirements on the handling of personal data.

Interoperability Standards: Interoperability standards define protocols and formats that enable different systems to exchange data and operate seamlessly together.

Multi-Stakeholder Collaboration: Multi-stakeholder collaboration involves cooperation among various stakeholders, including governments, businesses, and civil society, in shaping digital identity policies and standards.

Phishing: Phishing is a fraudulent attempt to obtain sensitive information, such as login credentials, by posing as a trustworthy entity in electronic communication.

Privacy by Design: Privacy by design is a framework that integrates privacy considerations into the design and architecture of systems, emphasizing proactive privacy protection.

Self-Sovereign Identity (SSI): Self-sovereign identity is a decentralized approach to digital identity that empowers individuals with control over their own identity data, reducing reliance on centralized authorities.

Tokenization: Tokenization is the process of substituting sensitive data with a non-sensitive equivalent, called a token, to enhance security.

Two-Factor Authentication (2FA): Two-factor authentication is a security method that requires users to provide two forms of verification before accessing an account or system.

Usability Testing: Usability testing involves evaluating the ease of use and user-friendliness of digital identity systems through user feedback and testing.

User-Centric Design: User-centric design is an approach that prioritizes user needs, preferences, and usability in the development of digital identity solutions.

Chapter 11
Navigating the Digital Era:
Exploring Privacy, Security, and Ownership of Personal Data

Atul Grover
DXC Technology, UK

ABSTRACT

In the midst of the 21st-century digital revolution, we find ourselves navigating a complex landscape where our analog instincts clash with our digital dependencies. Our smartphones, compact marvels of technology, house the entirety of our daily existence, from groceries to fashion and medicine. Yet, amidst this convenience, fundamental questions about data safety, security controls, and data ownership linger ominously in our minds. This chapter delves into the vital trifecta of privacy, security, and data ownership in our increasingly digitized world. In an era of ever-evolving technology, the ordinary user constantly frets about the sanctity of their data. While technology has undoubtedly bestowed countless benefits upon humanity, it also bears a dark side, epitomized by the surge in digital frauds and scams. Such concerns propel individuals to ponder deeply: Is my data truly private? and Who ultimately lays claim to, or safeguards, my data? This exploration ventures into the realms of artificial intelligence, virtual reality, the metaverse, online payments, and virtual classrooms.

1 INTRODUCTION

Digital, the word that changes every human's view begins on earth. Digital had no or very little existence 20-30 years ago but in the 21st century, this word carries everything in itself (Belk 2013). Whether is shopping payment gaming a music concert or travelling, from A to Z from 0 – to thousands, from anything to everything is part of it, contributing to it, and using these Digital words & world (Munir et al. 2015). As per the latest report 2/3 of adults in the world use digital payments for their day-to-day routine. 65% of the world's population uses the internet every day.

Before we step forward, let's spend time to understand why the word digital came into existence and would like to connect this with need, change, and solution. (Ashford 2000)

DOI: 10.4018/979-8-3693-1762-4.ch011

1.1 Need

A problem or opportunity to be addressed. Needs can cause changes by motivating stakeholders to act. Changes can also cause needs by eroding or enhancing the value delivered by existing solutions.

1.2 Change

The act of transformation in response to a need. Change works to improve the performance of an enterprise. These improvements are deliberate and controlled through business analysis activities.

1.3 Solution

A specific way of satisfying one or more needs in a context. A solution satisfies a need by resolving a problem faced by stakeholders or enabling stakeholders to take advantage of an opportunity.

But before we go deep, let's understand what exactly is the meaning of Digital and why we are using this (Conway 1996). Digital in layman's terms, is something that describes electronic technology that generates, stores, and processes data in terms of two states: positive and non-positive. Now when we add the word "lization" as a suffix with digital, it becomes digitalization which means something available or used via the internet. Now in the current era, everything is available on the Internet, whether it is grocery, education, medicine anything, or everything available on the internet. In a country like India, it is a very famous statement for digital payment "Paytm Karo".

With more & more involvement of technology or devices in the digital world, where a huge and huge personal data has been collected and there is always question related to ethics and legality around this data. Privacy and Security is one side where ownership of personal data underpins many points related to data management, its control, and usage example, the trust of users in the privacy of data, the uses of faith data to be secure, and more implications of this data in the future digital economy (Cummins & Schuller 2020).

Figure 1. Privacy, security, and ownership of personal data relationship

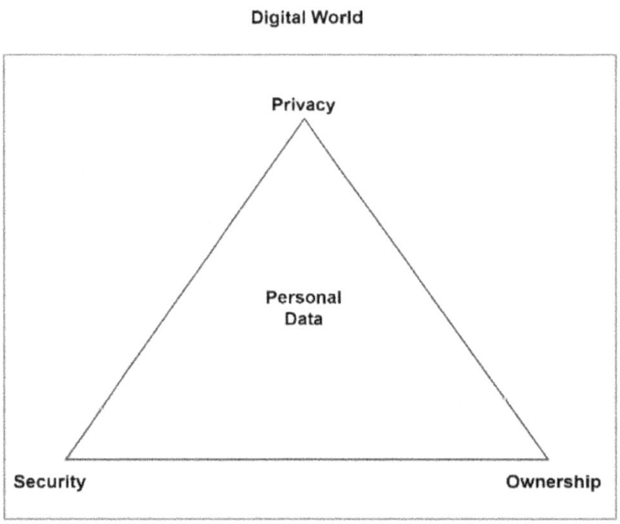

Figure 2. Privacy agreement

On one side digitalization is a plus for everyone because everything is available at the fingertips or on a small screen but it has some dark sides as well where users are still thinking to opt it fully or using it only in case of emergencies. Privacy is the biggest problem or threat. User data is secure in the digital world who owns the user's data means who is accountable if there is privacy leakage or if someone stole user data? (Spiekermann 2012) Let's discuss this in upcoming sections. Before that let's understand the meaning of these terms – Privacy, Security, and Ownership of Personal Data (Far & Rad 2022). Here the question comes what is personal data and what is nonpersonal data? It also raises the question of whether the law allows someone to own something especially the ownership of personal data.

1.4 Privacy

The word privacy is a combination of 7 alphabets but has a very deep meaning and users' points of concern. Privacy means something private and not shared with everyone or shared with authorized users only. In our context, user data like Age, Date of Birth, email ID, bank account, etc. are considered personal data and must be private, as anyone can miss it. The best example of privacy is cookie acceptance/rejection when the user opens any website.

1.5 Security

The term security, in layman's words secure/ safe. This means something or everything we as users want to keep safe or avoid any unauthorized access or usage. In the Digital world, security is also a concern for users because every day many users create accounts or register himself/herself in the cyber world and provide a lot of personal information. The security of this data is very important (Eskicioglu et al 2003).

1.6 Ownership

I am the owner of this house, I own this car and bank balance, etc. Ownership means owning something or in other words accountable for that particular stuff/thing/data. In the cyber world ownership is a big pain point. Like we said earlier, every day many users enter into digital worlds and provide data but at last, no one knows or is aware of who owns my data. This means to whom we as user reach out if the data get breached or stolen.

2 LITERATURE REVIEW

The above section covered the introduction and we discussed change, need and other core areas of this chapter. This section will cover the literature review, which supports the terminologies, challenges and implications. (Romansky 2017) wrote the paper and discussed the opportunities and challenges for users' privacy in the digital world. In this paper researcher wrote that the user data is collected in multiple ways and also stored in multiple places. The author concluded the opportunities by completing the survey. The emphasis of this paper is on PDP (Personal data protection) which means it is required that user must know about their rights in the digital world and control their data. The model of SISP (System for Information Security and Privacy) covers 4 sides including the Embedded tool, which cover the computer layer, 2nd is the physical layer, 3rd is external measures/ rule and the legislative layer. The author talked about the contemporary digital world which covers e-Governance and other components.

Security is another concern or area which always bothers users. (Minchev, 2017) talked about the security challenges in the dynamic digital world. The author called the digital world as Fourth Digital revolution. Along with this author considers the cultivation of expectations and mind models as a challenge in this digitized thinking. Minchev proposed a four-layered model, starting from context matrix establishing, scenario holistic risk assessment, Trend Dynamic validation & last interactive results verification. The author considered AI, and ML as major players in building this digital world and considered this a challenge from a security perspective.

Data leakage/ data breaches require strong data protection as well (Kaurin 2019). The theme of this paper about data protection for refuges digital entry. In 2015, Approximately 1.3 million asylum applications were submitted to European agencies & this is because of the crisis and unexpected experiences in Syria and other countries. Refugees need to provide a lot of information to submit these applications. In a research paper, the author described information – decision gap and considered a few bold questions like, how the data may be repurposed, how the data is stored and shared and more, along with this provided recommendations like, whenever data is collected explicitly consent needs to be taken, data must be covered under legal and regulatory frameworks, engaging user (Refugee) to design the solve the problem and there must be transparency and compliance meantime.

Ownership of personal data is always an issue. Who owns the data the user or organization? (Janeček 2018) The author analyses, defines and refines the ownership of data. In this paper, the author explained the personal and non-personal data. Not only this, the author covers that the boundary or extent of the ownership of data lies with the user or organization. Janecek showed 2 approaches concerning ownership of personal data and called these approaches as Top – the Down and bottom-up approaches. To draw the conclusion and approaches author considers IoT (Internet of Things) at large side. The definition of personal data varies from country to country and depends on the respective laws as well. The modern theory answers the question, why the law should allow someone to own something? To dig out this, refer to either the top-down (positivist approach) or the bottom-up (Natural Law approach).

Munir et al(2015), discussed the big challenges and data protection by considering the technology of big data. The author defined this world as big data because it is all about data nowadays. The research paper covers the definition of Big Data from multiple research papers. By considering multiple definitions, and different views of challenges respective to privacy, is privacy and data protection dead? By citing the words of Scott Mc, that "You have zero privacy anyway. Get over it." And famous words of Michael's death of privacy and much more. At the end of the paper, the researcher considered that privacy and data protection is still alive. Covered the challenge like the data collection is massive nowadays, collected from online platforms and devices.

3 DIGITAL TRANSFORMATION VALUE CREATION

In the last 2 sections, we learned about Digitalization & covered the literature review and now we talk about digital transformation value creation. What is digital transformation? It is a combination of 2 words Digital and transformation. Transformation is the buzzword in an era now, everyone, beginning in technology or finance or business or gaming or fashion everyone and anyone talks about this. But what does Transformation mean and how did this team come into the picture? In simple and layman's words, transformation means changing the form, nature, or appearance. For example, earlier in the 80s the mobile was so heavy and big, and now everyone knows this 5-6" devices changed everything and this is a transformation. Earlier doctors took a long time to operate but now with the best use of technology big operations are done quickly and successfully. The why we go far, India landed on the Moon at the South Pole and became the first country to land on the South Pole, this is the best example of technology and science transformation. Everything getting changed. The banking sector, medical science, and education are fully transformed and this COVID gives us a bad habit a bad or good tough to say but makes every human/individual/society use the digital world (Dove 2018).

Now the term comes value creation. Again, let's understand what is meaning of value here. The worth, importance, or usefulness of something to a stakeholder within a context. Value can be seen as potential or realized returns, gains, and improvements. It is also possible to have a decrease in value in the form of losses, risks, and costs. Value can be tangible or intangible if we go into the world of Service Management, where Value has a special meaning. As per ITIL4, Value means the perceived benefits, usefulness, and importance of something. For example, a gift that was given by my mother on my birthday is always valuable to me whether it is in good condition or bad but it always holds some importance in my life. If this is value then what is value creation? Easy to say, that creating the importance or usefulness of any product or service for the user is value creation. But is it so simple or any pillars or factors or components we need in this? ITIL4 says to create value these 8 pillars must exist also value creation is not a one-time work or activity; it is a journey and journey always. These 8 pillars include a few of these pillars can be clubbed together as well.

Figure 3. Eight pillars for value creation

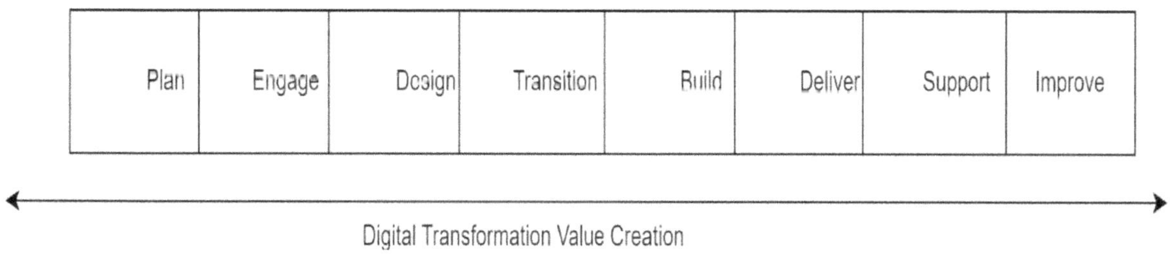

As these 8 pillars are important in the same manner, we need to understand how many types of values there are (Evanschitzky et al 2011).

3.1 Value-Add

Characteristics, features, and business activities for which the customer is willing to pay. For example, if on a website a company can provide subscription-based or multi-factor authorization to secure the data.

3.2 Non-Value-Add

Aspects and activities for which the customer is not willing to pay. For example, the company offers discounts or other benefits but the history says this company has no way to secure the data.

3.3 Business Non-Value-Add

Characteristics that must be included in the offering, activities performed to meet regulatory and other needs or costs associated with doing business, for which the customer is not willing to pay.

3.4 Relevancy

Explain how your product solves customers' problems or improves their situation. More education and benefit explanation of the product or service.

3.5 Quantified Value

Deliver specific benefits. For example, via Paytm or online banking how quickly the amount gets transferred to another's account.

3.6 Differentiation

Tell the ideal customer why they should buy from you and not from the competitor. Customer/society or individuals want to know what extra you as a company is giving which is not provided or given by other. Key highlights on Privacy of data/ownership make a great and positive impact on customers.

Till now we understood value creation and now it's time to combine digital transformation and value creation, which means let's read/understand what value this digital transformation brings and a few areas in depth.

3.7 For Customers

Digitalization brings more value for the customer, in the sense of new products, which give more choice, more convenience, every everything to their fingertips but this also brings the dark side, more users losing their data privacy, more and more searches happening which bring the infra cost high in terms of maintaining, etc.

3.8 For Companies

The biggest benefit of digital transformation to the companies is that it is easy for them to capture the market, opportunity, and new areas to create more value and co-creation of value and help to increase efficiency and effectiveness, but again companies also face risks like losing the legacy way of working or value chain, more and more competitions, less margins and faster innovative process.

3.9 For Society

Human is a social animal, a very old saying and it is always hence proved. Because a human can't survive without society or the society. This digital world/era brings many new things and benefits for a society like, information is available to everyone (in villages also, the internet is available now), public services have been improved and much more but along with these benefits few risks like regulation are on stake now, everyone wants to follow his way, more issue on ownership of data and privacy & security arises.

4 5W AND 1H PRINCIPLE

It is well said, where there is will, there is a way and in the current era it is important to understand why we are doing this, how it can be done, who will do and much more. It is important to understand the principle of 5W and 1H as these are the driver of the digital world and also has a direct relationship with privacy, Security, and Ownership (Knop & Mielczarek 2018).

Figure 4. 5W and 1H

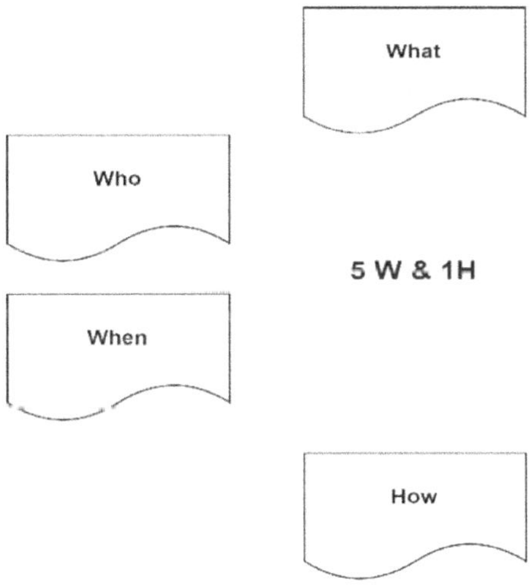

4.1 1W-What

There is a myth in the user's mind that every data they are providing or adding in the digital world is considered as personal and private data and the user asks for accountability for the same, so first W, clarify what is part or considered as Privacy. In other words, we can say it states what is the objective and what outcome we are expecting.

4.2 2W-Why

This word plays a crucial role. Why indicates the purpose or in simple words why state, why we want the user to know who owns the data why privacy is required, and why a digital world collects the data or asks for consent.

4.3 1H-How

The first and most important part of this journey. How a company /website collects the data, need to know or understand how my data is beneficial for others or how this data will be used for improvements. Due to a lack of clarity on the How part, users always think there is a privacy & security risk or threat.

4.4 3W-Where

The 3rd W in the sequence which states Where. It is important to understand where my data is impacted, where my data gets stored, or where this data will be used. Misconception or unavailability of all information makes users worried.

4.5 4W-Who

The 4th W, who which means who all are involved or who needs to take action in case of privacy/security and ownership of data. This W also plays a role like the RACI model which needs to be informed.

4.6 5W-When

The last and 5th W state, When. When this information is needed and also when a user needs to act like when logging in to the website privacy pop comes up and most importantly When to report when there is a threat.

In summary, 5W &1H make a clear picture for the user and support the user to clear all myths and provide a valid reason to collect data/ tell the user how this data will be used or stored and by whom it will be acted.

5. MOSCOW RULE

The most important and another rule that can play a vital role and open many perspectives in the digital world. Practically speaking MoSCoW is a method that is used in SDLC / Software Development, Project

Figure 5. MoSCoW rule

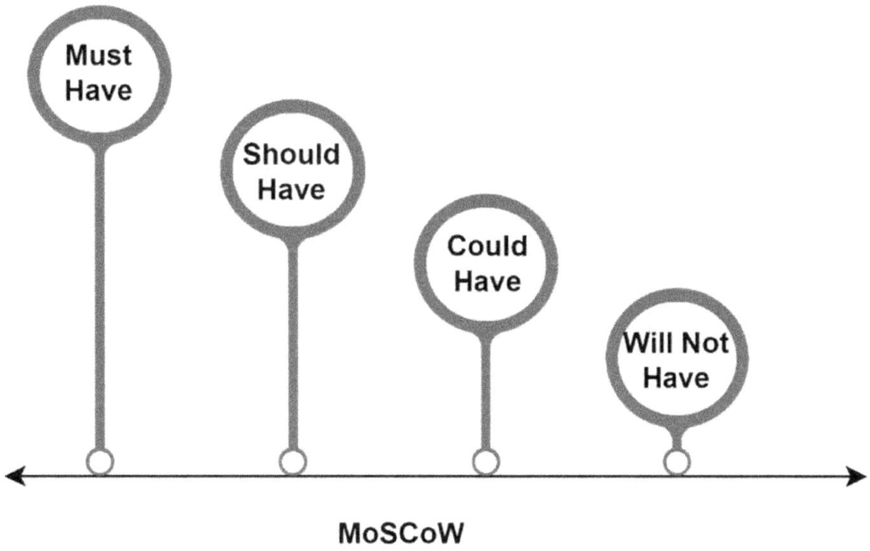

Management, or in the Business Analysis world/domain. Now the question comes can we relate or build a relationship of MoSCoW with Privacy, Security, and ownership of personal data? In simple words, we can say that the Moscow rule works like a framework or like a guideline for better understanding and taking care of privacy, security and ownership. (Miranda 2011).

5.1 Mo: Must Have

Meaning it is nonignorable or non-negotiable and without this feature or service the product or services can't be delivered or used. In the same manner in the domain of personal data, it is required that [personal data must be protected at all costs. Here personal data includes sensitive or confidential information like financial records, health data, personal identification and much more.

5.2 S: Should Have

Meaning that negotiable is possible but only up to some extent. It is good to have in a product or service. In the same context, the data which we are reciting is also easy to access by the authorized user if and when required. In this category, we can add preferences, habits, and other less critical personal data. Highlight the significance of securing this information and the potential risks if mishandled.

5.3 Co: Could Have

The 3rd one, say something that has a very small effect or impact on product sale/usage or service. Like – A mobile phone could show photo/memories highlights or a mobile have few apps pre-installed. The impact on services or usage without this option is very low. In this domain, the non-critical data could

Table 1. MoSCoW rule as per domains

Domains	Must	Should	Could	Will Not
Privacy and Security Implications	Robust Security Protocols Data Breach Handling	Ease of Access Privacy Impact Assessment	Impact of Privacy Breaches Cross Border Transfer	Third-Party Risk Management
Ownership of Personal Data	Individuals have control over their data	Right to know What data is collected, ability to request to delete the data	Where the data is stored	

be shared more openly. This might include non-sensitive preferences or information shared voluntarily by users. Emphasize the need for transparency and user consent even with seemingly less sensitive data.

5.4 W: Will Not Have

The last one, will not have. In simple words, it can be fully ignored and can be delayed or not required in products or services. Like A mobile phone doesn't have a pre-installed Radio. In this category or domain, we include the data which is not collected pt not kept or stored due to high risk in nature or not relevant to store.

At a glance what comes under M-S-C-W

6. TYPES AND CHALLENGES

After understanding the terms, it's time to understand the challenges or types of these issues. As per a survey conducted 20% of adult in the US is victims of privacy threat which is a loss of $17 billion in monetary form. It is not one thing, 5-7% of users deactivated their accounts in the digital world due to security issues and ownership unclear (Kaurin 2019).

Before we see the types /challenges in Privacy, Security, and Ownership, need to understand why this data is so important. We are in the Digital age and data plays a very vital role in capturing the market and user base. A very basic example is if a user check something on the internet about Flight details or may be interested in pursuing some product, and as soon as the user search once over the internet user start getting ads/spam call or highlights or even in the news and start getting similar stuff. So, to be precise a company needs data for:

- To capture the user base or market
- To facilitate the selling
- For better services
- Enhance user experience
- And to improve the products/ services and many more.

This data has been pooled in a big platform and then with the help of different algorithms it has been proceeding and further improvement has been done. Data collection is not only a one-time activity, it is a regular activity, a user login from a web browser App or email, and many other ways to collect the data. (Minchev 2017).

Figure 6. Wisdom process

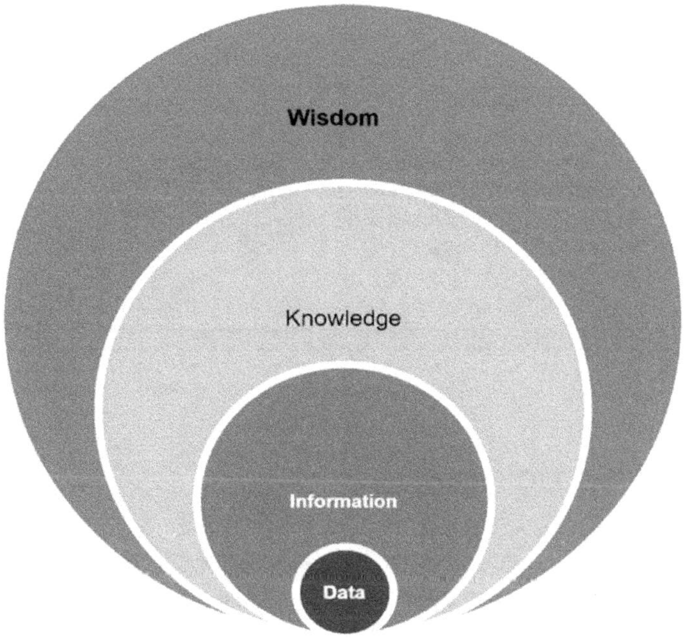

But the question is why a user is worried about data sharing.

- In 2021, a $6.9 Billion financial loss was recorded which happened due to internet crime.
- In 2021, 23% of Americans lost their money in Spam Calls only.
- 51% of the population in the UK had Impersonation fraud.
- Nearly 43 million UK adult internet users have encountered suspected scams online.

These are a few facts and many others which are still uncovered. By seeing these figures daily in the news or newspaper users are getting worried and raising their voices for better Privacy, security, and Accountability of their data (Munir et al. 2015). As per the survey,

- 80% of users are not interested in sharing their data.
- 76% of users said they don't want marketing ads based on their data collection.
- 51% of users said data security is the biggest challenge.

These figures support or are inclined that yes there is a need for strong Privacy, robust security, and ownership of data (Raina & Palaniswami 2021). Now, it's time to understand the different types of these terms. (Olphert et al., 2005).

Figure 7. Privacy threats

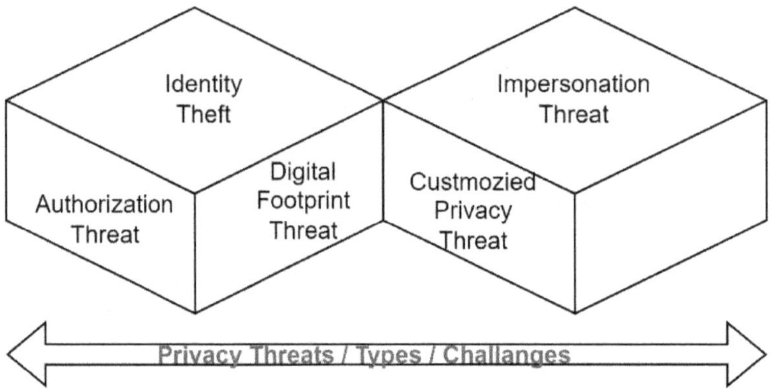

6.1 Privacy Threats/Types/Challenges

First and foremost, privacy is the biggest concern. The leakage of private data or data breaches has a direct impact on the user, as this data can be used for bad causes as well. Here are a few challenge types below. (Kernighan 2021). It is not easy to add every privacy issue/threat or classify them in the same category so for better understanding, we have divided privacy threats below

6.1.1 Identity Theft

In layman's language, it states, that a user loses his identity or he /she can be recognized based on his/ her data which was only with him/her but now is available publicly or on the internet. This includes the full names/ secret asset keys or bank details. This means if this data has been stolen or hacked, has a bad impact on the users. The only Bank scams are famous for this.

6.1.2 Impersonation Threat

Someone able to work/act/behave/authorize an actual user. In simple words nowadays it is very popular and technology plays the biggest role in the same. An unauthorized user creates the same image/face and calls to victim knowns by describing some worst situation and asking for money. This means in the image it was not you but someone who acted on behalf of you and did the wrong stuff (Regan 2002).

6.1.3 Authorization Threat

Nowadays we are in the habit of storing every credit and arrestive its banking or social media login or any e-commerce site or app on local devices and knowingly or unknowing this data has been saved which we consider as private but captured by the best use of technology and used by unauthorized user to make transactions.

Figure 8. Security threat

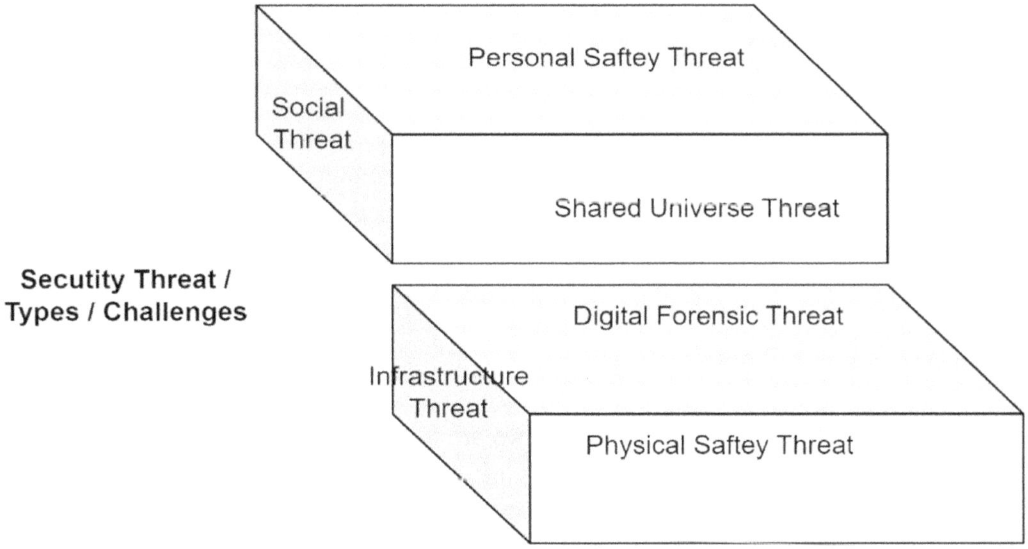

6.1.4 Digital Footprint Threat

Yes, we are in the digital world, and technologies like Metaverse / Virtual reality /Artificial intelligence are the future (Far, 2022). To build a good footprint in the digital world, the user preferences/habits/ activities are getting captured and attackers can use this information to exploit human beings or even use these in illegal activity.

6.1.5 Customized Privacy Threat

Every internet platform has its way of maintaining privacy and collecting data. Finally, websites are more eager to know the user's financial status rather than name, and few are interested in age and like to build some adult content rather than gender. And few are only interested in collecting social data. Later all these data were put in platforms and conclusions were drawn.

6.2 Security Threats/Type/Challenges

Security is another major concern for users. Because if the data is secure users still feel comfortable and happy to share data up to some extent. Here are a few types of threats under security (Romansky & Noninska 2020).

6.2.1 Personal Safety Threat

The major threat is physical and social. To use some latest technologies users, use the devices and hackers can attack on same. By using the devices, hackers / unauthorized users can get the exact location or behaviour of the user and perform some illegal activity. For example- A digital watch keeps track of the location and if a hacker gets data, the user is not home and can do illegal acts. Social media especially vlogging plays a major role. In a Vlog user is sharing all information without knowing who is watching the content (Romansky 2017).

6.2.2 Social Threat

Recently a case happened, a couple's video went viral as they captured some intimate movements on their mobile phone and it got viral on the internet and now it has a very bad impact on society. Not only this, Twitter is another example where the message gets leaked and has a very bad impact on users. Raised in child pornography / cyberbullying & biased outcomes come under this threat.

6.2.3 Shared Universe Threat

The way we are progressing in the digital world, we are mixing the Virtual world Real world. We are emerging the both worlds and this is a major threat to security. Technology like Metaverse is aggravating a lot of other stuff and, due to this we are losing our natural/real-world interest and sharing more and more data on digital platforms. As per research, this concept is called the risks of polarization and radicalization. The real world is different than the virtual world. Physical Safety Threat.

There is a very thin line between personal safety and physical safety. Recently a user lost 10,000 Rs from his bank due to a Facebook photo. The hacker takes the user's photo from Facebook which has a bank card in her hand and with the dark side if this technology issued a new card and did this illegal activity. Web users post everything on social media without knowing who is seeing that content and what his/her intention and it directly impacts human beings.

6.2.4 Infrastructure Threat

Building this big complex and user-friendly system called the digital world needs a good infrastructure and software. Maybe an end user doesn't have to know how this infrastructure has been built or maintained but it is a security threat. The data is passed and stored in this hardware infra (servers etc.) and this can be easily compromised by the hackers who take the data.

6.2.5 Digital Forensics Threat

The last threat added to the list means trying to build the same crime in the digital world that happens in the real world. For example, we heard a lot of rape cases/eve teasing, and money scams, and now similar is happening in digital work. One of the metaverse users said in an interview, she was raped by another avatar in the metaverse world and there is no way to complain because there is no such rule or punishment that exists in the digital world (Minchev 2017).

6.3 Ownership Threats/Types/Challenges

The word ownership is very specific in meaning. It has different meanings for different users and has been different in different laws & countries. In one of the papers, the researcher asked if The Ownership

Figure 9. Ownership threats

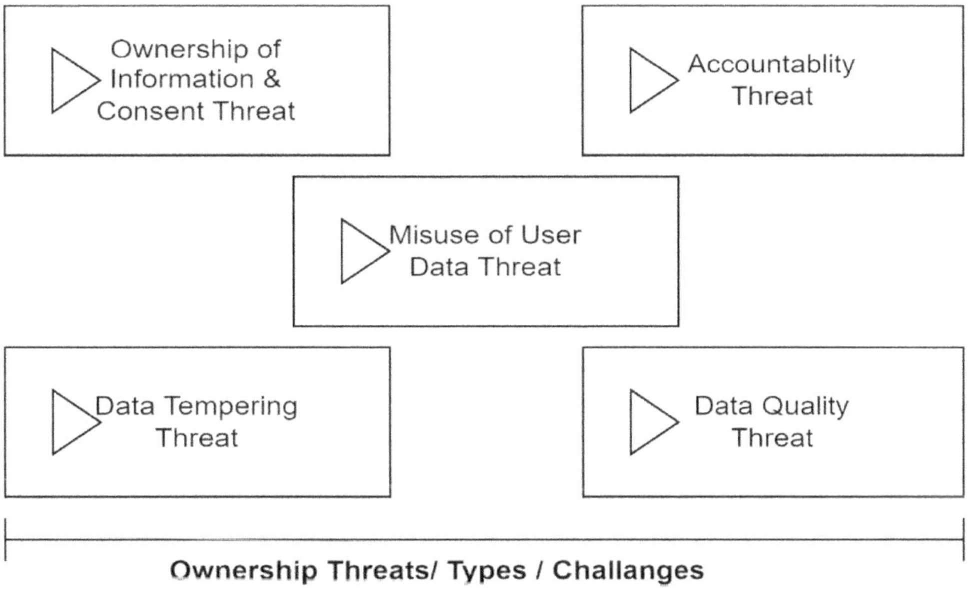

of Personal Data is the same as Personal Data Ownership. The answer is NO (Watkins et all 2016) & these are two different concepts.

Many new rules/standards are coming that support users and companies to protect their data. The GDPR (General Data Protection Regulation) is one of the latest standards applied by government bodies along with this there are many data protection rules/rights like the fundamental right to respect for private life and the fundamental right to protection of personal data (Lynch 2001). It is tough to imagine any data ownership without thinking about the ownership of personal data (Pagallo2017). Personal data ownership is always a concern and in many companies and countries, personal data is considered an economic asset. But the question remains the same Who owns the personal data (Dove 2018)?

6.3.1 Ownership of Information and Consent Threat

The first and foremost is ownership of confirmation and consent threat. What does this mean? This means even user provided the agreement to save the personal data but this mean it can be considered used anywhere or only for a specific purpose. There are many cases and situations where organizations collect personal data without the consent of the user.

6.3.2 Accountability Threat

The second threat is accountability. What is the meaning of accountability here? If we go into the ITIL / ITSM world accountability means a user or we can say a final user who finally reviews the task and gives confirmation. But in the cyber world or digital world, accountability means the principle that an individual is entrusted to safeguard and control equipment, keying material, and information and is answerable to proper authority for the loss or misuse of that equipment or information. Now the question is why it is a threat or challenge. There are cases where an organization said that your data would be safe with us but still leaked. Now in this case who is accountable?

6.3.3 Misuse of User Data Threat

Data has been processed or follows a complete lifecycle that passes through many stages/vendors/tools/platforms & here this data can be misused as well. Like with collected data the wrong conclusion can be derived for profitable purposes or misused to target the audience for the wrong product. Also, this is a threat where due to switching multiple apps / due to interlinking of user accounts or profiles and not using the right cookies the data can be gathered and then misused.

6.3.4 Data Tempering Threat

The 4th in the queue is the Data tempering threat. The words tempered indicate a change in the actual or real facts. Like, a survey said 70% of users think security is the biggest concern and to prove or to create false hopes/ information the data has been tempered, and put like 95% of users said security is the biggest issue. The main purpose of data tempering is to make hype of any product/service or any party. Where the facts/figures say something different.

6.3.5 Data Quality Threat

Last but not least data quality threat, which is on similar lines as above. Sometimes many users provide the wrong or low-quality data or even correct data is provided by the user but a company or organization converts them into low-quality.

7. DATA PROTECTION

Technology plays a vital role in today's world each and everything is driven by technology and this digital world is one of the best examples of the same. But the question arises, for privacy or security or ownership only the user's responsibility or anyone else is responsible or playing a role? Building or protecting both are equally important for Users and technology and here we will discuss a few more about the role of technology in the digital world (Weber 2015).

Ethics is one of the terms that need to be followed by everyone, irrespective of which domain you are working or on which organization you are working & here we are talking about Digital Ethics (Kaurin 2019).

Privacy has a special place in digital ethics for many reasons.

- Privacy is an integral part of any economic and dignity of individuals
- Controlling and safeguarding personal data is every user's fundamental right and it supports or provides the ability to make a decision.

Figure 10. Data protection

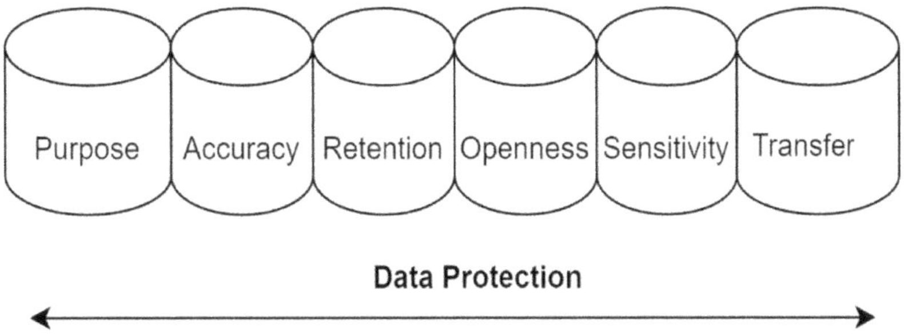

Security is the next phase in the same sequence. Privacy and Security work hand in hand and share a similar reason:

- Securing personal information/data is the user's right.
- Data security ensures the user and enhances/increases the faith in the brand.

Ownership

- The user who shares the data is the actual owner but using the same data unauthorized way is problematic.
- Misuse of the data.

As long as the ethics concerns, now need to discuss data protection. Yes. Till now we discussed Privacy and its threats, security and challenges, and issues in the Ownership of personal data. Now the question is how to protect this data or in simple words need for Data Protection or Protection of Data (Osburg 2017).

If there is a breach of privacy it has a direct impact on trust and risk to security and puts a question on ownership of Data. We can say, that privacy, Security, and ownership are tightly coupled, and a breach of any of these leads to a loss of monetary or reputation. There are laws to protect these and violations of these rules/principles lead the roads to privacy and security threats. The law is there to protect these and has underlined punishments as well.

Why do we need data protection? Because of daily changes in technology or ubiquity of the technology and greed and need for an information-intensive environment. There is a need for data privacy and security on an urgent basis but it is not that easy to implement any rule/standard or principle. We can't solve the breach in privacy or security completely but we can reduce the chances of happening these breaches. There are a few principles that need to be followed:

7.1 Data Collection Purpose

The first and most important step is that whenever any organization collects the data for any purpose must mention the purpose for data collection as per the law defined.

7.2 Accuracy

The 2nd one is accuracy. This means collecting only the data that is required for the purpose and data must be provided that is accurate.

7.3 Retention

The 3rd one, is how long an organization keep the user's data, if we say there are law is that followed? That is the question. So, it is required whenever data is collected, the retention period must be mentioned.

7.4 Openness

This is another step where the organization needs to tell the user how this data will be processed or what personal data they are holding & data collection is necessary but not excessive.

7.5 Sensitivity

Many users don't want to share data that is very sensitive like sex choices/religion or medical-related data. In this case, the organization must explicitly tell users what data is collected and need to take individual consent.

7.6 Transfer

In any case, it is not recommended to transfer the used data to another organization or vendor unless the 3rd parties also comply with the data protection rules/ standards.

These are a few principles/ rules for data protection. But the question is this only the organization's responsibility to protect the data or is there any recommendation for the user to protect the data?

- Before accepting the web cookies, the user must read the terms and conditions very carefully.
- Avoid linking multiple accounts.
- Avoid logging on public networks or public systems desktops etc.
- Don't save information unnecessary
- Avoid unknown or unsecured websites
- Configure the privacy settings

8. IMPLICATIONS AND REAL-WORLD EXAMPLES

This in chapter, we discussed about digital world, the definition of privacy, Security and ownership of data along with the 5 W – 1H and we discussed threats in these domains. In this section, we are going to discuss the implications that the Digital Era brought regarding privacy, security and ownership of personal data. Here are a few key and summary points.

In this section, we are going to discuss the uses case and implications of digital worlds in the area of Privacy, security and ownership of personal data. Due to the unchanging nature of attacks in the digital world things are tough and tough. If cheating or fund fraud happens in the real world, can happen in same in the digital world. If a bank gets robbed, in the same manner in the digital world data can be robbed and much more. In simple or easy words, we can, what can happen in the real world or the physical world can same in the digital world too.

We are progressing well and we as humans want everything to automate and we can say, automation is the key. In countries like the USA / UK, every payment from small to big is done online. The concept of Auto Debit is famous and this is the place where crime happens. Attackers / unauthorised users insert the code in these ways and rob the money of people.

More digitalization is not good for humans to begin as this interconnected world of computers so-called internet has no boundaries or we can say the sky is the limit. This means, that to do the cyber attack attacker does not need to be present or near the place/venue. By sitting far and far these attacks can take place.

We saw that in the domain of privacy concerns companies or organisations collect user data, especially personal data and in many instances, it is not known to users explicitly which raises an alarming concern for users. Also, companies use user behaviour to manipulate or capture the set of audience. Not only this in the current era of cyber-attacks and data breaches which directly hit users' data and lead to theft including financial loss and reputation damage, the implication is that personal information, even seemingly innocuous data, can be used to paint a comprehensive picture of an individual. A few examples in 2019, Facebook's 500 million users' data including phone numbers was found on the website, also recently during COVID-19, due to security vulnerabilities the user data exposed included the test reports and health information. In 2021, T- T-Mobile was affected by a data breach of 50 million users.

If we move to the next domain, like security challenges it includes cyber threats and these are not 1 or 2 many formats/patterns like Malware a common way, phishing or ransomware which directly pose a risk to personal and company data security. 24*7 connection on the internet which bring IoT vulnerability threat and causes and directly impact the user's personal and data security. Yamaha Motor's Philippines motorcycle manufacturing subsidiary confirmed the latest ransomware attack on a server which was accessed without authorization by a third party and partial leakage of employee personal information. Not only this, the British Library which has more than 11 million visitors annually and 16000 users who use this site daily confirms the ransomware attack which caused the major outages. Also, Toyota Financial Services confirmed the latest ransomware attack that they also detected unauthorised access on some of the systems based in Europe & Africa. The Medusa ransomware gang listed TFS (Toyota Financial Services) demanded the payment of $8000000 to delete their data from their site called Dark Web.

In the last let's talk about the ownership of personal data, this is again a big concern for users. The first and foremost issue is there is a lack of control over data, which means it's not clear who is collecting, who is storing and who is processing the data in big organisations. It is not only the case here, even organizations or companies sell this data or we can monetise users' data without providing compensation or no transparency and earn millions and billions of profits on this.

8.1 Implications and Mitigation

8.1.1 Legal and Ethical Controls

Agree, that we need innovation and we are on the way to building an innovative world but we need to get the balance between innovation and convincing ourselves on ethical grounds again that using personal data and price is important.

8.1.2 User Education and Training

We know, what rights a human holds or provides by our organisation, in the same manner, we need to educate each one of us regarding digital rights, and best practices for keeping safe online security and privacy. Sharing data over social media brings security to par.

8.1.3 Regulation and Compliance

If there is no punishment, crimes will happen, so to make it stronger, government bodies and other regulatory firms play a vital and significant role in developing ad creating & enforcing the laws to protect privacy and security and define laws about ownership of personal data as well. The evolving legal landscape globally attempts to address data ownership and protection through regulations like GDPR and CCPA.

8.1.4 Technology Development

The success of innovation depends on technology or we can say innovation and technology work hand in hand. Innovations like decentralized systems and enhanced privacy-preserving technologies aim to give users more control over their data.

8.1.5 Change in Business Mindset

To run a safe and secure business it is important that businesses adapt their practices or improve or enhance their way of working to ensure compliance with new regulations, prioritize data protection, and build trust with their consumers.

8.2 Real-World Examples

After discussing about implications, it's time to look at some real-world examples that how personal data gets attacked and ways.

8.2.1 Social Media/Public Access

The first and foremost way to get or collect user data is this social media app. Everyone either a teenager or a young guy, creates an account on platforms like Facebook, Instagram or TikTok and gives access to their phone memory and uploads all the data in the format of images or video which is drawing a trend or showing the habits of individual.

8.2.2 External Apps/Application

We as humans want to track everything, a heartbeat/walking steps running kilometres and much more and use the app so-called Fitness or Healthcare apps where we intentionally or unintentionally provide very sensitive data including medical history. The challenge here is ensuring this data remains secure and confidential.

8.2.3 Internet All Over the World

Technology is so vast that we are addicted to technology that we want everything can be done by technology and automation. Devices like Bluetooth speakers Thermostats or security cameras either at home to in the office capture GBs of data and problem arises when this data is vulnerable to hacking or unauthorized access, potentially compromising home security and personal privacy.

8.2.4 Online Shopping and Recommendations

COVID-19 bring online, the buzz words in the limelight. From real world or retail shopping, we moved to online shopping and using an online platform like Amazon / Myntra or entertainment app like Netflix / Prime is running with a solid log, if you as a user search about one topic or outfit it starts recommending the same, not only this, with support of browsing history utilized for a recommendation. The issue arises about how much data is being collected and how it's utilized.

8.2.5 Selling the Data

Data is key in today's IT world. Companies collect and sell the data for profits like sales to the Loan Department/e-commerce Website / Medical departments and much more. & The surprise is that users often have little or no control over how this data is used or shared.

Figure 11. Real-world example

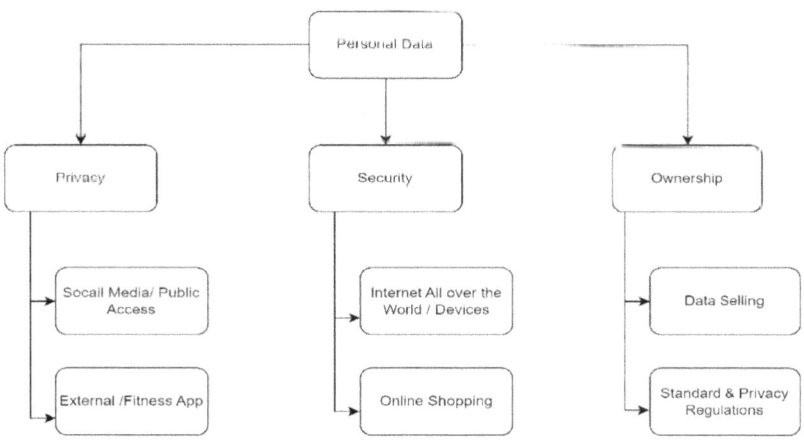

8.2.6 Standard and Regulations

The implementation of laws like the European Union's General Data Protection Regulation (GDPR) aims to give individuals more control over their data. It sets rules on data protection, consent, and user rights, impacting how companies handle and process user information.

The digital era or digital world or digitalisation offers great and immense benefits and conveniences but also poses and brings significant challenges/impacts which are concerning privacy, security, and ownership of personal data. Keeping a good balance between technological advancements and protecting individuals' rights and data is essential for a safer and more ethical digital landscape.

9. ETHICAL CONSIDERATIONS

Ethics play a vital role in human life. Whether it is the real world the digital world or the virtual world ethical consideration is required and crucial in the digital world especially in and around privacy, security and ownership of personal data. Below are a few key highlights of ethical aspects:

9.1 Be Clear and Provide Consent

Transparency is key in the digital world and it is required that it is mentioned what data is collected and who and how it is going to be used, also not only this, clear consent from individuals before collecting data and using the data. Individuals must know how their data is going to be used.

9.2 Collect as per Need

Data collection is required but is this required to collect all types of data? So ethically need to collect only the required or necessary data should be less from the personal side. Organizations think that over collection of data beyond's essential for the intended purpose only.

9.3 Secure the Data

Ethically it is the responsibility of each and everyone to safely guard the personal data to avoid or stop unauthorised access breaches or misuse. It is possible and required to build the robot's security measures and laws and regulations.

9.4 Ownership and Control

Ethically, it is required that there must be a framework that more ownership and control of data must be given to individuals only. This does not include the right to access but also correct or delete the information if and then required.

9.5 Be Equal for Everyone

Automation is key in today's world. Machine learning along with automation helps to draw the trends and patterns based on users' data but on ethical grounds, these patterns or algorithms should not be biased or lead to any discriminatory outcome or results. The pattern/ also must be fair and built by considering sensitive personal data equally.

9.6 Accountability and Responsibility (R&A)

RACI, which stands for Responsible, Accountable, Consulted and Informed. Organizations or companies must have a clear-cut segregation of Resposniobel and Account concerning privacy, security and ownership of personal data. On ethical grounds, companies or clients handle and take responsibility in case of any data breach or are accountable for their actions.

9.7 Rules Across Globel

Across the globe, multiple cultures, multi-languages and multiple legal frameworks exist. And it's time to extend the ethical discussion globally. Respect the cultures and make sure that the ethical consideration upholds the universal human rights.

9.8 Respect for Individuals

Individuals should have the autonomy to control their data. Respecting their choices regarding what data is collected, how it's used, and with whom it's shared is essential for ethical data handling.

9.9 Impact Assessment and Sustainability

It is required that the ethical consideration consider the long-term benefits or the impact assessment of personal data collection and usage also, needs to consider the ethical point of view to use data which can be sustained for upcoming generations but must be handled responsibly.

9.10 Respect for Privacy Rights

Recognizing and respecting the fundamental right to privacy is paramount. Upholding privacy rights involves balancing the need for data usage with the protection of an individual's privacy and freedoms.

10. WHAT'S NEXT NOW IN DIGITAL TRANSFORMATION

The above sections summarize a lot about Privacy, Security, and Ownership of Data but the question remains, are we good to move deeply in Digitalization or shall we take a step back and re-analyze or rethink? It is well said everything comes with some cost and risk.

Now the question is how to survive as there is no reverse gear or no go back to the old state/stage, the prime reason is COVID. The 2 years of lockdown changed the perspective of human nature towards technology. From online classes to online grocery to online medicines and consultation everything available on mobile phones. The cost is in the form of privacy breaches, leakage of data, security attacks, and unaccountability of ownership of personal data. A few highlights or points to move forward as below:

10.1 Transform Yourself

It is time to move on. Nokia and HMT are examples of whose products were best in their days but now they do not exist in the market because these brands have not moved as per time. So it is important if an organization wants to exit it needs to move as per time and transform itself. Netflix and Amazon are many more examples that transform themselves on time and lead the market (Reddy & Reinartz 2017).

10.2 Innovative Mindset and Actions

The next step or possible way to survive is innovation. SDLC waterfall model is now in history and agile is a new way of working. Let's make minds accept the changes and we need to change our mindset not only for new thoughts but also to change our mindset for actions. New ways to make decisions, to implement them, and also new ways to learn from mistakes and experiences.

10.3 Charge and Speed Up

Time and tide wait for none. The digital world is changing so fast that if you missed out 1 step you will be 100 steps away. Industries need to charge themselves as a customer, society needs to understand that the benefits of digital transformation bring a lot of changes and we need to speed up. Also, speeding up doesn't mean starting to change or transforming everything, need to plan like we start with Process or People or Product first, what is a priority for us then speed up the plan on the same.

10.4 Right People Right Time Right Skill

The last but important point is time to put the right people with the right skills and the right time. This is need of an hour. If either a customer or a company, society if anyone or everyone wants to move forward or use digital transformation must have to bring or put the right people. Many companies try to build digital transformation but they fail because lack of the right skills not using the right people or not at the right time. So, the key to progress is Right – People, Skills, and Time.

11. SUMMARY

It is very clear from the above sections that there are issues and challenges in the digital world concerning Privacy, security, and Ownership of personal data but it's not the organization's responsibility for these, we users are also equally responsible for the same.

Rules/ principles are there that we need to follow to protect our data, to secure the data, but this is also clear that these rules also need upgrades and more collaboration with technology. This is also true that we are new in this digital world of technology and we are still not aware of the impact of these data collection or breaches in the long run. Someone said Digital Literacy = Digital Equity. The determination, dedication, and new innovative minds who are helping to build this digital world also working to make data protection robust and standard. For data ownership, we can follow two approaches the bottom-up or Top-down approach. Both are good approaches to make the ownership of personal data term stable and bring positive change. Last but not least, we are humans and we learn from our mistakes. We experiment, we fail and we improve and the same goes and mature with time and experience.

Experimenting with the right attitude or must say with an open mind, with an innovative mind with the right people at the right time, and with the right skills is the key factor in the digital era. If we human begin do not move forward with time or as individuals I will not update myself with time, I am losing the race. It's time to work as WE not as I. Let's understand the cost we are paying to use this digital world and try to build a safe, secure world to live and live in for the coming generations.

REFERENCES

Ashford, N. A. (2000). An innovation-based strategy for a sustainable environment. In *Innovation-oriented environmental regulation: theoretical approaches and empirical analysis* (pp. 67–107). Physica-Verlag HD. doi:10.1007/978-3-662-12069-9_5

Belk, R. W. (2013). Extended self in a digital world. *The Journal of Consumer Research*, *40*(3), 477–500. doi:10.1086/671052

Conway, P. (1996). *Preservation in the digital world*. Council on Library and Information Resources.

Cummins, N., & Schuller, B. W. (2020). Five crucial challenges in digital health. *Frontiers in Digital Health*, *2*, 536203. doi:10.3389/fdgth.2020.536203 PMID:34713029

Dove, E. S. (2018). The EU general data protection regulation: Implications for international scientific research in the digital era. *The Journal of Law, Medicine & Ethics*, *46*(4), 1013–1030. doi:10.1177/1073110518822003

Eskicioglu, A. M., Town, J., & Delp, E. J. (2003). Security of digital entertainment content from creation to consumption. *Signal Processing Image Communication*, *18*(4), 237–262. doi:10.1016/S0923-5965(02)00143-1

Evanschitzky, H., Wangenheim, F. V., & Woisetschläger, D. M. (2011). Service & solution innovation: Overview and research agenda. *Industrial Marketing Management*, *40*(5), 657–660. doi:10.1016/j.indmarman.2011.06.004

Far, S. B., & Rad, A. I. (2022). Applying digital twins in metaverse: User interface, security, and privacy challenges. *Journal of Metaverse*, *2*(1), 8–15.

Janeček, V. (2018). Ownership of personal data in the Internet of Things. *Computer Law & Security Report*, *34*(5), 1039–1052. doi:10.1016/j.clsr.2018.04.007

Kaurin, D. (2019). *Data protection and digital agency for refugees*. Academic Press.

Kernighan, B. W. (2021). *Understanding the digital world: What you need to know about computers, the internet, privacy, and security*. Princeton University Press.

Knop, K., & Mielczarek, K. (2018). Using 5W-1H and 4M Methods to Analyze and Solve the Problem with the Visual Inspection Process Case Study. In *MATEC Web of Conferences* (Vol. 183, p. 03006). EDP Sciences.

Lynch, C. (2001). The battle to define the future of the book in the digital world. *First Monday*, *6*(6). Advance online publication. doi:10.5210/fm.v6i6.864

McConaghy, M., McMullen, G., Parry, G., McConaghy, T., & Holtzman, D. (2017). Visibility and digital art: Blockchain as an ownership layer on the Internet. *Strategic Change*, *26*(5), 461–470. doi:10.1002/jsc.2146

Minchev, Z. (2017). Security challenges to digital ecosystems dynamic transformation. *Proc. of BISEC*, 6-10.

Miranda, E. (2011). Timeboxing planning: Buffered Moscow rules. *Software Engineering Notes*, *36*(6), 1–5. doi:10.1145/2047414.2047428

Munir, A. B., Mohd Yasin, S. H., & Muhammad-Sukki, F. (2015). Big data: big challenges to privacy and data protection. *International Scholarly and Scientific Research & Innovation, 9*(1).

Olphert, C. W., Damodaran, L., & May, A. J. (2005, August). Towards digital inclusion–engaging older people in the 'digital world'. In *Accessible Design in the Digital World Conference 2005* (pp. 1-7). Academic Press.

Osburg, T. (2017). *Sustainability in a digital world needs trust.* Springer International Publishing. doi:10.1007/978-3-319-54603-2

Pagallo, U. (2017). The legal challenges of big data: Putting secondary rules first in the field of EU data protection. *Eur. Data Prot. L. Rev.*, *3*(1), 36–46. doi:10.21552/edpl/2017/1/7

Raina, A., & Palaniswami, M. (2021). The ownership challenge in the Internet of Things world. *Technology in Society*, *65*, 101597. doi:10.1016/j.techsoc.2021.101597

Reddy, S. K., & Reinartz, W. (2017). Digital transformation and value creation: Sea change ahead. *GfK Marketing Intelligence Review*, *9*(1), 10–17. doi:10.1515/gfkmir-2017-0002

Regan, P. M. (2002). Privacy as a common good in the digital world. *Information Communication and Society*, *5*(3), 382–405. doi:10.1080/13691180210159328

Romansky, R. (2017). A survey of digital world opportunities and challenges for user's privacy. *International Journal on Information Technologies and Security*, *9*(4), 97–112.

Romansky, R. P., & Noninska, I. S. (2020). Challenges of the digital age for privacy and personal data protection. *Mathematical Biosciences and Engineering*, *17*(5), 5288–5303. doi:10.3934/mbe.2020286 PMID:33120553

Spiekermann, S. (2012). The challenges of privacy by design. *Communications of the ACM*, *55*(7), 38–40. doi:10.1145/2209249.2209263

Spremić, M., & Šimunic, A. (2018, July). Cyber security challenges in the digital economy. In *Proceedings of the World Congress on Engineering* (Vol. 1, pp. 341-346). International Association of Engineers.

Watkins, R. D., Denegri-Knott, J., & Molesworth, M. (2016). The relationship between ownership and possession: Observations from the context of digital virtual goods. *Journal of Marketing Management*, *32*(1-2), 44–70. doi:10.1080/0267257X.2015.1089308

Weber, R. H. (2015). The digital future–A challenge for privacy? *Computer Law & Security Report*, *31*(2), 234–242. doi:10.1016/j.clsr.2015.01.003

Chapter 12

Ethics and Artificial Intelligence:
A Theoretical Framework for Ethical Decision Making in the Digital Era

Yashpal Azad
https://orcid.org/0000-0003-2957-8917
Eternal University, India

Amit Kumar
https://orcid.org/0000-0002-1686-3279
Eternal University, India

ABSTRACT

In the rapidly evolving technological landscape, the pervasive integration of artificial intelligence (AI) has brought to light the pressing need to address the ethical dimensions associated with its widespread adoption. This study comprehensively explores AI ethics for ethical decision-making in the digital era. It offers a structured guide for aligning AI with ethical principles, emphasizing transparency, bias mitigation, and interdisciplinary collaboration in AI deployment. Additionally, it delves into the evolving AI landscape, highlighting potential societal impacts. It calls upon policymakers and stakeholders to engage in persistent dialogue and to remain adaptable in the face of a continuously transforming technological environment, advocating for the continuous refinement and adaptation of regulatory frameworks. This framework acts as a compass for ethically sound AI decisions, fostering a responsible, human-centric approach. It aims to forge a symbiotic relationship where AI uplifts society while upholding ethical values, making it a tool for societal betterment.

INTRODUCTION

The term Artificial Intelligence (AI) was first coined by John McCarthy in a 1956 conference (Kaplan & Haenlein, 2019). Alan Turing, even earlier, had envisioned machines emulating human behavior through the Turing test (Hodges, 2009). Over time, advancements in computational power enable instant calculations and real-time evaluation of new data based on past information. The development and advancement

DOI: 10.4018/979-8-3693-1762-4.ch012

of Artificial Intelligence (AI) represent a significant milestone in the field of computer science, aiming to emulate human intelligence in machines. This involves enabling machines to perform tasks that typically require human cognitive abilities, such as reasoning, voice recognition, learning, problem-solving, and quick decision-making (Kaushik et al., 2022). AI operates through rule-based algorithms and has the capacity to acquire new skills through machine learning, where it learns from exposure to computer data (Webber & Nilsson, 2014). Notably, AI can be classified into three main types: Artificial Narrow Intelligence (ANI), which excels in specific functions like face recognition; Artificial General Intelligence (AGI), which aspires to replicate broader human intelligence and problem-solving capabilities; and Artificial Superintelligence (ASI), a concept envisioning machines surpassing human intelligence and attaining self-awareness, albeit still confined to the realm of science fiction (Russell & Norvig, 2016).

In the broader landscape, AI encompasses the replication of human intelligence, with machine learning being a subset that facilitates autonomous decision-making processes (Webber & Nilsson, 2014). The integration of AI, including machine learning, has witnessed significant strides in various aspects of daily life. Through data analysis and pattern recognition, AI enhances system performance across diverse applications. Natural Language Processing (NLP) plays a crucial role in facilitating communication between humans and machines, while robotics introduces autonomous decision-making capabilities. The potential of these technologies to revolutionize industries is vast, with applications spanning healthcare, finance, transportation, and entertainment (Gongane et al., 2022; Wu et al., 2022). As AI continues to evolve, its impact on these industries and society at large is poised to be transformative, ushering in a new era of technological innovation and integration.

UNDERSTANDING THE CONCEPT OF AI

The comprehensive overview of artificial intelligence (AI) begins by elucidating the broad capabilities of AI systems, which can mimic human cognitive functions such as interpreting speech, playing games, and pattern recognition. The learning process of AI involves massive data processing, enabling the system to make decisions based on patterns it identifies. While some AI systems require human supervision during their learning phase, others are designed to learn autonomously, like mastering a video game through repeated plays (Mitchell, 2019).

The Concept of Strong AI vs. Weak AI

AI experts distinguishes between strong AI and weak AI, to define the machine intelligence, strong AI refers to the artificial general intelligence, which aspiring to replicate human-like problem-solving across various tasks (such as learning from a novel situation, learn from experience, and perform any intellectual task that human can do), while on the other hand, weak AI sometimes referred to narrow AI or specialized AI, operates within specific contexts, excelling in narrowly defined problems (such as voice activated personal assistants, providing weather updates, setting reminders, or answering general knowledge questions etc.) (Flowers, 2019).

The Concept of Machine Learning and Deep Learning

The differentiation between machine learning (ML) and deep learning is important to understand, emphasizing that deep learning is a subset of machine learning and machine learning is a sub-field of AI. Machine learning involves algorithms learning from data without explicit programming, encompassing supervised and unsupervised learning. On the other hand, deep learning employs biologically inspired neural network architectures with hidden layers, allowing for in-depth learning and complex pattern recognition (Janiesch, Zschech & Heinrich, 2021).

Types of AI

There are broadly the four types of AI, categorized based on task complexity, viz a) Reactive machines, b) Limited memory, c) Theory of mind, and d) Self-awareness. The machine based on reactive machines model operates based on fundamental AI principles, focusing on perceiving and reacting to its immediate environment without the ability to store or rely on past experiences. This limited memory capacity prevents it from informing real-time decision making based on historical data. These machines are intentionally designed to have a narrow worldview, performing specific, specialized duties. Despite their inability to store memories, this design choice contributes to their reliability and consistency, ensuring that the AI reacts consistently to new stimuli each time, Examples include, Deep Blue, developed by IBM in 1990, designed to play chess in real time as per chess rules (Newborn, 2012; Chung, Thaichon & Quach, 2022).

Secondly, limited memory systems have the capability to retain past data and predictions, enabling it to make informed decisions by referencing historical information. It surpasses the simplicity of reactive machines, offering greater complexity and potential. To develop limited memory AI, continuous training is essential, either by training the model to analyze and utilize new data or by constructing an AI environment that automatically trains and renews models. The process involves establishing training data, creating the machine learning model, ensuring predictive abilities, receiving feedback, storing feedback as data, and repeating these steps cyclically for ongoing improvement, example include, Autonomous Vehicles (Ma et al., 2020).

Further, the concept of theory of mind in AI remains theoretical and has not yet been achieved due to current technological limitations. The concept involves imbuing AI machines with the psychological capacity to understand that living entities including humans, animals and machines, have thoughts and emotions influencing their behavior. This would enable AI to engage in self-reflection, comprehend human emotions and thought processes, and use that understanding to make informed decision autonomously. For this to occur, machines must me capable of real time processing such as the "mind" and the role of emotions in decision-making. Ultimately, this approach envisions a bidirectional relationship, where machines and humans engage in a mutual understanding of emotions and decision-making processes (Cuzzolin et al., 2020).

Finally, the concept of self-awareness in AI is aspired for the future, if in near future the goal of establishing theory of mind is established the final step will be the establishing AI to become self-aware. A future stage of AI development envisions machines with human-level consciousness. Self-aware AI would comprehend its own existence and the emotional states of others, interpreting not just explicit communication but also in the nuances in how it is conveyed. Achieving this model will require researchers to grasp the intricacies of human consciousness and then replicate and integrate this understanding into machines. This transformative step in AI development aims to create entities that not only understand themselves but also process an awareness of the world and the emotions of those around them (Chatila et al., 2018; Bellman et al., 2020; Chung, Thaichon & Quach, 2022).

Therefore, this comprehensive overview of artificial intelligence (AI) covers key aspects, beginning with the broad capabilities of AI systems, including mimicking human cognitive functions through data processing. The distinction between strong AI, aspiring to replicate human-like problem-solving, and weak AI, excelling in specific tasks, is emphasized. The differentiation between machine learning (ML) and deep learning, where deep learning is a subset of ML, is highlighted. The four types of AI—reactive machines, limited memory, theory of mind, and self-awareness—are discussed, with examples illustrating their functionalities. The concept of theory of mind remains theoretical, involving the understanding of thoughts and emotions, while self-aware AI, possessing human-level consciousness, is an aspirational future development requiring a profound understanding of human consciousness for integration into machines.

The Ethical Dilemma in the Age of AI

The swift advancement of AI technology raises critical ethical concerns, encompassing decision-making in learning, human rights, data ownership, privacy, consent, and algorithmic biases (Giuffrida, 2019; Ungerer & Slade, 2022). Privacy, especially in the digital age, is a paramount worry due to the extensive collection and storage of personal information online. Robust data protection laws and transparent regulations, empowering individuals to control their data usage, are pivotal for mitigating these concerns (Finn, Wright & Friedewald, 2013).

Security is another vital ethical concern in this digital era, given the evolving cyber threats (Smith et al., 2019; Formosa, Wilson & Richards, 2021). To protect personal and financial data, as well as intellectual property, implementing strong cybersecurity measures such as encryption, two-factor authentication, and regular security audits is imperative. Data protection, closely intertwined with privacy and security, hinges on responsible handling, secure storage, clear guidelines for retention and disposal, and granting individuals access and control over their data (Romanski & Noninska, 2020).

Ethical concerns tied to AI encompass bias, accountability, and transparency (Gunning & Aha, 2019). Biased AI algorithms can result in discriminatory outcomes across various domains, underscoring the necessity for fairness and accountability. Implementing mechanisms that ensure transparency in AI decision-making processes and hold AI systems accountable is crucial. Simultaneously, addressing online harassment, especially on social media platforms, is a significant ethical concern in today's digital age (Davis & Koepke, 2016; Kiritchenko, Nejadgholi & Fraser, 2021). Cyberbullying, hate speech, and online harassment profoundly impact mental health and well-being, necessitating stringent policies, regulations, and advanced moderation tools for the detection and removal of harmful content (Abaido, 2020; Popat & Tarrant, 2023).

Therefore, to address digital age ethical concerns effectively, a comprehensive strategy is required to tackle ethical challenges in the digital age, including the enforcement of robust privacy laws to protect individuals' personal information, thereby granting them control over their data. Addressing the escalating cyber threats involves bolstering cybersecurity measures through the implementation of advanced security protocols and encryption to safeguard digital systems. Mitigating biases in Artificial Intelligence is imperative, necessitating proactive efforts to identify and rectify unfair outcomes, fostering fairness and equity. The prevalence of online harassment in communication platforms requires measures for prevention and response to ensure a safer digital space. The overarching objective is to cultivate a secure and fair digital environment by not only addressing specific concerns like privacy, cybersecurity, and bias but also by promoting a culture of respect and fairness online, all while respecting individuals' rights and well-being.

SIGNIFICANCE OF PRESENT STUDY AND RESEARCH GAP

While ethical considerations in AI are gaining importance, a notable research gap exists in a comprehensive analysis and comparison of existing theoretical models guiding ethical decision-making in AI. Individual studies are present, but a consolidated systematic review delineating these models, along with their strengths, weaknesses, and practical applicability, is lacking. Moreover, understanding the practical implementation and adaptation of these theoretical models across diverse AI contexts, domains, and cultures is crucial. Bridging this gap is vital for a holistic understanding of theoretical frameworks and their effective application in promoting ethical AI development and decision-making.

OBJECTIVES OF THE STUDY

This chapter seeks to explore the intricate domain of Artificial Intelligence (AI), a groundbreaking field in computer science dedicated to emulating human intelligence in machines. Our examination will encompass the core principles of AI, including its subfields such as Machine Learning (ML), and their pervasive integration into daily life. The chapter will also scrutinize the ethical challenges stemming from the swift progress in AI, focusing on critical issues like privacy, security, bias, accountability, and transparency. Furthermore, the study aims to conduct a comprehensive analysis of theoretical frameworks and literature to propose a robust conceptual model that prioritizes ethical considerations in AI decision-making for development and deployment in the digital age. This model will discern key issues influencing ethical decision-making in AI systems, encompassing aspects like bias, transparency, accountability, and human intervention in managing AI. Additionally, the study will delve into the broader societal impact and implications of AI decision-making. However, the specific objectives of this study are:

Objective-1: To conduct the comprehensive literature review and analyze the existing theoretical frameworks concerned with ethical considerations in AI development and deployment.

Objective-2: To propose a conceptual model emphasizing the key factors such as bias mitigation, transparency, accountability, and the importance of human-AI collaboration.

Objective-3: To analyze the impact of ethical decision-making of AI systems in a broad spectrum of socio-cultural, ethnic, and demographic aspects of human life and suggest practical implications for organizations, policymakers, AI developers, ethicists, and future research.

Literature Review and Conceptual Framework

Integration of AI Into Ethical Decision Making

The integration of AI in ethical decision-making is increasingly receiving significant attention, with a comprehensive exploration of both theoretical underpinnings and practical implications. Designing AI systems to align with human values is crucial, employing value alignment and value-sensitive design approaches. Challenges involve translating abstract ethical principles into computationally tractable AI behavior guidelines (Lipton, 2018; Ballard, Chappell & Kennedy, 2019; Friedman et al., 2021; Hen et al., 2021; Gan & Moussawi, 2022; Umbrello & Yampolskiy, 2022).

Furthermore, addressing biases and ensuring fairness in AI decision-making is vital for ethical integration. Research focuses on reducing biases in training data and mitigating unfair outcomes, especially in domains like hiring and criminal justice (Barocas & Hardt, 2019). Governance and policies surrounding ethical AI integration are equally crucial (Thuraisinghan, 2019). An emerging aspect is the integration of AI into ethical decision-making across diverse cultures. Research explores how ethical considerations and AI frameworks require adaptation to respect cultural variations and norms (Kalpan & Haenlein, 2019).

Additionally, De Moura et al., (2020) studied ethical challenges in autonomous driving, introducing a decision-making algorithm based on a Markov Decision Process (MDP) to navigate ethical dilemmas during normal driving conditions. This study explicitly considers collision scenarios, evaluating harm to different road users and prioritizing safety. This model incorporate various moral approaches based on Rawlsian, utilitarianism, contractarianism, and egalitarianism for making ethical decision. The study suggested mitigating challenge from the inherent subjectivity in defining ethical principles and the diversity of different approaches which can complicate achieving consensus on right behavior in autonomous vehicles.

Therefore, these findings emphasize the complexity of integrating Artificial Intelligence (AI) into ethical decision-making, covering key dimensions such as value alignment, bias mitigation, explainability, governance, and cross-cultural considerations. Ensuring AI systems align with human values and ethical principles is crucial to avoid conflicts with societal norms. Addressing inherent biases in AI algorithms requires proactive measures for identification and rectification. The transparency of AI systems, known as explain ability, is vital for building trust and accountability among stakeholders. Governance plays a pivotal role in establishing responsible frameworks and regulations for AI development and deployment. Acknowledging the diversity of ethical perspectives across societies, especially in cross-cultural contexts, ensures AI systems are adaptable and sensitive to cultural nuances. Together, these dimensions contribute to a comprehensive understanding of how AI can drive ethical outcomes across various societies, recognizing the challenges involved in integrating AI into ethical decision-making processes.

The Significance of Value System Design Approach in Shaping AI to Align Human Values and Ethics

Designing AI systems aligned with human values and ethics is paramount for responsible and beneficial AI deployment. A value system design approach plays a crucial role in achieving this alignment, integrating ethical considerations into the core of AI development (Winkler & Speikermann, 2019). Similarly, Mittelstadt et al. (2016) provide a comprehensive view of algorithm ethics, emphasizing the need for ethical considerations and a value-driven design approach to ensure alignment with human values. Likewise, Boddington and Boddington (2017) advocate for a virtue ethical approach, embedding virtuous values in AI systems to promote ethical behavior.

In a similar vein, Dignum (2022) highlights the need for a value-centric approach in responsible AI development, asserting that integrating human values into AI design is fundamental for ethical behavior. Furthermore, Jobin, Lenca, and Vayena (2019) analyze AI ethics guidelines globally, emphasizing the necessity of a value-oriented approach and the incorporation of diverse societal values into AI system design.

AI, Machine Learning, and Ethical Decision Making

Machine learning (ML) has rapidly advanced and become prevalent in various sectors, from healthcare to finance and many more (Pramod, Naicker & Tyagi, 2021). However, the usage of ML algorithms raises ethical concerns about bias, fairness, transparency, and accountability in decision-making (Danks & London, 2017; Wang et al., 2023). Algorithmic bias is a critical ethical concern in ML, stemming from skewed or prejudiced training data, leading to unintentional discriminatory behavior towards certain groups (Diakopoulos, 2016). This bias can perpetuate societal inequalities, resulting in unfair outcomes.

Furthermore, few studies suggest that to counter algorithmic bias and promote fairness and inclusivity, proactive measures throughout the model development process are crucial. Issues such as transparency and interpretability are fundamental in ethical ML practices, ensuring comprehension of the model's operations and factors influencing its decisions. Furthermore, explainable AI models play a pivotal role in achieving transparency, enabling stakeholders to understand ML model decisions, identify biases, and enhance trust (Zliobaite, 2015; Rudin, 2019, Morse et al., 2021; Kordzadeh & Ghasemaghaei, 2022, Zowghi & da Rimini, 2023).

Further, background studies while addressing the ethical implications of ML necessitate clear guidelines and regulations to protect individuals' privacy. Additionally, interdisciplinary collaboration involving ethicists, social scientists, and technologists is essential to comprehensively address ML's ethical challenges and develop robust ethical frameworks for responsible ML development and deployment (Goodman & Flaxman, 2017; Villaronga, Kieseberg & Li, 2018; Gunning & Aha, 2019; Gebru et al., 2021).

The Development of AI Ethics in Finance

Cao (2020) in his review offers a through exploration of AI's recent surge in FinTech, highlighting decades of significant developments and its potential impact on shaping a smart economy. The review recognizes AI empowered finance as a dynamic and crucial field, spanning AI, data science, economics and finance. The review emphasizes the transformative influence of new-generation AI, data science and machine learning on the vision, mission, and societal aspects of economics and finance. The work underscores how these advancements drive the evolution of smart FinTech, facilitating the creation of personalized, advanced and secure economic and financial mechanisms.

The recent strides in AI ethics within the finance sector represent a commendable shift towards addressing the ethical implications of artificial intelligence (AI) (Dwivedi et al., 2021). This evolving landscape underscores the industry's growing recognition of the imperative for responsible and transparent use of AI technologies. The emphasis on ethical considerations brings attention to potential risks such as biased algorithms and data privacy concerns, significantly influencing financial decision-making processes. To mitigate these challenges, financial institutions are increasingly adopting ethical frameworks and guidelines to ensure fairness, accountability, and transparency in AI applications (Fleming & Morgan, 2012). This development is crucial for fostering trust among consumers and stakeholders as AI becomes more integrated into various financial processes. The concurrent focus on ethical standards, including human involvement, is a positive step towards creating a more responsible and sustainable financial ecosystem (Brusseau, 2023). Staying abreast of these developments and understanding the evolving ethical landscape in AI remains paramount for professionals, policymakers, and researchers in the finance sector.

The literature collectively illuminates vital aspects at the intersection of AI and finance. Cath (2018) provides a comprehensive analysis of the ethical, legal, and technical challenges in governing AI, particularly within the finance sector. Mhlanga's (2020) exploration of AI's impact on digital financial inclusion within Industry 4.0 underscores its transformative role in enhancing accessibility to financial services. Aitken et al. (2020) delve into establishing a social license for financial technology, exploring the private sector's adoption of ethical data practices to build societal trust and contribute to innovation. Truby et al. (2020) suggest the imperative for proactive regulatory measures in governing AI in banking, highlighting both opportunities and challenges within the financial industry's regulatory landscape. Buckley et al.'s (2021) examination emphasizes the necessity of human oversight in AI-driven financial processes, advocating for regulatory considerations to put humans in the loop. These insights collectively contribute to a nuanced understanding of the multifaceted relationship between AI and finance.

Fritz-Morgenthal, Hein & Papenbrock (2022) investigated the managing model risk in banks and financial institutions. The research, based on insights from Round Table AI members and external speakers, covers a range of models from simple rules-based to advanced AI/ML models, with a focus on credit, insurance, and various financial risks. The findings foresee the rise of complex models in anti-money laundering and derivatives valuation. It recognizes the success of similar models in areas like sales optimization and robo-advisory, offering practical guidance for establishing a governance framework for responsible AI/ML deployment.

On the other hand, Qiang, Rhim and Moon (2023) address the challenge of selecting and implementing AI ethics frameworks (AIEFs) in the industry, despite the proliferation of AIEFs over the last decade, their adoption remail unclear. The study assesses the four existing frameworks to assess the AI ethics concerns of a real-world system. The study compares the experiences of a third-party auditor and the audited company, revealing a common positive sentiments towards the assessment process, despite varying completion limits. Importantly each framework offers distinct benefits, suggesting their potential synergetic use at different stages of AI development. The study emphasizes the need for the Ai ethics community to specify framework suitability and expected benefits, facilitating more effective adoption in the industry.

In conclusion the synthesis of extensive literature on integration of AI into ethical decision-making necessitates a holistic approach, considering dimensions such as value alignment, bias mitigation, explainability, governance, and cross-cultural considerations. Scholars emphasized the importance of value-sensitive design approaches and the translation of abstract ethical principles into practical AI behavior guidelines. Addressing biases, ensuring fairness, and implementing proactive measures in machine learning development are vital for ethical AI practices. Transparency and interdisciplinary collaboration involving ethicists, social scientists, and technologists are crucial for building trust and accountability. The evolving field of AI ethics in finance underscores the need for responsible and transparent need use of AI technologies, emphasizing ethical frameworks and guidelines. Ultimately, aligning AI systems with human values and navigating diverse ethical perspectives are essential for deriving ethical outcomes and avoiding conflicts with societal norms.

CASE STUDIES AND PRACTICAL APPLICATIONS

Integrating AI into ethical decision-making is a nuanced task that demands careful analysis. Here are illustrative case studies showcasing AI's integration into ethical contexts:

Case Study 1: Healthcare Decision Support

In the case study exploring the integration of artificial intelligence (AI) in healthcare, the focus is on utilizing AI to facilitate ethical decision-making in patient care. The application of AI algorithms involves the comprehensive analysis of vast patient datasets, enabling the generation of personalized treatment recommendations. This process is guided by ethical principles such as patient autonomy, beneficence, and non-maleficence (Obermeyer et al., 2019). A specific example of AI implementation is demonstrated through IBM Watson for Oncology, a system designed to support oncologists in treatment planning. By leveraging AI to sift through extensive medical literature, patient records, and clinical trial data, healthcare professionals can make more informed and ethically sound decisions, ultimately improving the quality of care provided to patients. The convergence of AI and healthcare not only showcases technological advancements but also underscores the importance of aligning technological applications with ethical considerations in the medical field (Kreps & Neuhauser, 2013).

Case Study 2: Criminal Justice Fairness

The case study delves into the integration of artificial intelligence (AI) within the criminal justice system, specifically focusing on aiding judges in making equitable decisions related to bail, sentencing, and parole. Utilizing AI algorithms, the system endeavors to predict the likelihood of reoffending by analyzing historical data, with the overarching goal of mitigating decision-making biases (Chouldechova, 2017). One prominent example is the "COMPAS" system, which has been widely employed to forecast recidivism risk among individuals within the U.S. criminal justice system. By harnessing AI's analytical capabilities, the criminal justice system aims to enhance the objectivity and fairness of decisions pertaining to individuals' legal outcomes, ultimately contributing to a more just and unbiased judicial process (Bellamy et al., 2019).

Case Study 3: Ethical Investment Decisions

Within the realm of finance, the case study explores the integration of artificial intelligence (AI) to analyze market trends and financial indicators, facilitating ethical investment decisions characterized by regulatory compliance and transparency. Notably, the Robo-Advisor Wealthfront leverages AI-driven algorithms to construct personalized and diversified investment portfolios based on clients' financial situations and objectives. This approach ensures a tailored and ethically sound investment strategy, aligning with regulatory standards and emphasizing transparency in the decision-making process. This application of AI in finance not only optimizes investment portfolio construction but also underscores the potential of technology to contribute to ethical and transparent financial practices, catering to the diverse needs and preferences of investors (Max, Kriebitz, & Von Websky, 2021).

Case Study 4: Responsible AI-Powered Chatbots

The case study explores the widespread adoption of AI-powered chatbots for customer service interactions by companies worldwide, emphasizing the importance of ethical implementation to align with company values and principles. In this context, the ethical considerations encompass ensuring that chatbots not only accurately represent the company's ethos but also provide unbiased information, respect user privacy, and adhere to robust data protection principles. Google's AI-powered chatbot, "Duplex," serves as a notable example, demonstrating transparency in its functionalities, particularly in making reservations and appointments. The case study focuses on how the ethical guidelines governing Duplex contribute to a responsible and trustworthy interaction between the chatbot and users, highlighting the pivotal role of ethical considerations in shaping the deployment of AI technologies for customer service applications (Leviathan & Matias, 2018).

Case Study 5: Autonomous Vehicles and Ethical Decision Making

AI algorithms in autonomous vehicles face complex challenges, especially in ethical dilemmas involving prioritizing occupant or pedestrian safety. An example is the "trolley problem," where the vehicle must choose between potential harm to pedestrians and risks to its occupants. This highlights the need for clear guidelines in programming autonomous vehicles, addressing ethical considerations like minimizing harm, legal issues, and transparency. Engaging in open discussions with the public, policymakers, and ethicists is crucial for developers to establish ethical frameworks that balance safety, transparency, and fairness, building trust in this advancing technology (Martinho et al., 2021).

Case Study 6: The Application of AI In Finance

The integration of AI in financial sector has transformed various aspects of operations, including risk assessment, fraud detection, customer service, and investment strategies. One notable application is the use of AI algorithms in robo-advisors for personalized financial advice. Wealthfront, a robo advisor platform was founded in 2011, exemplifies the application of AI in finance. Using Ai algorithm Wealthfront assesses investment risk tolerance, dynamically adjusts portfolios based on market conditions, employ's tax-loss harvesting strategies, and automates rebalancing. The outcomes include cost-effective investing with lower fees, personalized financial planning, aligned with individual goals, and efficient tax management, showcasing how Ai in robo-advisors transforms finance by offering accessible, tailored, and efficient investment solutions (Bandi & Kothari, 2022).

Case Study 7: The Application of AI in Remote Sensing in Earth Science

AI in remote sensing has revolutionized Earth Science Applications, enhancing the efficiency and precision of satellite and aerial data analysis. Convolutional neural network (CNNs) is a real-world example that play crucial role in land cover classification, distinguishing forests, urban areas, water bodies, and agricultural land for improved land management (Janga, 2023; Atik & Ipbuker, 2021). In deforestation monitoring AI detects changes in forest cover, classifies deforested areas, and assesses biodiversity and carbon storage impacts. Ai-driven remote sensing aids crop monitoring, disease detection, and yield prediction in agriculture, supporting data-driven decisions for precision farming (Zulfiqar et al., 2021). Wildlife conservation benefits from AI as image recognition algorithms track and monitor animal, providing crucial data for understanding migration pattern and population dynamics (Nazir & Kaleem, 2021). Climate research uses AI to analyze satellite data, monitoring sea surface temperatures, ice covers, and atmospheric conditions, contributing to a better understanding of climate change (Huntingford et al., 2019). AI powered remote sensing is essential for early detection and response to natural disasters, identifying terrain changes after events like earthquakes or floods (Tan et al., 2021). Additionally, AI analyses satellite imagery for urban growth and planning, offering valuable insights for infrastructure development and assessing the environmental impact of expanding urban areas (Yigitcanlar et al., 2020).

Therefore, in summary, the mentioned cases highlight the diverse ethical applications of Artificial Intelligence (AI), underscoring the imperative to confront biases and adhere to ethical principles within AI development and deployment. In the evolving landscape of AI, it is crucial to acknowledge the potential biases embedded in algorithms, data, and decision-making processes. Addressing biases is vital to ensure that AI systems do not perpetuate or amplify existing inequalities, discrimination, or unfair practices. Additionally, upholding ethical principles in AI involves considering factors such as transparency, accountability, and privacy to safeguard individuals and communities affected by AI technologies. The continuous refinement of AI systems is essential to align with evolving ethical standards and norms. Researchers and developers must remain vigilant, regularly assessing and updating AI models to mitigate biases and enhance ethical considerations. This ongoing commitment to ethical AI practices is necessary for fostering trust among users, promoting responsible AI innovation, and maximizing the positive impact of AI on society as a whole

THEORETICAL FRAMEWORK FOR ETHICAL DECISION-MAKING IN AI

This section outlines a theoretical model that integrates key factors influencing ethical decision-making in AI, incorporating established ethical frameworks like utilitarianism, deontology, virtue ethics, and rights-based ethics. By amalgamating these ethical perspectives, the model aims to comprehensively address the intricate challenges posed by AI (see details below):

UTILITARIANISM

Utilitarianism, an ethical theory by Bentham and Mill, gauges an action's morality by its utility in maximizing overall happiness or well-being. In AI, this means optimizing systems for society's maximum benefit, enhancing well-being, happiness, or utility while minimizing harm, and aligning with the greater good. For example, AI in healthcare improves medical care and aligns with utilitarian principles, potentially saving lives (see Figure 1).

Ethical concerns arise in utilitarianism regarding AI. The risk lies in AI inadvertently perpetuating biases or inequalities while optimizing for the majority, potentially neglecting or harming minority groups and causing an unequal benefit distribution. Striking a balance to minimize negative consequences is crucial in applying utilitarianism to AI (Bentham & Mill, 2004).

Advantages of Utilitarianism in the Context of AI

Maximization of Societal Welfare:

Focus on Majority Benefit: Utilitarianism in AI emphasizes solutions that benefit the majority, aiming to maximize overall happiness and well-being. This approach can lead to advancements in various sectors, contributing to societal welfare.

Figure 1. Utilitarian ethic

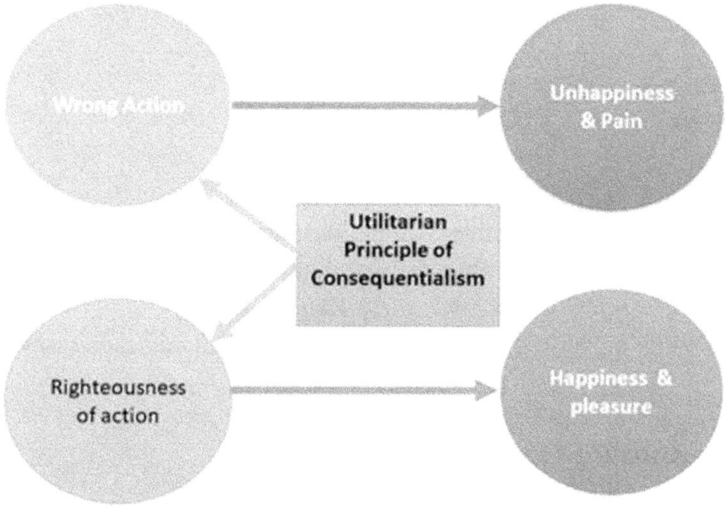

Potential for Significant Positive Impact: Optimizing Critical Areas: Utilitarianism can be particularly impactful in critical areas such as healthcare. Optimizing AI for medical diagnoses and treatment can have significant positive consequences, potentially saving lives on a large scale.

Efficient Resource Allocation: Directed Efforts: Utilitarianism aids in efficiently allocating resources by directing efforts towards AI applications with the most substantial positive impact. This ensures that resources are used optimally for maximum societal benefit.

Adaptability to Varied Contexts: Flexible Ethical Framework: Utilitarianism is adaptable to different AI contexts and domains. This flexibility allows for the optimization of benefits across diverse areas, including the environment, education, and social services.

Disadvantages of Utilitarianism in the Context of AI

Risk of Minority Neglect: The focus on maximizing benefits for the majority may inadvertently lead to neglecting the needs and interests of minority groups. This can perpetuate biases and inequalities in AI systems.

Ethical Trade-offs and Dilemmas: Utilitarianism can create ethical dilemmas in AI decision-making, where achieving the greatest good for the majority might involve making difficult choices that harm or neglect certain individuals or groups. Balancing competing interests becomes challenging.

Subjectivity in Defining "Utility": Defining and measuring utility or well-being in AI is subjective, with differing interpretations among stakeholders. This subjectivity adds complexity to the task of objectively maximizing happiness or well-being for the majority.

Difficulty in Qualifying Consequences: Evaluating the consequences of AI actions in terms of happiness or well-being is intricate. It can be challenging to accurately qualify the potential impact on overall societal welfare, leading to unintended negative consequences.

In conclusion, while Utilitarianism provides a framework for maximizing AI benefits for societal welfare, it is crucial to address biases, ethical trade-offs, and challenges in utility measurement. Balancing the pursuit of welfare maximization with harm mitigation is essential for the responsible application of utilitarian principles in the development and deployment of AI systems.

DEONTOLOGY

Deontology, linked to Kant, centers on action morality tied to rules and duties, regardless of outcomes. In AI ethics, it underlines adherence to ethical guidelines during system development, deployment, and use. The emphasis is on AI operating within defined moral boundaries, following societal, cultural, or ethical standards. The categorical imperative, a key Kantian concept, suggests an AI action or system is morally acceptable when universally applicable without ethical contradictions (Huang et al., 2022). Refer to Figure 2.

Advantages of Deontology in the Context of AI

Clear Ethical Guidelines: Deontology provides clear ethical guidelines for AI development and deployment. It establishes a structured framework that developers and practitioners can follow when making ethical decisions, contributing to responsible AI practices.

Emphasis on Moral Duties: Deontology places a strong emphasis on moral duties and obligations, fostering trust in AI systems. By aligning with societal, cultural, and ethical standards, it helps build a foundation of responsibility in the development and deployment of AI technologies.

Figure 2. Deontological ethic

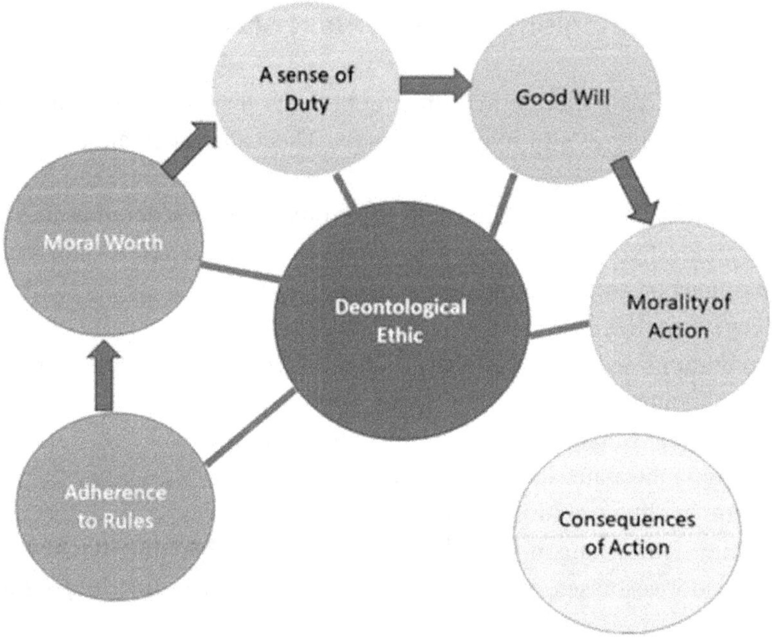

Protection of Fundamental Values: Deontology prioritizes ethical values such as fairness, autonomy, transparency, and privacy. This emphasis ensures that AI applications align with fundamental ethical principles, promoting ethical and equitable use.

Universal Applicability: The concept of the categorical imperative in deontology enhances ethical applicability and relevance. By emphasizing universal acceptance without causing contradictions, it provides a framework that can be applied across diverse cultural and societal contexts.

Disadvantages of Deontology in AI

Rigidity in Adherence to Rules: Strict adherence to rules in deontology may lead to inflexibility when dealing with complex and evolving ethical situations in AI. This rigidity could impede adaptive responses to emerging challenges, hindering the ability to address novel ethical dilemmas.

Conflict Resolution Dilemmas: Deontology lacks explicit guidance on resolving conflicts when multiple ethical principles or rules conflict. This poses challenges in decision-making, especially in the dynamic AI domain where conflicting principles may need to be balanced.

Difficulty in Application and Interpretation: There may be varying interpretations of moral duties, and disagreements may arise regarding the correct application of deontological principles to specific AI contexts. This ambiguity can lead to challenges in consistent and universally accepted ethical standards.

Lack of Considerations for Consequences: Deontology focuses on adherence to rules regardless of consequences. This approach may overlook the overall societal benefits or harms that AI systems could generate. It hinders a holistic evaluation of AI's impact, potentially missing the broader consequences of ethical decisions.

In conclusion, while deontology offers a compact ethical framework and emphasizes adherence to ethical principles, it must strike a balance with flexibility and considerations of consequences in the dynamic field of AI. Finding this balance is crucial for ensuring responsible AI development and deployment that aligns with ethical standards and addresses the evolving challenges in the AI landscape.

VIRTUE ETHICS (ARISTOTELIAN ETHICS)

Aristotle, a foundational figure in virtue ethics, emphasized virtue's role in moral decisions in his work "Nicomachean Ethics" (Aristotle, 350 BCE). Unlike consequentialist (e.g., utilitarianism) and deontological theories (e.g., Kantian ethics), virtue ethics prioritizes moral character development through practicing virtues. This approach leads to ethical conduct and the pursuit of the common good (Quinn, 2007). See Figure 3.

In AI ethics, virtue ethics promotes designing AI systems that embody virtues such as honesty, transparency, fairness, and accountability. For example, transparent AI systems align with honesty, while fair treatment of all users embodies fairness.

Advantages of Virtue Ethics in AI

Character Development Focus: Virtue ethics emphasizes moral character development, encouraging the embodiment of virtues in AI systems. This focus on character contributes to the promotion of ethical behavior in AI.

Figure 3. Virtue ethics

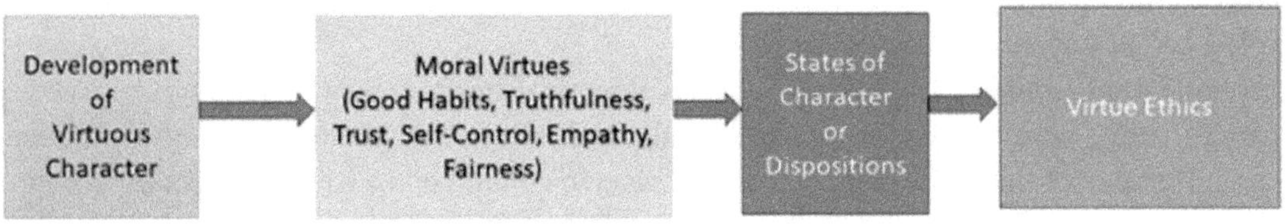

Holistic Ethical Perspective: Virtue ethics considers not only individual virtues but also the overall character and virtues that contribute to the common good. This holistic perspective encourages responsible AI behavior that aligns with broader ethical considerations.

Adaptability and Flexibility: Virtue ethics allows for adaptability and flexibility, enabling the incorporation of virtues into AI systems according to diverse stakeholder needs. This adaptability ensures that ethical considerations are contextually relevant.

Promotes Positive AI Impact: Designing AI systems with virtues such as fairness promotes a positive societal impact. It ensures that AI applications are equitable, trustworthy, and contribute to the overall well-being of society.

Disadvantages of Virtue Ethics in AI

Subjectivity and Interpretation: Virtue ethics relies on subjective interpretation, leading to potential disagreements on prioritizing and applying virtues in AI development. Different stakeholders may have varying perspectives on what virtues should be emphasized.

Lack of Clear Guidance in Conflicts: Virtue ethics does not offer clear guidance during conflicts between virtues, challenging decision-making in complex AI scenarios. Resolving conflicts between virtues may require nuanced considerations.

Potential Overlooking of Consequences: Emphasizing virtues may lead to overlooking potential consequences of AI actions. While virtues are essential, considering the broader implications and consequences of AI decisions is crucial for a comprehensive ethical evaluation.

Difficulty in Measurement and Evaluation: Measuring virtues in AI systems is challenging. Unlike more tangible metrics, virtues are abstract qualities that can be difficult to quantify and evaluate, making it hard to assess their genuine embodiment in AI systems.

In conclusion, virtue ethics in AI underscores the importance of character development and virtues. To ensure responsible applications, addressing subjectivity, conflicts, consequences, and practical virtue measurement within AI systems is vital. Striking a balance between virtues and consideration of consequences and societal impact is crucial for responsible AI development. The ongoing ethical discourse in AI should aim to navigate these challenges and refine virtue ethics principles in the context of evolving technology.

Figure 4. Right-based justice

FAIRNESS OR RIGHTS-BASED ETHICS

Rights-based ethics emphasizes upholding fundamental rights and freedoms. Applied to AI, it advocates prioritizing human rights, including privacy, freedom of expression, non-discrimination, and equal treatment. Surveillance, ensures privacy while maximizing benefits, pushing for unbiased AI designs that promote fairness and inclusivity (Reamer, 2019). See Figure 4.

Advantages of Rights-Based AI Ethics

Fundamental Rights Protection: Rights-based AI ethics places a strong emphasis on protecting fundamental human rights. This ensures that AI development and deployment align with principles that respect and uphold the rights of individuals.

Equitable Treatment and Inclusivity: This framework focuses on providing equal opportunities and treatment, reducing biases in AI systems. It promotes inclusivity across demographics, contributing to fair and just outcomes.

Preservation of Privacy and Freedom: Rights-based AI ethics safeguards privacy and freedom, particularly important in AI applications like surveillance. It seeks to balance the benefits of AI with the preservation of individual privacy and freedom.

Promotion of Fairness and Justice: Rights-based ethics aims to mitigate biases and societal inequalities, promoting fairness and just outcomes. This contributes to the creation of a more equitable and just society through AI applications.

Disadvantages of Rights-Based AI Ethics

Conflict with Other Ethical Frameworks: Balancing rights-based ethics with other ethical frameworks, such as utility or virtue ethics, can be challenging when conflicting principles arise in specific AI applications. Finding a harmonious integration may require careful consideration.

Complex Application: The application of conflicting rights or determining the extent of a right's application in AI can be complex. Resolving conflicts between rights may require nuanced and context-specific decision-making.

Figure 5. Principled artificial intelligence

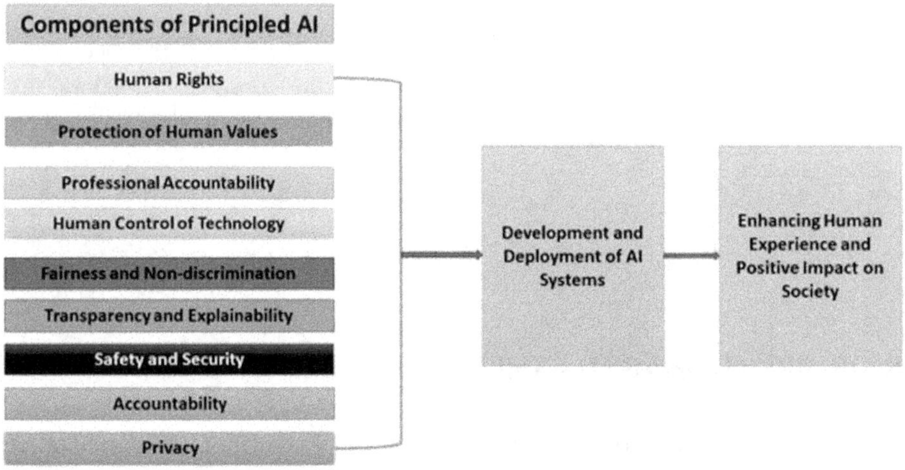

Interpretation of Rights: The interpretation of rights can vary across cultures, societies, and individuals. Diverse perspectives on applying rights-based ethics in AI may lead to challenges in creating universally accepted standards.

Technological Limitations: Current technological capabilities may limit the complete realization of respecting and upholding fundamental rights in AI. Technical constraints may hinder the full integration of rights-based principles into AI systems.

In conclusion, integrating rights-based AI ethics is crucial for human rights and fairness. Harmonizing with other ethical frameworks, addressing complexity in the application of conflicting rights, navigating diverse interpretations across cultures, and overcoming technological limitations are vital for effective integration. Balancing fundamental rights with broader ethical implications is key to ensuring responsible and ethical AI innovation that aligns with human values.

PRINCIPLED ARTIFICIAL INTELLIGENCE (PAI)

Principled Artificial Intelligence (PAI) advocates seamlessly integrating ethics into the AI life cycle, aligning AI with moral and societal values from the start. Core principles include Ethical Foundation Integration, Transparency, Accountability, and Human-Centric Design. These principles emphasize ethics integration, elucidating AI processes, ensuring traceability, and prioritizing human well-being (Fjeld et al., 2020). See Figure 5.

Advantages of Ethical Integration in AI

Ethical Integration: Ethical integration ensures that ethical principles are embedded from the start, fostering the development of responsible and morally aligned AI systems. This proactive approach contributes to ethical AI development and deployment.

Transparency: Ethical integration focuses on making AI processes and decisions interpretable, enhancing understanding among stakeholders. Transparency contributes to trust in AI systems and facilitates informed decision-making.

Accountability: Ethical integration allows for the tracing and comprehension of AI algorithms' decision-making. This addresses biases and fosters accountability, making it clear who is responsible for AI outcomes.

Human-Centric Design: Ethical integration prioritizes human well-being, safety, and societal benefits in AI technology. This human-centric design enhances the user experience and ensures that AI serves the best interests of individuals and society.

Interdisciplinary Involvement: Ethical integration encourages diverse stakeholder involvement, incorporating various perspectives and insights. This interdisciplinary approach helps identify and address ethical concerns comprehensively.

Ethical Impact Assessments: Ethical integration includes ethical impact assessments, allowing the identification and addressing of ethical concerns early in the development process. This proactive approach promotes responsible AI deployment.

Public Trust and Acceptance: By adhering to ethical standards and involving stakeholders, ethical integration helps build public trust, which is crucial for the responsible adoption of AI technologies. Trust fosters acceptance and positive reception of AI applications.

Disadvantages of Ethical Integration in AI

Implementation Challenges: Integrating ethics into AI development can be resource-intensive and challenging for organizations. Allocating the necessary resources, expertise, and time is crucial for effective ethical integration.

Complexity and Time: Ethical integration may increase development complexity and time. Balancing ethical considerations with technical requirements adds layers of complexity to the development process.

Subjectivity in Ethical Assessment: Ethical considerations can be subjective, leading to potential disagreement among stakeholders. Differing perspectives on what is ethically acceptable may result in delays or conflicts during the development process.

Balancing Ethical Goals: Balancing conflicting ethical goals during AI development can be challenging. Different ethical principles may sometimes compete, requiring careful consideration and decision-making.

Constraints on Innovation: Overemphasis on ethics may inadvertently stifle innovation in AI technologies. Striking a balance between ethical considerations and innovation is essential for the continued advancement of AI.

Resource Intensiveness: Involving diverse stakeholders and conducting ethical assessments may require substantial resources. Ensuring comprehensive input from various perspectives demands time and effort.

In conclusion, ethical integration in AI offers notable gains in ethics, transparency, accountability, and public trust. However, organizations must navigate challenges such as resource intensiveness, complexity, subjectivity, and potential impacts on innovation. The effective integration of ethical principles into AI systems requires careful consideration and a commitment to balancing ethical goals with practical development requirements.

Figure 6. The principle of double effect

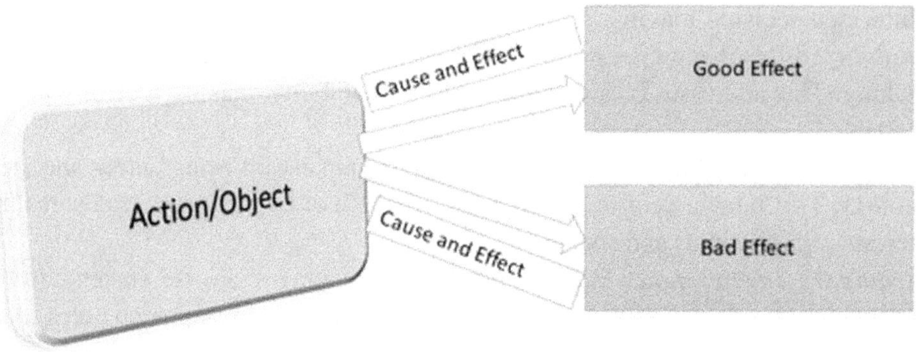

THE PRINCIPLE OF DOUBLE EFFECT

The Principle of Double Effect (PDE), rooted in ancient philosophy and refined by modern ethicists like Philippa Foot and Joseph M. Boyle Jr., evaluates actions with dual intended outcomes. It has applications in ethics, law, medicine, and increasingly in AI ethics. In AI ethics, PDE helps distinguish intended positive outcomes (e.g., efficiency, better diagnostics) from foreseeable negative consequences (e.g., biased decision-making, job displacement). This differentiation aids in assessing proportionality and necessity to achieve the desired goal (Moreland, 2012, Bentzen, 2016). Refer to Figure 6.

Advantages of the Principle of Double Effect (PDE) Model

Ethical Analysis Framework: PDE provides a structured ethical analysis framework for evaluating actions with dual effects, facilitating a systematic assessment of the morality of AI actions. This framework aids in organizing ethical considerations and promoting thorough analysis.

Balanced Decision-Making: PDE encourages a balanced consideration of both intended positive outcomes and potential negative consequences. This promotes thoughtful and balanced ethical deliberations in AI development and deployment.

Intention Focus: PDE underscores the significance of analyzing the intention behind actions. In the context of AI, this prompts a critical evaluation of AI systems' goals and motives, ensuring that the intentions align with ethical principles.

Alignment with Ethical Theories: Applying ethical theories within the PDE framework allows for a comprehensive analysis of the moral implications of AI actions. This aids in a thorough ethical assessment, considering various ethical perspectives and principles.

Clarity and Differentiation: PDE helps in clearly defining and differentiating between intended positive outcomes and foreseeable negative consequences of AI actions. This enhances decision-making clarity, making it easier to identify and assess the ethical implications of actions.

Holistic Assessment: PDE enables a holistic evaluation of AI actions by considering their intended goals, potential benefits, and risks. This holistic approach leads to more informed ethical judgments that take into account the broader context of AI applications.

Adaptability to Various Domains: The flexibility of PDE makes it adaptable across different domains, including law, medicine, finance, and AI. This adaptability allows for a consistent ethical evaluation approach across diverse fields.

Disadvantages of the Double Effect Model

Complexity and Interpretation: PDE can be complex to interpret and apply, especially in the rapidly evolving field of AI. This complexity may lead to misinterpretations or disagreements on the ethical assessment of AI actions.

Subjectivity and Value Differences: Evaluating the proportionality and necessity of AI actions can be subjective, with differing interpretations of an acceptable balance between intended and unintended effects. Value differences among stakeholders may lead to ethical disagreements.

Difficulty in Predicting Consequences: Accurately predicting all possible consequences of an AI action can be challenging. This makes the effective application of PDE in practice and foreseeing all potential negative effects difficult.

Time-Consuming Analysis: Applying PDE, especially in complex AI systems, may be time-consuming. This can potentially delay decision-making processes and hinder the timely development and deployment of AI technologies.

Resource Intensiveness: Comprehensive application of PDE may require significant resources, including expert input and computational capabilities. This resource intensiveness can be a limitation for some organizations.

Incorporating Unintended Consequences: Fully accounting for all unintended consequences, especially those that may only become apparent after AI system deployment, can be challenging. This poses a threat to the effectiveness of PDE frameworks in addressing unforeseen ethical issues.

In conclusion, while PDE provides a crucial ethical lens for evaluating the dual impacts of AI actions, it poses challenges such as complexity, subjectivity, prediction difficulty, resource needs, and adapting to evolving tech landscapes. Thoughtful consideration and critical evaluation are essential when using the PDE model in the ethical analysis of AI systems.

MACHINE LEARNING FAIRNESS MODELS

Machine Learning Fairness Models play a critical role in addressing bias and promoting equity in AI decision-making. They employ various techniques and algorithms, like fairness indicators and fairness-aware algorithms, to rectify biases within AI systems. The primary objective is to integrate fairness metrics and constraints into AI system design to ensure sensitivity to and mitigation of biases, ultimately resulting in equitable and just AI outcomes. These models highlight the significance of fairness indicators and fairness-aware algorithms in promoting equitable AI decision-making. Fairness indicators provide a measurable assessment of fairness in AI system outcomes across different demographic or sensitive attribute groups, allowing a quantitative evaluation of biases and areas for improvement. Fairness-aware algorithms are purpose-built to consider fairness throughout model training and decision-making processes (Selbst et al., 2019). See Figure 7.

Figure 7. Machine learning fairness model

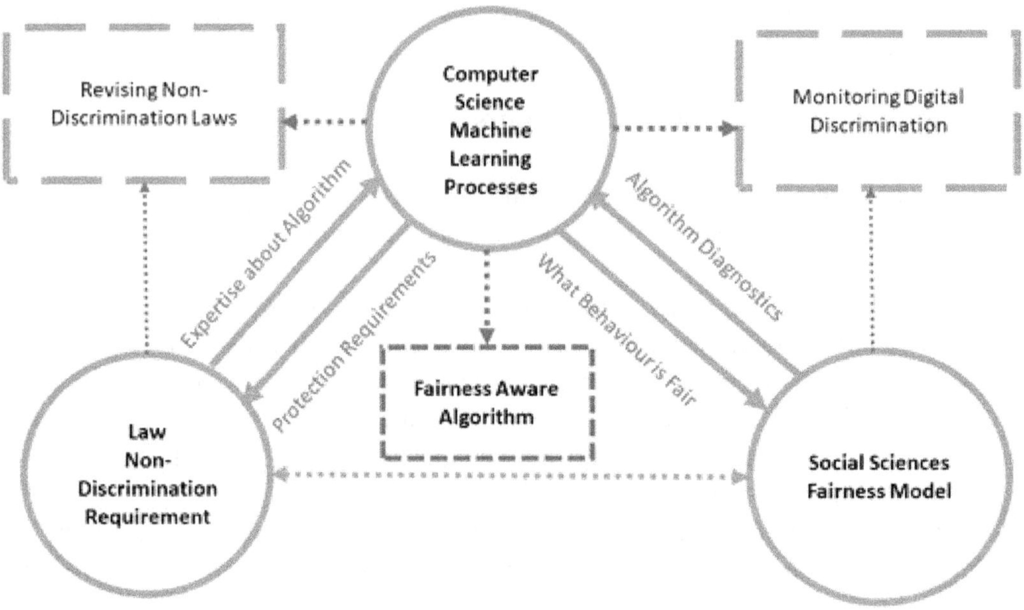

Advantages of ML Fairness Models

Bias Mitigation: ML fairness models effectively mitigate biases in AI systems by incorporating fairness metrics and constraints during design. This promotes more equitable decision-making and reduces the impact of biased algorithms.

Promoting Equity and Justice: By reducing biases, ML fairness models promote equity and justice, especially in domains where biased decisions can have significant societal impacts. This contributes to a fairer distribution of resources, opportunities, and benefits.

Quantifiable Fairness Assessment: Fairness indicators provide a measurable, quantitative assessment of AI system fairness across various demographic or sensitive attribute groups. This aids in improvement and progress measurement, allowing for ongoing refinement.

Transparent Decision-Making: The inclusion of fairness indicators and fairness-aware algorithms enhances transparency in AI decision-making. This fosters trust and accountability, as stakeholders can understand and assess the fairness of the AI system.

Alignment with Ethical Principles: ML fairness models align with ethical principles of fairness and non-discrimination. They contribute to the development of AI systems that treat all individuals fairly, irrespective of their characteristics.

Social Acceptance and Adoption: Addressing bias and promoting equity through fairness models enhances societal acceptance and adoption of AI technologies. Building trust in AI systems encourages widespread usage across various sectors.

Disadvantages of ML Fairness Models

Challenges in Fairness Definition: Defining fairness universally is challenging due to varying stakeholder perspectives. Different individuals and communities may have different notions of fairness, making it difficult to design models that satisfy all definitions.

Algorithmic Trade-Offs: Incorporating fairness constraints can result in trade-offs with other model performance aspects, such as accuracy. Achieving a delicate balance between fairness and accuracy is challenging, and optimizing for one may compromise the other.

Data Quality and Bias: Fairness models heavily rely on the quality and representativeness of training data. Biased data may perpetuate or reinforce biases inadvertently, impacting the effectiveness of fairness interventions.

Computational Complexity: Implementing fairness-aware algorithms may introduce additional computational complexity. This can potentially limit real-time applications and require significant computational resources.

Overfitting to Fairness Constraints: Fairness-aware algorithms might overfit to fairness constraints, impacting system robustness and effectiveness. Overemphasis on fairness may compromise the system's ability to adapt to diverse and dynamic environments.

Socio-Technical Challenges: Addressing fairness is a socio-technical challenge involving the navigation of complex societal issues. Understanding the broader fairness implications in AI requires consideration of historical and cultural contexts, adding complexity to the technical aspects.

In conclusion, ML fairness models, while beneficial for bias mitigation and promoting equity, present challenges including defining fairness, algorithmic trade-offs, data quality, computational complexity, and socio-technical considerations. Overcoming these challenges is crucial to maximize the potential of fairness models in AI systems and to ensure that AI technologies are developed and deployed ethically and responsibly.

EXPLAINABILITY AND TRANSPARENCY MODELS

These models improve AI interpretability, addressing the 'black-box' nature of many ML models. Understanding AI decision-making is crucial for ethics, aligning AI with ethical norms and societal expectations. Examples include Local Interpretable Model-agnostic Explanations (LIME) and Shapley Additive Explanations (SHAP). LIME provides a local understanding of AI predictions, approximating the model's behavior near a specific instance, and simplifying the model while preserving interpretability. This helps users comprehend prediction rationale, revealing how input features influence outcomes (Ribeiro et al., 2016).

Conversely, SHAP, based on cooperative game theory, computes 'Shapley values' to allocate each feature's contribution to a prediction. It offers a more comprehensive and theoretically grounded understanding of feature importance and contribution to the prediction process, attributing outcomes to specific features and shedding light on their role and impact on AI decision-making (Lundberg & Lee, 2017). See Figure 8.

Utilizing these models enhances understanding of AI, building trust and confidence. Furthermore, it can result in improved transparency, aligning with ethics and societal expectations—critical for the responsible deployment of AI and the establishment of an ethical framework in AI applications.

Figure 8. Explainability and transparency model

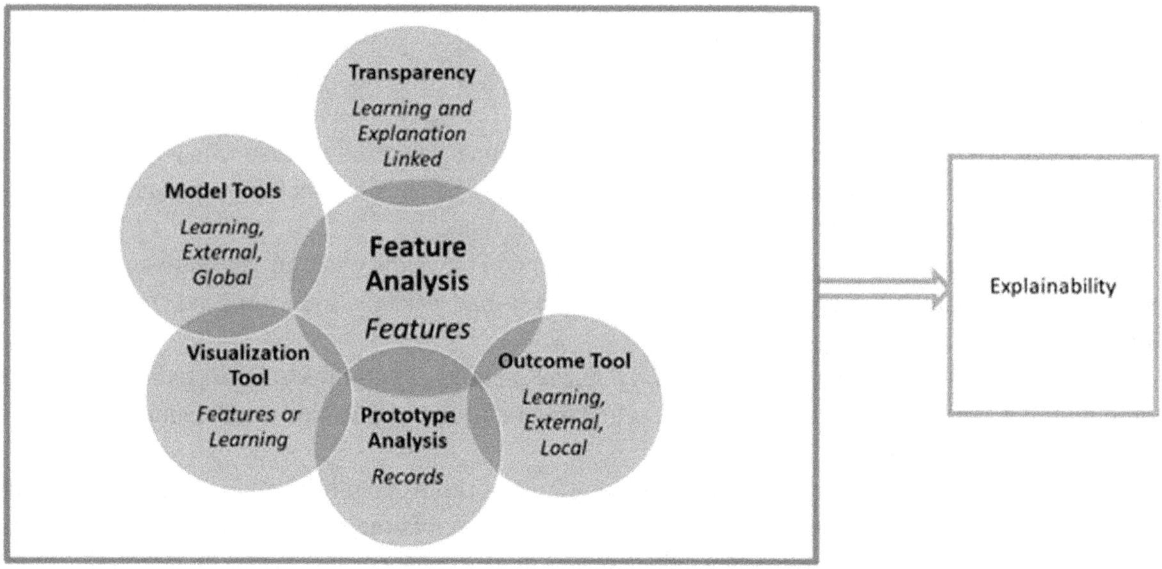

Advantages of Explainability Models in AI

Enhanced Transparency: Explainability models address the "black-box" nature of AI, offering insights into AI decisions. This enhanced transparency allows stakeholders to understand the inner workings of the model, making AI systems more accountable.

Facilitates Ethical Decision-Making: Understanding AI decision processes is vital for ethical decision-making. Explainability models ensure that decisions comply with ethical norms and societal expectations, fostering responsible AI practices.

Local Understanding of Predictions: Techniques like Local Interpretable Model-agnostic Explanations (LIME) offer local interpretations, aiding users in understanding why a specific prediction was made for a particular instance. This provides context-specific insights.

Feature Importance Insights: Shapley Additive explanations (SHAP) provides comprehensive insights into feature importance and contribution to predictions. This allows stakeholders to understand which features influence model outcomes the most.

Builds Trust and Confidence: By providing explanations and insights, explainability models foster trust and confidence in AI systems. Users and stakeholders can have a clearer understanding of how and why AI systems make specific decisions.

Responsible AI Deployment: The transparency provided by explainability models supports responsible AI deployment. It aligns with ethical frameworks and promotes responsible usage, ensuring that AI systems operate within ethical boundaries.

Alignment with Ethical Frameworks: Explainability aligns AI actions with ethical frameworks, ensuring that AI systems adhere to ethical principles. This alignment is crucial for maintaining societal trust and meeting ethical expectations.

Disadvantages of Explainability Models in AI

Simplification and Accuracy Trade-off: Some models simplify AI models for better understanding, potentially trading off accuracy for explainability. Achieving a balance between accuracy and interpretability is a challenge.

Complexity of Interpretations: Interpretations can be complex, hindering effective use and understanding, especially for non-experts. This complexity may limit the practical utility of explanations for a broader audience.

Dependence on Model Type: The effectiveness of explainability models varies based on the AI model used. Certain models may not be well-suited for specific interpretation techniques, limiting the applicability of explainability.

Interpretability-Performance Trade-offs: Balancing interpretability and high predictive performance is crucial for effective model deployment. Some highly complex models may sacrifice interpretability, while simpler models may not achieve the desired predictive accuracy.

Inherent Bias in Interpretations: Models may inadvertently introduce biases in explanations, influencing fairness perceptions and contributing to decision-making biases. Careful consideration is needed to avoid reinforcing or introducing biases through interpretability models.

In conclusion, explainability models significantly contribute to transparency, ethical decision-making, and trust in AI systems. However, careful consideration of trade-offs, complexity, model dependence, and potential biases is necessary to maximize their benefits in AI. Striking a balance between accuracy and interpretability is essential for effective and responsible use of explainability models in diverse AI applications.

CONCEPTUAL MODEL

The literature underscores the importance of ethical decision-making in AI. Our comprehensive review highlights diverse theoretical frameworks and studies essential for ethical AI decision-making. These frameworks prioritize fairness, transparency, explainability, trust, and societal benefit in AI model development and deployment. Addressing fraud and bias, they integrate principles such as virtue ethics, right-based justice, and moral worth. Ethical decision-making in AI is complex, necessitating ongoing improvement to mitigate framework disadvantages in AI design, development, and deployment. Our proposed model aims to enhance credibility and trust in AI systems during decision-making (See Figure 9).

The comprehensive conceptual framework guides responsible AI development, integrating diverse ethical perspectives to prioritize societal welfare, ethical adherence, transparency, fairness, and individual rights. Responsible AI aims to enhance human well being while upholding ethical values, and fostering an equitable and inclusive future.

To ensure fairness, equity, and responsible decision-making in AI, bias mitigation is crucial. Strategies encompass inclusive data collection, fairness-aware algorithms, transparent AI, Human-in-the-Loop approaches, ethical impact assessments, continuous monitoring, stakeholder engagement, and bias detection tools. These strategies, informed by various ethical perspectives, promote the development and deployment of fair and responsible AI systems aligned with societal values and ethics.

The proposed conceptual framework draws theoretical support from key ethical theories and governance perspectives in the realm of AI ethics. It combines elements from utilitarianism, deontology, transparency,

Figure 9. Conceptual model of ethical frameworks in developing and deploying responsible AI systems

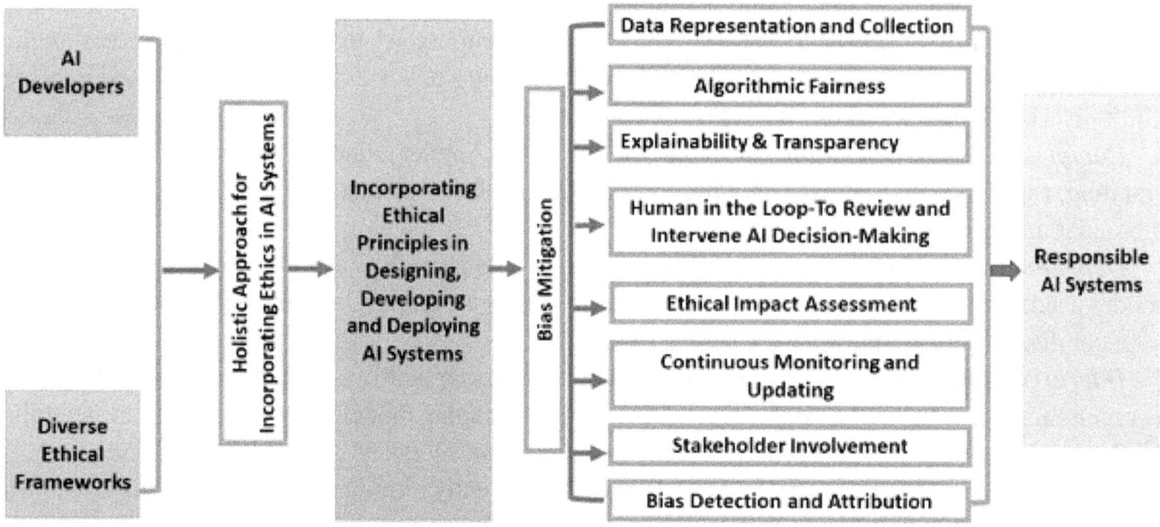

and accountability theories, fairness and justice theories, virtue ethics, and ethical pluralism to guide the development of fair, transparent, and responsible Ai systems aligned with societal values and ethics.

REVIEW METHODOLOGY

Study Selection and Screening

To identify relevant literature, we initially utilized major databases including Scopus, Web of Science, and IEEE Xplore. Subsequently, we complemented this search with Google Scholar and other data repositories (refer to section 3.2).

Selection and Inclusion of Studies

In our second step, we established search terms and criteria for inclusion and exclusion (as outlined in Table 1). These searches were limited to journal articles, conference proceedings, book chapters, published books, and policy documents in the English language from the last decade. Databases were limited to related research disciplines for relevance (Philosophy and Ethics, Law and Policy, Sociology, and Anthropology, Psychology and Cognitive Science, Human-Computer Interaction (HCI), Engineering and Technology Ethics, Bioethics, Communication Studies, Interdisciplinary Studies, Management, Business and Accounting). Applying this strategy yielded 217 search results, which were refined to 110 unique, relevant studies (research papers, published books, book chapters, and conference proceedings) after removing duplicates.

Table 1. Search terms for the literature review

Databases	Search Terms
Scopus, IEEE Xplore, Web of Science, Springer Link, Google Scholar and Other Sources	"Artificial Intelligence", "Machine Learning", "Ethical decision-making", "Transparency", "Responsible AI", "Societal Impact", "Regulatory Frameworks", Natural Language Processing", "Ethics and AI", "Ethical Frameworks for AI", "Ethical Decision Making in Digital Era", "AI Ethics Theoretical Frameworks", "AI and Ethics Decision Support System", "AI and Moral Reasoning".

Evaluating Relevance and Quality of Studies

In the third step of our methodology, we evaluated the relevance and quality of 90 refined studies extracted after removing duplicate records, based on predefined inclusion and exclusion criteria (Tranfield, Denyer & Smart, 2003; Grant & Booth, 2009). Inclusions encompassed articles directly linked to identified themes. The screening process involved evaluating titles and abstracts to identify relevant articles. Defined criteria were applied to filter out irrelevant studies, resulting in 43 records being excluded. A thorough examination of the remaining 67 articles' full texts for alignment with research objectives led to 50 articles for a comprehensive review of the present study after eliminating 17 records due to quality issues (See Fig 10).

Data Mining

For data extraction, we employed a spreadsheet to meticulously record metadata for each study selected in our systematic literature review. Table 2 illustrates the collected metadata for the 31 chosen studies. These categories have been thoroughly derived from the literature review to ensure cohesion and clarity.

Figure 10. Selection of the studies, assessment, and inclusion (presented using PRISMA flow diagram)

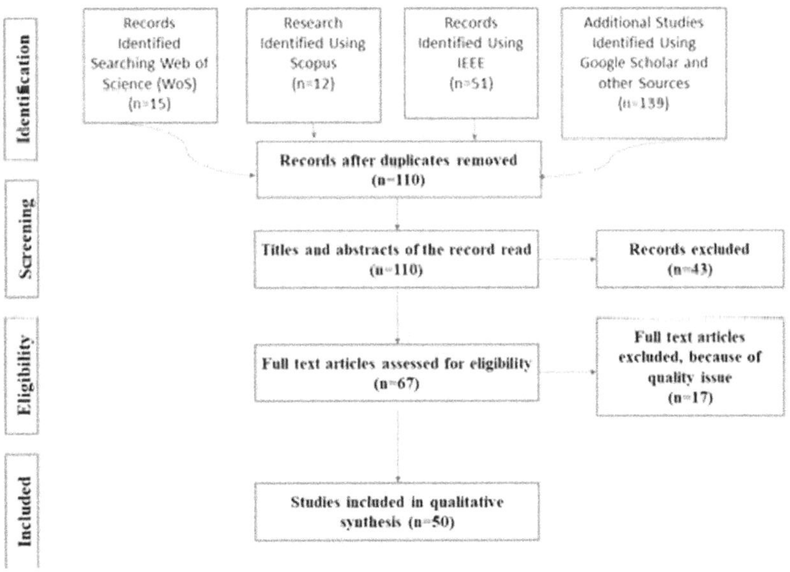

Data Synthesis and Analysis

We employed structured data extraction and meticulous qualitative analysis to systematically capture, analyze, and synthesize essential information from the literature. Themes were identified and findings interpreted, sparking discussions on AI integration implications. Transparency, value alignment, bias mitigation, governance, and practical applications were central in these discussions.

Table 2. Overview of information collected about each of the selected articles

Sr No	Study Title	Authors	Year	Source	Study Objective
1	Judgment Call the Game: Using Value Sensitive Design and Design Fiction to Surface Ethical Concerns Related to Technology	Ballard, S., Chappell, K. M., & Kennedy, K.	2019	In *Proceedings of the 2019 on Designing Interactive Systems Conference*	The study aims to address AI's complexity, recognizing its potential and societal harm, and promote responsible AI through collaboration with corporations, non-profits, and academic researchers.
2	Fairness and Abstraction in Sociotechnical Systems	Selbst, A. D., Boyd, D., Friedler, S. A., Venkatasubramanian, S., & Vertesi, J.	2019	In Proceedings of the conference on fairness, accountability, and transparency	The objective is to evaluate the drawbacks of using computer science fundamentals (abstraction, modular design) to define and implement fairness in machine learning within societal contexts.
3	Think Your Artificial Intelligence Software Is Fair? Think Again	Bellamy, R. K., Dey, K., Hind, M., Hoffman, S. C., Houde, S., Kannan, K., & Zhang, Y.	2019	*IEEE Software*	The study aims to debunk the misconception that machine-learning software is inherently unbiased and always leads to fair decisions. It underscores that machine-learning software can possess biases, akin to or distinct from human biases.
4	Does AI Raise Any Distinctive Ethical Questions?	Boddington, P., & Boddington	2017	Book (Towards a code of ethics for artificial intelligence)	The chapter explores AI's ethical dimensions in relation to advancing technologies, underlining the necessity of integrating AI ethics with broader ethical concerns. It covers human understanding, societal integration, economic impact, cultural implications, predictive challenges, and data analysis.
5	Aligning artificial intelligence with human values: reflections from a phenomenological perspective	Han, S., Kelly, E., Nikou, S., & Svee, E. O	2021	AI & SOCIETY	The aim of the study is to address the critical concern of ensuring that Artificial Intelligence (AI) is directed towards humane objectives, emphasizing AI value alignment with human values throughout its design and use.
6	Algorithmic Bias in Autonomous Systems	Danks, D., & London, A. J.	2017	IJCAI	The paper aims to provide a taxonomy that categorizes different types and sources of algorithmic bias, emphasizing their impacts on the proper functioning of autonomous systems.
7	Accountability in algorithmic decision making	Diakopoulos, N.	2016	Communications of the ACM	The paper aims to provide insights into the challenges and considerations surrounding accountability when automated algorithms are involved in decision-making processes.
8	Responsible Artificial Intelligence – from Principles to Practice	Dignum, V.	2022	arXiv preprint arXiv	The goal is to underscore that responsible AI goes beyond reliability—it includes design, purpose, and inclusivity. The ultimate aim is to create comprehensive methods and tools supporting AI practitioners and facilitating involvement from all stakeholders, ensuring AI systems align with societal principles and values.
9	Eight grand challenges for value sensitive design from the 2016 Lorentz workshop	Friedman, B., Harbers, M., Hendry, D. G., van den Hoven, J., Jonker, C., & Logler, N.	2021	Ethics and Information Technology	The study aims to outline and delve into eight crucial challenges in value sensitive design, presenting key questions for each challenge. These challenges encompass a wide array of issues, from power dynamics to value tensions, necessitating dedicated research and design efforts in the field.
10	A Value Sensitive Design Perspective on AI Biases	Gan, I., & Moussawi, S.	2022	Proceedings of the 55th Hawaii International Conference on System Sciences	It aims to classify AI-related biases into distinct categories and propose effective strategies to mitigate these biases.

continued on following page

Table 2. Continued

Sr No	Study Title	Authors	Year	Source	Study Objective
11	Datasheets for datasets	Gebru, T., Morgenstern, J., Vecchione, B., Vaughan, J. W., Wallach, H., Iii, H. D., & Crawford, K.	2021	Communications of the ACM	Datasheets for datasets aim to serve dataset creators and consumers. Creators are encouraged to reflect on dataset creation processes and implications, while consumers gain information to make informed usage decisions and prevent unintentional misuse.
12	European Union Regulations on Algorithmic Decision-Making and a "Right to Explanation"	Goodman, B., & Flaxman, S.	2017	AI magazine	The article aims to analyze the potential impact of the EU's General Data Protection Regulation on machine learning algorithms. It focuses on the restrictions it imposes on automated decision-making and the introduction of a "right to explanation" for users affected by algorithmic decisions. The objective is to highlight challenges for the industry and opportunities for computer scientists to design algorithms and frameworks that ensure fairness and enable explanation.
13	Explainable artificial intelligence (xai)	Gunning, D	2017	Defense advanced research projects agency (DARPA)	XAI aims to develop ML techniques that produce understandable models without compromising performance. It strives to ensure users can comprehend, trust, and manage the new AI systems effectively.
14	The global landscape of AI ethics guidelines	Jobin, A., Ienca, M., & Vayena, E.	2019	Nature machine intelligence	The study aims to analyze recent ethical AI principles and guidelines from various entities to understand if a global consensus on 'ethical AI' is emerging and to map the necessary ethical requirements, technical standards, and best practices for its realization.
15	Siri, Siri, in my hand: Who's the fairest in the land? On the interpretations, illustrations, and implications of artificial intelligence	Kaplan, A., & Haenlein, M.	2019	Business horizons	The objective of this article is to analyze the distinctions between AI and related concepts like the Internet of Things and big data. It emphasizes that AI should not be seen as a monolithic term but understood in a nuanced manner.
16	Algorithmic bias: review, synthesis, and future research directions	Kordzadeh, N., & Ghasemaghaei, M.	2022	European Journal of Information Systems	The objective is to review, summarize, and synthesize current literature on algorithmic bias and provide recommendations for future research in information systems.
17	New directions in eHealth communication: opportunities and challenges	Kreps, G. L., & Neuhauser, L.	2013	Patient education and counseling	The article reviews communication issues in designing effective eHealth applications, guiding strategic development and implementation of health technologies.
18	The mythos of model interpretability: In machine learning, the concept of interpretability is both important and slippery	Lipton, Z. C.	2018	Queue	The objective of this article is to analyze and refine the discourse on interpretability in machine learning.
19	Ethical considerations about the implications of artificial intelligence in finance	Max, R., Kriebitz, A., & Von Websky, C.	2021	Handbook on Ethics in Finance	This chapter outlines AI's ethical impact on finance, encompassing responsibility, legal accountability, and broader ethical concerns.
20	The ethics of algorithms: Mapping the debate	Mittelstadt, B. D., Allo, P., Taddeo, M., Wachter, S., & Floridi, L.	2016	Big Data & Society	The primary goal is to elucidate the ethical significance of algorithmic mediation.
21	Do the ends justify the means? Variation in the distributive and procedural fairness of machine learning algorithms	Morse, L., Teodorescu, M. H. M., Awwad, Y., & Kane, G. C.	2021	Journal of Business Ethics	This paper aims to gain insight into fairness perceptions regarding five algorithmic criteria, specifically focusing on distributive fairness and procedural fairness dimensions.
22	Human values as the basis for sustainable information system design	Winkler, T., & Spiekermann, S.	2019	IEEE Technology and Society Magazine	The article contends that companies should reassess this perspective, highlighting that both sustainability and a business model are fundamentally grounded in human values.

continued on following page

Table 2. Continued

Sr No	Study Title	Authors	Year	Source	Study Objective
23	Dissecting racial bias in an algorithm used to manage the health of populations	Obermeyer, Z., Powers, B., Vogeli, C., & Mullainathan, S.	2019	Science	The article aims to dissect racial bias within an algorithm employed for population health management.
24	Artificial intelligence and data science governance: Roles and responsibilities at the c-level and the board	Thuraisingham, B.	2020	IEEE	The objective is to explore corporate governance and the roles of corporate officers and the board of directors, particularly emphasizing the heightened interest following the Enron scandal in the early 2000s.
25	Machine learning and deep learning: Open issues and future research directions for the next 10 years.	Pramod, A., Naicker, H. S., & Tyagi, A. K.	2021	Computational Analysis and Deep Learning for Medical Care	This work aims to comprehensively cover ML and deep learning, spanning their evolution, widespread applications, societal benefits, and associated challenges and limitations.
26	Stop explaining black box machine learning models for high stakes decisions and use interpretable models instead	Rudin, C.	2019	Nature Machine Intelligence	The objective is to stress the importance of creating interpretable machine learning models initially, rather than relying on explanations for opaque models, to mitigate issues and prevent potential harm to society.
27	Designing AI for explainability and verifiability: a value sensitive design approach to avoid artificial stupidity in autonomous vehicles	Umbrello, S., & Yampolskiy, R. V.	2022	International Journal of Social Robotics	The objective is to explore designing decision matrix algorithms utilizing the belief-desire-intention model for autonomous vehicles, aiming to minimize the risks associated with opaque architectures.
28	Humans forget, machines remember: Artificial intelligence and the right to be forgotten	Villaronga, E. F., Kieseberg, P., & Li, T.	2018	Computer Law & Security Review	The article aims to explore AI memory and the Right to Be Forgotten, assessing their connection within the legal framework. It particularly focuses on understanding the interplay between privacy and transparency in current E.U. privacy law.
29	Ethical Considerations of Using ChatGPT in Health Care	Wang, C., Liu, S., Yang, H., Guo, J., Wu, Y., & Liu, J.	2023	Journal of Medical Internet Research	The objective of this article is to address the potential ethical challenges associated with utilizing ChatGPT in healthcare.
30	A survey on measuring indirect discrimination in machine learning	Zliobaite	2015	ArXiv preprint arXiv	The objective is to review and organize discrimination measures used for assessing discrimination in data and discrimination-aware predictive models.
31	Diversity and Inclusion in Artificial Intelligence	Zowghi, D., & da Rimini, F.	2023	arXiv preprint arXiv	The objective is to examine the interaction between humans and technology within the larger societal context, emphasizing a socio-technological approach over purely technical or human factors.
32	Ethical decision making for autonomous vehicles.	De Moura, N., Chatila, R., Evans, K., Chauvier, S., & Dogan, E.	2020	IEEE	The objective is to address ethical dilemma situations that may arise during autonomous driving.
33	AI in Finanee: A Review	Cao, L	2020	SSRN	To conduct a comprehensive review and future direction for AI application in FinTech sector in Finance.
34	Artificial Intelligence (AI): Multidisciplinary perspectives on emerging challenges, opportunities, and agenda for research, practice and policy.	Dwivedi, Y. K., Hughes, L., Ismagilova, E., Aarts, G., Coombs, C., Crick, T., ... & Williams, M. D.	2021	ScienceDirect (Elsevier)	The study was focused on drawing on expert insights, examines the significant opportunities, assesses impact, identifies challenges, and proposes a research agenda stemming from the rapid integration of AI in diverse domains.
35	Standards for Financial Decision-Making: Legal, Ethical, and Practical Issues.	Fleming, R. B., & Morgan, R. C.	2012	HeinOnline	This article reviews the established standards governing guardians of the estate in their decision-making processes concerning financial matters.
	AI human impact: toward a model for ethical investing in AI-intensive companies.	Brusseau, J.	2023	Taylor & Francis Online	This paper asserts that traditional ESG (Environmental, social and governance) frameworks fall short in addressing companies at the forefront of AI integration. It argues that effectively assessing these entities, heavily reliant on contemporary big data, predictive analytics, and machine learning, demands specialized metrics derived from established AI ethics principles.

continued on following page

Table 2. Continued

Sr No	Study Title	Authors	Year	Source	Study Objective
37	Governing artificial intelligence: ethical, legal and technical opportunities and challenges.	Cath, C.	2018	The Royal Society Publishing	This paper conducts a comprehensive examination of the ethical, legal-regulatory, and technical challenges inherent in crafting governance regimes for AI Systems.
38	Industry 4.0 in finance: the impact of artificial intelligence (ai) on digital financial inclusion.	Mhlanga, D.	2020	MDPI	The study aims to explore the influence of AI on digital financial inclusion.
39	Establishing a social licence for Financial Technology: Reflections on the role of the private sector in pursuing ethical data practices.	Aitken, M., Toreini, E., Carmichael, P., Coopamootoo, K., Elliott, K., & van Moorsel, A.	2020	Sage	The article leverages existing literature on the Social License in extractive industries to examine potential approaches for establishing a Social License in emerging data-intensive industries. It also draws on established trust literature from psychology and organizational science to assess the relevance and trustworthiness in evolving practices within data-intensive industries.
40	Banking on AI: mandating a proactive approach to AI regulation in the financial sector.	Truby, J., Brown, R., & Dahdal, A.	2020	Taylor & Francis Online	This article contends that fostering a sustainable future for AI innovation in the financial sector requires advocating a proactive regulatory approach, pre-empting potential harm.
41	Regulating artificial intelligence in finance: Putting the human in the loop.	Buckley, R. P., Zetzsche, D. A., Arner, D. W., & Tang, B. W.	2021	Informit	This article proposes a framework to comprehend and tackle the growing influence of AI in Finance, with a central emphasis on human responsibility to address the AI "Black Box" problem.
42	Ethical issues in focus by the autonomous vehicles industry. Transport reviews, 41(5), 556-577.	Martinho, A., Herber, N., Kroesen, M., & Chorus, C.	2021	Taylor & Francis Online	The study delves into the ethical issues central to the autonomous vehicle industry.
43	Artificial Intelligence: An Asset for the Financial Sector.	Bandi, S., & Kothari, A. (2022).	2022	Wiley Online Library	The article concentrates on elucidating a paradigm shift within the finance sector through the application of AI, specifically examining its impact on insurance, stock exchanges, and mutual funds.
44	A Review of Practical AI for Remote Sensing in Earth Sciences. Remote Sensing, 15(16), 4112.	This	2023	MDPI	The article aims to synthesize the current literature on AI applications in remote sensing, providing a comprehensive analysis of methodologies, outcomes, and limitations in the field.
45	Integrating convolutional neural network and multiresolution segmentation for land cover and land use mapping using satellite imagery.	Atik, S. O., & Ipbuker, C.	2021	MDPI	This study aim to conduct land cover and land use mapping utilizing the proposed CNN-MRS model.
46	AI-Forest Watch: semantic segmentation based end-to-end framework for forest estimation and change detection using multi-spectral remote sensing imagery.	Zulfiqar, A., Ghaffar, M. M., Shahzad, M., Weis, C., Malik, M. I., Shafait, F., & Wehn, N.	2021	SPIE Digital Library	This study introduces a novel approach employing deep convolutional neural network-based semantic segmentation for processing multi-spectral space-borne images.
47	Advances in image acquisition and processing technologies transforming animal ecological studies. Ecological Informatics, 61, 101212.	Nazir, S., & Kaleem, M.	2021	Elsevier	The article reviews the progress in image acquisition and processing technologies applied in animal ecological studies.
48	Machine learning and artificial intelligence to aid climate change research and preparedness.	Huntingford, C., Jeffers, E. S., Bonsall, M. B., Christensen, H. M., Lees, T., & Yang, H.	2019	IOP Science	The study focuses on concurrently leveraging ML and AI to enhance understanding and capitalize more effectively on existing data and simulations.

continued on following page

Table 2. Continued

Sr No	Study Title	Authors	Year	Source	Study Objective
49	Can we detect trends in natural disaster management with artificial intelligence? A review of modelling practices.	Tan, L., Guo, J., Mohanarajah, S., & Zhou, K.	2021	Springer Link	This paper conducts a systematic review of the application of AI models across various stages of Natural Disaster Management (NDM).
50	Contributions and risks of artificial intelligence (AI) in building smarter cities: Insights from a systematic review of the literature.	Yigitcanlar, T., Desouza, K. C., Butler, L., & Roozkhosh, F.	2020	MDPI	This paper aims to provide insights into the contribution of AI to the development of smarter cities.

RESULTS AND DISCUSSION

Discussion

In this section, we have summarized the key findings from the existing literature on the problem under investigation. The results and discussion delve into the multidimensional aspects of integrating artificial intelligence (AI) into ethical decision-making. The key themes explored include transparency and explainability, value alignment and ethical AI design, bias mitigation and fairness, governance and policy frameworks, and case studies/practical applications. Additionally, theoretical, and conceptual frameworks highlighting various ethical perspectives in AI are discussed.

The findings of our systematic review stress that transparency and explainability are crucial in AI decision-making, especially in critical sectors like healthcare and criminal justice, fostering trust and accountability. Providing meaningful explanations for AI decisions is vital (Lipton, 2018). Aligning AI with human values and ethics is essential for responsible deployment (Winkler & Spiekermann, 2019), minimizing conflicts and ethical dilemmas. Integrating diverse cultural values into AI frameworks promotes inclusivity and sensitivity to various contexts (Kaplan & Haenlein, 2019). These pillars are integral for ethically integrating AI across domains and global communities.

Our literature review emphasizes addressing biases and ensuring fairness in AI decision-making, forming the foundation for ethical AI integration. Evolving strategies like fairness-aware algorithms are crucial in reducing biases and promoting equitable outcomes in AI systems (Selbst et al., 2019; Zliobaite, 2015). Additionally, governance and policy frameworks are vital for responsible AI development and deployment, safeguarding individual rights and privacy. Multidisciplinary collaboration involving ethicists, social scientists, technologists, and policymakers is key to comprehensively addressing AI's ethical challenges (Gunning & Aha, 2019; Goodman & Flaxman, 2017).

Insights gained from real-world case studies highlight the diverse applications of AI in ethical decision-making across healthcare, criminal justice, and finance. AI enhances decision-making while upholding ethical principles, aiding in disease diagnosis, optimizing resource allocation, and improving risk assessment. Continuous improvements and vigilance are essential to align AI with ethical standards, necessitating ongoing research, regulation, and collaboration for responsible integration (Obermeyer et al., 2019; Chouldechova, 2017).

The review highlights the value of various ethical frameworks, such as utilitarianism, deontology, virtue ethics, and rights-based ethics, in guiding AI ethics. Each framework offers unique advantages and challenges, emphasizing the need to balance these perspectives in the dynamic AI field. Utilitarianism aims at maximizing societal welfare, guiding AI development towards the greatest benefit for the majority. Deontology focuses on rules and duties, setting clear guidelines and restrictions in AI applications. Virtue ethics prompts considering the character and virtues of AI developers and users, fostering responsible AI practices. Rights-based ethics emphasizes individual rights, ensuring AI respects fundamental human rights in design and deployment (Russel, 2022).

The proposed conceptual model underscores the critical need to mitigate biases while integrating these diverse ethical perspectives into AI. Strategies like careful data representation, algorithmic fairness measures, transparent decision-making processes, and active stakeholder involvement are essential. Ethical impact assessments and continuous monitoring are crucial practices to ensure unbiased AI systems aligned with ethical principles. This approach enhances trust, accountability, and responsible development and deployment of AI technologies.

CONCLUSION

In conclusion, the integration of AI into ethical decision-making signifies a multifaceted endeavor demanding ongoing research, interdisciplinary collaboration, and an unwavering dedication to harmonize AI technologies with human values and ethics. It is paramount to address critical aspects such as transparency, bias reduction, fairness, governance, and practical applications to ensure the responsible and impactful integration of AI. By embracing a variety of ethical frameworks and deploying effective strategies to mitigate biases, we can endeavor towards the development of responsible and equitable AI systems that positively influence society. This approach not only enhances the credibility and trust in AI technologies but also amplifies their potential to contribute to a better and more just world. The journey towards responsible AI integration is a shared responsibility, one that necessitates active engagement from stakeholders across domains and the concerted efforts of researchers, developers, policymakers, and communities at large.

Implications of the Study

The integration of artificial intelligence (AI) into ethical decision-making carries profound implications for key stakeholders, such as organizations, policymakers, AI developers, ethicists, and future researchers. This comprehensive examination, grounded in identified literature review themes, unveils critical considerations.

Organizations: For responsible deployment across sectors, organizations must prioritize transparency, explainability, and value alignment in AI systems. Vital for inclusivity and contextual sensitivity, the incorporation of diverse cultural values into the AI framework demands strategic attention (Ring, Maker & Sarazen, 2023). Ethical AI design, coupled with robust bias-mitigation and fairness strategies, necessitates organizational investment (Selbst et al., 2019).

Policymakers: Playing a pivotal role in regulating AI development, policymakers are urged to shape governance and policy frameworks. Effective response to ethical challenges requires collaborative efforts with multidisciplinary experts (Cabrera et al., 2023; Gunning & Aha, 2019). Legislative focus should be on enacting laws safeguarding individual rights, privacy, and preventing biases in AI systems.

AI Developers: Transparency, explainability, and value alignment are imperatives for AI developers during system design (Lipton, 2018). Actively mitigating biases through fairness-aware algorithms and integrating diverse ethical perspectives into AI development are crucial (Zliobaite, 2015). The incorporation of ethical impact assessment and continuous monitoring throughout the development lifecycle is paramount (Obermeyer et al., 2019

Ethicists: Ethicists play a proactive role in shaping AI ethics by advocating for the integration of various ethical frameworks. Drawing on utilitarianism, deontology, virtue ethics, and right-based ethics is essential (Russel & Norving, 2016). Engaging in interdisciplinary collaboration, ethicists provide ethical guidance, ensuring that AI aligns with fundamental human rights and values (Azenberg & Van Den Hovan, 2020). This collaborative effort contributes to the ethical evolution of AI, reinforcing its responsible integration into decision-making processes.

Future Research: Future research should focus on advancing fairness-aware algorithms and exploring novel ways to address biases in AI designing and development (Helberger, Araujo & de Vrees, 2020; Selbst et al., 2019). Moreover, researchers should continue to investigate the practical applications of AI in ethical decision-making across different domains, identifying challenges and opportunities for responsible integration (Dwivedi et al., 2021; Chouldechova, 2017).

The seamless integration of artificial intelligence (AI) into ethical decision-making demands a persistent commitment to ongoing research, interdisciplinary collaboration, and the steadfast alignment of AI technologies with human values and ethics. This ethical framework is not just a requisite but a fundamental necessity for guiding future AI research (Boddington, P., & Boddington, 2017; Han et al., 2021).

Ensuring the responsible and impactful incorporation of AI entails addressing pivotal aspects, including transparency, bias reduction, fairness, governance, and practical applications (Villaronga, E Kieseberg & Li, 2018; Selbst et al., 2019; Thuraisingham, 2020; Morse et al., 2021; Janga et al., 2023). Ethical frameworks act as a guiding compass, directing researchers and practitioners towards developing AI systems that are not only technically advanced but also socially responsible (Aitken et al., 2020; Brusseau, 2023).

By embracing diverse ethical frameworks and implementing effective strategies to mitigate biases, the research community can significantly contribute to the creation of responsible and equitable AI systems (Gan & Moussawi, 2022). This approach enhances the credibility and trustworthiness of AI technologies, thereby fostering a positive societal impact (Thiebes, Lins & Sunyaev, 2021).

This ethical framework plays a crucial role in navigating the delicate balance between technological innovation and ethical considerations, ensuring that AI systems align with societal values, promote inclusivity, and foster fairness (Hakami & Hernandez, 2020). As the pace of AI development accelerates, a steadfast commitment to ethical principles becomes increasingly vital to prevent unintended consequences and ensure positive contributions to society.

The journey towards responsible AI integration is a collective responsibility, necessitating active engagement from stakeholders across domains. Researchers, developers, policymakers, and communities must collaborate to establish and uphold ethical standards, fostering an environment where AI advances in a beneficial, transparent, and just manner. This collaborative effort paves the way for a future where AI technologies positively shape our world and contribute to a more ethical and equitable society (Leslie, 2019; Schiff et al., 2020; Kamila & Jasrotia, 2023).

LIMITATIONS OF THE STUDY

While this study provides valuable insights into the integration of AI into ethical decision-making and its implications for various stakeholders, it is important to acknowledge its limitations:

Scope and Depth

The study may not comprehensively cover all potential aspects and implications of AI integration into ethical decision-making due to limitations in the scope and depth of the research.

Generalizability

The findings and implications discussed in this study may be context-dependent and may not be universally applicable across all organizational settings, domains, or regions.

Research Bias

The study's findings may be influenced by any inherent biases in the selected literature, data sources, or methodologies used for synthesizing information.

Evolution of AI Landscape

The rapidly evolving nature of AI technologies and methodologies might mean that some insights presented in this study could become outdated or require revision as new advancements emerge.

Availability of Literature

The limitations of available literature, especially regarding certain aspects of AI integration or specific domains, may have constrained the depth of analysis and coverage in this study.

Time Constraints

The study's timeframe and resource limitations may have restricted the depth of analysis and the ability to delve deeply into certain topics or conduct primary research.

Declaration of Conflicting Interests

The authors affirm no conflicts of interest related to the research, writing, or publication of this article.

REFERENCES

Abaido, G. M. (2020). Cyberbullying on social media platforms among university students in the United Arab Emirates. *International Journal of Adolescence and Youth, 25*(1), 407–420. doi:10.1080/02673843.2019.1669059

Aitken, M., Toreini, E., Carmichael, P., Coopamootoo, K., Elliott, K., & van Moorsel, A. (2020). Establishing a social licence for Financial Technology: Reflections on the role of the private sector in pursuing ethical data practices. *Big Data & Society, 7*(1). doi:10.1177/2053951720908892

Aizenberg, E., & Van Den Hoven, J. (2020). Designing for human rights in AI. *Big Data & Society, 7*(2), 2053951720949566. doi:10.1177/2053951720949566

Atik, S. O., & Ipbuker, C. (2021). Integrating convolutional neural network and multiresolution segmentation for land cover and land use mapping using satellite imagery. *Applied Sciences (Basel, Switzerland), 11*(12), 5551. doi:10.3390/app11125551

Ballard, S., Chappell, K. M., & Kennedy, K. (2019). Judgment call the game: Using value sensitive design and design fiction to surface ethical concerns related to technology. In *Proceedings of the 2019 on Designing Interactive Systems Conference* (pp. 421-433). 10.1145/3322276.3323697

Bandi, S., & Kothari, A. (2022). Artificial Intelligence: An Asset for the Financial Sector. *Impact of Artificial Intelligence on Organizational Transformation*, 2.

Bellamy, R. K., Dey, K., Hind, M., Hoffman, S. C., Houde, S., Kannan, K., Lohia, P., Mehta, S., Mojsilovic, A., Nagar, S., Ramamurthy, K. N., Richards, J., Saha, D., Sattigeri, P., Singh, M., Varshney, K. R., & Zhang, Y. (2019). Think your artificial intelligence software is fair? Think again. *IEEE Software, 36*(4), 76–80. doi:10.1109/MS.2019.2908514

Bellman, K., Landauer, C., Dutt, N., Esterle, L., Herkersdorf, A., Jantsch, A., ... Tammemäe, K. (2020). Self-aware cyber-physical systems. *ACM Transactions on Cyber-Physical Systems, 4*(4), 1-26.

Bentham, J., & Mill, J. S. (2004). Utilitarianism and other essays. Academic Press.

Bentzen, M. M. (2016). The Principle of Double Effect Applied to Ethical Dilemmas of Social Robots. In Robophilosophy/TRANSOR (pp. 268-279). Academic Press.

Boddington, P., & Boddington, P. (2017). Does AI Raise Any Distinctive Ethical Questions? *Towards a code of ethics for artificial intelligence*, 27-37.

Brusseau, J. (2023). AI human impact: Toward a model for ethical investing in AI-intensive companies. *Journal of Sustainable Finance & Investment, 13*(2), 1030–1057. doi:10.1080/20430795.2021.1874212

Buckley, R. P., Zetzsche, D. A., Arner, D. W., & Tang, B. W. (2021). Regulating artificial intelligence in finance: Putting the human in the loop. *The Sydney Law Review, 43*(1), 43–81.

Cabrera, J., Loyola, M. S., Magaña, I., & Rojas, R. (2023, June). Ethical dilemmas, mental health, artificial intelligence, and llm-based chatbots. In *International Work-Conference on Bioinformatics and Biomedical Engineering* (pp. 313–326). Springer Nature Switzerland. doi:10.1007/978-3-031-34960-7_22

CaoL. (2020). AI in finance: A review. Available at SSRN 3647625.

Cath, C. (2018). Governing artificial intelligence: Ethical, legal and technical opportunities and challenges. *Philosophical Transactions. Series A, Mathematical, Physical, and Engineering Sciences*, *376*(2133), 20180080. doi:10.1098/rsta.2018.0080 PMID:30322996

Chatila, R., Renaudo, E., Andries, M., Chavez-Garcia, R. O., Luce-Vayrac, P., Gottstein, R., Alami, R., Clodic, A., Devin, S., Girard, B., & Khamassi, M. (2018). Toward self-aware robots. *Frontiers in Robotics and AI*, *5*, 88. doi:10.3389/frobt.2018.00088 PMID:33500967

Chouldechova, A. (2017). Fair prediction with disparate impact: A study of bias in recidivism prediction instruments. *Big Data*, *5*(2), 153–163. doi:10.1089/big.2016.0047 PMID:28632438

Chung, K., Thaichon, P., & Quach, S. (2022). Types of artificial intelligence (AI) in marketing management. In *Artificial intelligence for marketing management* (pp. 29–40). Routledge. doi:10.4324/9781003280392-4

Cuzzolin, F., Morelli, A., Cirstea, B., & Sahakian, B. J. (2020). Knowing me, knowing you: Theory of mind in AI. *Psychological Medicine*, *50*(7), 1057–1061. doi:10.1017/S0033291720000835 PMID:32375908

Danks, D., & London, A. J. (2017). Algorithmic Bias in Autonomous Systems. In Ijcai (Vol. 17, No. 2017, pp. 4691-4697). doi:10.24963/ijcai.2017/654

De Moura, N., Chatila, R., Evans, K., Chauvier, S., & Dogan, E. (2020). *Ethical decision making for autonomous vehicles. In 2020 IEEE intelligent vehicles symposium (iv)*. IEEE.

Diakopoulos, N. (2016). Accountability in algorithmic decision making. *Communications of the ACM*, *59*(2), 56–62. doi:10.1145/2844110

Dignum, V. (2022). *Responsible Artificial Intelligence—from Principles to Practice.* arXiv preprint arXiv:2205.10785.

Dwivedi, Y. K., Hughes, L., Ismagilova, E., Aarts, G., Coombs, C., Crick, T., Duan, Y., Dwivedi, R., Edwards, J., Eirug, A., Galanos, V., Ilavarasan, P. V., Janssen, M., Jones, P., Kar, A. K., Kizgin, H., Kronemann, B., Lal, B., Lucini, B., ... Williams, M. D. (2021). Artificial Intelligence (AI): Multidisciplinary perspectives on emerging challenges, opportunities, and agenda for research, practice and policy. *International Journal of Information Management*, *57*, 101994. doi:10.1016/j.ijinfomgt.2019.08.002

Dwivedi, Y. K., Hughes, L., Ismagilova, E., Aarts, G., Coombs, C., Crick, T., Duan, Y., Dwivedi, R., Edwards, J., Eirug, A., Galanos, V., Ilavarasan, P. V., Janssen, M., Jones, P., Kar, A. K., Kizgin, H., Kronemann, B., Lal, B., Lucini, B., ... Williams, M. D. (2021). Artificial Intelligence (AI): Multidisciplinary perspectives on emerging challenges, opportunities, and agenda for research, practice and policy. *International Journal of Information Management*, *57*, 101994. doi:10.1016/j.ijinfomgt.2019.08.002

Finn, R. L., Wright, D., & Friedewald, M. (2013). Seven types of privacy. *European data protection: Coming of age*, 3-32.

Fjeld, J., Achten, N., Hilligoss, H., Nagy, A., & Srikumar, M. (2020). *Principled artificial intelligence: Mapping consensus in ethical and rights-based approaches to principles for AI.* Berkman Klein Center Research Publication.

Fleming, R. B., & Morgan, R. C. (2012). Standards for Financial Decision-Making: Legal, Ethical, and Practical Issues. *Utah L. Rev.*, 1275.

Flowers, J. C. (2019, March). Strong and Weak AI: Deweyan Considerations. In AAAI spring symposium: Towards conscious AI systems (Vol. 2287, No. 7). AAAI.

Formosa, P., Wilson, M., & Richards, D. (2021). A principlist framework for cybersecurity ethics. *Computers & Security*, *109*, 102382. doi:10.1016/j.cose.2021.102382

Friedman, B., Harbers, M., Hendry, D. G., van den Hoven, J., Jonker, C., & Logler, N. (2021). Eight grand challenges for value sensitive design from the 2016 Lorentz workshop. *Ethics and Information Technology*, *23*(1), 5–16. doi:10.100710676-021-09586-y

Fritz-Morgenthal, S., Hein, B., & Papenbrock, J. (2022). Financial risk management and explainable, trustworthy, responsible AI. *Frontiers in Artificial Intelligence*, *5*, 779799. doi:10.3389/frai.2022.779799 PMID:35295866

Gan, I., & Moussawi, S. (2022). A Value Sensitive Design Perspective on AI Biases. *Proceedings of the 55th Hawaii International Conference on System Sciences.* 10.24251/HICSS.2022.676

Gebru, T., Morgenstern, J., Vecchione, B., Vaughan, J. W., Wallach, H., Iii, H. D., & Crawford, K. (2021). Datasheets for datasets. *Communications of the ACM*, *64*(12), 86–92. doi:10.1145/3458723

Giuffrida, I. (2019). Liability for AI decision-making: Some legal and ethical considerations. *Fordham Law Review*, *88*, 439.

Gongane, V. U., Munot, M. V., & Anuse, A. D. (2022). Detection and moderation of detrimental content on social media platforms: Current status and future directions. *Social Network Analysis and Mining*, *12*(1), 129. doi:10.100713278-022-00951-3 PMID:36090695

Goodman, B., & Flaxman, S. (2017). European Union regulations on algorithmic decision-making and a "right to explanation". *AI Magazine*, *38*(3), 50–57. doi:10.1609/aimag.v38i3.2741

Grant, M. J., & Booth, A. (2009). A typology of reviews: An analysis of 14 review types and associated methodologies. *Health Information and Libraries Journal*, *26*(2), 91–108. doi:10.1111/j.1471-1842.2009.00848.x PMID:19490148

Gunning, D., & Aha, D. (2019). DARPA's explainable artificial intelligence (XAI) program. *AI Magazine*, *40*(2), 44–58. doi:10.1609/aimag.v40i2.2850

Hakami, E., & Hernandez Leo, D. (2020). How are learning analytics considering the societal values of fairness, accountability, transparency and human well-being? A literature review. *LASI-SPAIN 2020: Learning Analytics Summer Institute Spain 2020: Learning Analytics. Time for Adoption? 2020 Jun 15-16; Valladolid, Spain. Aachen: CEUR; 2020,* 121-41.

Han, S., Kelly, E., Nikou, S., & Svee, E. O. (2021). Aligning artificial intelligence with human values: Reflections from a phenomenological perspective. *AI & Society*, 1–13.

Helberger, N., Araujo, T., & de Vreese, C. H. (2020). Who is the fairest of them all? Public attitudes and expectations regarding automated decision-making. *Computer Law & Security Report*, *39*, 105456. doi:10.1016/j.clsr.2020.105456

Hodges, A. (2009). *Alan Turing and the Turing test*. Springer Netherlands. doi:10.1007/978-1-4020-6710-5_2

Huang, C., Zhang, Z., Mao, B., & Yao, X. (2022). An overview of artificial intelligence ethics. *IEEE Transactions on Artificial Intelligence*.

Huntingford, C., Jeffers, E. S., Bonsall, M. B., Christensen, H. M., Lees, T., & Yang, H. (2019). Machine learning and artificial intelligence to aid climate change research and preparedness. *Environmental Research Letters*, *14*(12), 124007. doi:10.1088/1748-9326/ab4e55

Janga, B., Asamani, G. P., Sun, Z., & Cristea, N. (2023). A Review of Practical AI for Remote Sensing in Earth Sciences. *Remote Sensing (Basel)*, *15*(16), 4112. doi:10.3390/rs15164112

Janiesch, C., Zschech, P., & Heinrich, K. (2021). Machine learning and deep learning. *Electronic Markets*, *31*(3), 685–695. doi:10.100712525-021-00475-2

Jobin, A., Ienca, M., & Vayena, E. (2019). The global landscape of AI ethics guidelines. *Nature Machine Intelligence*, *1*(9), 389–399. doi:10.103842256-019-0088-2

Kamila, M. K., & Jasrotia, S. S. (2023). Ethical issues in the development of artificial intelligence: recognizing the risks. *International Journal of Ethics and Systems*.

Kaplan, A., & Haenlein, M. (2019). Siri, Siri, in my hand: Who's the fairest in the land? On the interpretations, illustrations, and implications of artificial intelligence. *Business Horizons*, *62*(1), 15–25. doi:10.1016/j.bushor.2018.08.004

Kaushik, M., Pathak, S., Bhatt, P. K., Singh, S. K., Tripathi, A., & Pandey, P. K. (2022). Artificial Intelligence (AI). *Intelligent System Algorithms and Applications in Science and Technology*, 119-133.

Kiritchenko, S., Nejadgholi, I., & Fraser, K. C. (2021). Confronting abusive language online: A survey from the ethical and human rights perspective. *Journal of Artificial Intelligence Research*, *71*, 431–478. doi:10.1613/jair.1.12590

Kordzadeh, N., & Ghasemaghaei, M. (2022). Algorithmic bias: Review, synthesis, and future research directions. *European Journal of Information Systems*, *31*(3), 388–409. doi:10.1080/0960085X.2021.1927212

Kreps, G. L., & Neuhauser, L. (2010). New directions in eHealth communication: Opportunities and challenges. *Patient Education and Counseling*, *78*(3), 329–336. doi:10.1016/j.pec.2010.01.013 PMID:20202779

Leslie, D. (2019). *Understanding artificial intelligence ethics and safety*. arXiv preprint arXiv:1906.05684.

Leviathan, Y., & Matias, Y. (2018). *Google Duplex: An AI system for accomplishing real-world tasks over the phone*. Academic Press.

Lipton, Z. C. (2018). The mythos of model interpretability: In machine learning, the concept of interpretability is both important and slippery. *ACM Queue; Tomorrow's Computing Today*, *16*(3), 31–57. doi:10.1145/3236386.3241340

Lundberg, S. M., & Lee, S. I. (2017). A unified approach to interpreting model predictions. *Advances in Neural Information Processing Systems*, 30.

Ma, Y., Wang, Z., Yang, H., & Yang, L. (2020). Artificial intelligence applications in the development of autonomous vehicles: A survey. *IEEE/CAA Journal of Automatica Sinica, 7*(2), 315-329.

Martinho, A., Herber, N., Kroesen, M., & Chorus, C. (2021). Ethical issues in focus by the autonomous vehicles industry. *Transport Reviews*, *41*(5), 556–577. doi:10.1080/01441647.2020.1862355

Max, R., Kriebitz, A., & Von Websky, C. (2021). Ethical considerations about the implications of artificial intelligence in finance. *Handbook on ethics in finance*, 577-592.

Mhlanga, D. (2020). Industry 4.0 in finance: The impact of artificial intelligence (ai) on digital financial inclusion. *International Journal of Financial Studies*, *8*(3), 45. doi:10.3390/ijfs8030045

Mitchell, M. (2019). Artificial intelligence: A guide for thinking humans. Academic Press.

Mittelstadt, B. D., Allo, P., Taddeo, M., Wachter, S., & Floridi, L. (2016). The ethics of algorithms: Mapping the debate. *Big Data & Society*, *3*(2). doi:10.1177/2053951716679679

Moreland, M. P. (2012). Mistakes about Intention in the Law of Bioethics. *Law and Contemporary Problems*, *75*, 53.

Morse, L., Teodorescu, M. H. M., Awwad, Y., & Kane, G. C. (2021). Do the ends justify the means? Variation in the distributive and procedural fairness of machine learning algorithms. *Journal of Business Ethics*, 1–13.

Müller, V. C., & Bostrom, N. (2016). Future progress in artificial intelligence: A survey of expert opinion. *Fundamental issues of artificial intelligence*, 555-572.

Nazir, S., & Kaleem, M. (2021). Advances in image acquisition and processing technologies transforming animal ecological studies. *Ecological Informatics*, *61*, 101212. doi:10.1016/j.ecoinf.2021.101212

Obermeyer, Z., Powers, B., Vogeli, C., & Mullainathan, S. (2019). Dissecting racial bias in an algorithm used to manage the health of populations. *Science*, *366*(6464), 447–453. doi:10.1126cience.aax2342 PMID:31649194

Popat, A., & Tarrant, C. (2023). Exploring adolescents' perspectives on social media and mental health and well-being–A qualitative literature review. *Clinical Child Psychology and Psychiatry*, *28*(1), 323–337. doi:10.1177/13591045221092884 PMID:35670473

Pramod, A., Naicker, H. S., & Tyagi, A. K. (2021). Machine learning and deep learning: Open issues and future research directions for the next 10 years. *Computational analysis and deep learning for medical care: Principles, methods, and applications*, 463-490.

Pramod, A., Naicker, H. S., & Tyagi, A. K. (2021). Machine Learning and Deep Learning: Open Issues and Future Research Directions for the Next 10 Years. *Computational Analysis and Deep Learning for Medical Care*, 463–490.

Qiang, V., Rhim, J., & Moon, A. (2023). No such thing as one-size-fits-all in AI ethics frameworks: A comparative case study. *AI & Society*, 1–20. doi:10.100700146-023-01653-w

Quinn, A. (2007). Moral virtues for journalists. *Journal of Mass Media Ethics*, *22*(2-3), 168–186. doi:10.1080/08900520701315764

Reamer, F. G. (2019). Ethical theories and social work practice. The Routledge handbook of social work ethics and values, 15-21.

Ribeiro, M. T., Singh, S., & Guestrin, C. (2016, August). " Why should i trust you?" Explaining the predictions of any classifier. In *Proceedings of the 22nd ACM SIGKDD international conference on knowledge discovery and data mining* (pp. 1135-1144). 10.1145/2939672.2939778

Ring, M., Ai, D., Maker-Clark, G., & Sarazen, R. (2023). Cooking up Change: DEIB Principles as Key Ingredients in Nutrition and Culinary Medicine Education. *Nutrients*, *15*(19), 4257. doi:10.3390/nu15194257 PMID:37836541

Romansky, R. P., & Noninska, I. S. (2020). Challenges of the digital age for privacy and personal data protection. *Mathematical Biosciences and Engineering*, *17*(5), 5288–5303. doi:10.3934/mbe.2020286 PMID:33120553

Rudin, C. (2019). Stop explaining black box machine learning models for high stakes decisions and use interpretable models instead. *Nature Machine Intelligence*, *1*(5), 206–215. doi:10.103842256-019-0048-x PMID:35603010

Russell, S. (2022). Artificial intelligence and the problem of control. *Perspectives on Digital Humanism*, 19.

Russell, S., & Norvig, P. (2016). *Artificial Intelligence: A Modern Approach*. Pearson.

Schiff, D., Rakova, B., Ayesh, A., Fanti, A., & Lennon, M. (2020). *Principles to practices for responsible AI: closing the gap*. arXiv preprint arXiv:2006.04707.

Selbst, A. D., Boyd, D., Friedler, S. A., Venkatasubramanian, S., & Vertesi, J. (2019, January). Fairness and abstraction in sociotechnical systems. In *Proceedings of the conference on fairness, accountability, and transparency* (pp. 59-68). 10.1145/3287560.3287598

Smith, K. T., Jones, A., Johnson, L., & Smith, L. M. (2019). Examination of cybercrime and its effects on corporate stock value. *Journal of Information, Communication and Ethics in Society*, *17*(1), 42–60. doi:10.1108/JICES-02-2018-0010

Tan, L., Guo, J., Mohanarajah, S., & Zhou, K. (2021). Can we detect trends in natural disaster management with artificial intelligence? A review of modeling practices. *Natural Hazards*, *107*(3), 2389–2417. doi:10.100711069-020-04429-3

Thiebes, S., Lins, S., & Sunyaev, A. (2021). Trustworthy artificial intelligence. *Electronic Markets*, *31*(2), 447–464. doi:10.100712525-020-00441-4

Thuraisingham, B. (2020). Artificial intelligence and data science governance: Roles and responsibilities at the c-level and the board. In *2020 IEEE 21st international conference on information reuse and integration for data science (IRI)* (pp. 314-318). IEEE.

Tranfield, D., Denyer, D., & Smart, P. (2003). Towards a methodology for developing evidence-informed management knowledge by means of systematic review. *British Journal of Management*, *14*(3), 207–222. doi:10.1111/1467-8551.00375

Truby, J., Brown, R., & Dahdal, A. (2020). Banking on AI: Mandating a proactive approach to AI regulation in the financial sector. *Law and Financial Markets Review*, *14*(2), 110–120. doi:10.1080/17521440.2020.1760454

Umbrello, S., & Yampolskiy, R. V. (2022). Designing AI for explainability and verifiability: A value sensitive design approach to avoid artificial stupidity in autonomous vehicles. *International Journal of Social Robotics*, *14*(2), 313–322. doi:10.100712369-021-00790-w

Ungerer, L., & Slade, S. (2022). Ethical considerations of artificial intelligence in learning analytics in distance education contexts. In *Learning Analytics in Open and Distributed Learning: Potential and Challenges* (pp. 105–120). Springer Nature Singapore. doi:10.1007/978-981-19-0786-9_8

Villaronga, E. F., Kieseberg, P., & Li, T. (2018). Humans forget, machines remember: Artificial intelligence and the right to be forgotten. *Computer Law & Security Report*, *34*(2), 304–313. doi:10.1016/j.clsr.2017.08.007

Wang, C., Liu, S., Yang, H., Guo, J., Wu, Y., & Liu, J. (2023). Ethical Considerations of Using ChatGPT in Health Care. *Journal of Medical Internet Research*, *25*, e48009. doi:10.2196/48009 PMID:37566454

Webber, B. L., & Nilsson, N. J. (Eds.). (2014). *Readings in artificial intelligence*. Morgan Kaufmann.

Winkler, T., & Spiekermann, S. (2019). Human values as the basis for sustainable information system design. *IEEE Technology and Society Magazine*, *38*(3), 34–43. doi:10.1109/MTS.2019.2930268

Wu, C., Li, X., Guo, Y., Wang, J., Ren, Z., Wang, M., & Yang, Z. (2022). Natural language processing for smart construction: Current status and future directions. *Automation in Construction*, *134*, 104059. doi:10.1016/j.autcon.2021.104059

Yigitcanlar, T., Desouza, K. C., Butler, L., & Roozkhosh, F. (2020). Contributions and risks of artificial intelligence (AI) in building smarter cities: Insights from a systematic review of the literature. *Energies*, *13*(6), 1473. doi:10.3390/en13061473

Zliobaite, I. (2015). *A survey on measuring indirect discrimination in machine learning*. arXiv preprint arXiv:1511.00148.

Zowghi, D., & da Rimini, F. (2023). *Diversity and Inclusion in Artificial Intelligence*. arXiv preprint arXiv:2305.12728.

Zulfiqar, A., Ghaffar, M. M., Shahzad, M., Weis, C., Malik, M. I., Shafait, F., & Wehn, N. (2021). AI-ForestWatch: Semantic segmentation based end-to-end framework for forest estimation and change detection using multi-spectral remote sensing imagery. *Journal of Applied Remote Sensing*, *15*(2), 024518–024518. doi:10.1117/1.JRS.15.024518

Chapter 13
The Intersection of Ethics and Big Data:
Addressing Ethical Concerns in Digital Age of Artificial Intelligence

Divya Goswami
https://orcid.org/0009-0004-6210-1568
Chitkara Business School, Chitkara University, Punjab, India

Balraj Verma
https://orcid.org/0000-0002-6542-3261
Chitkara Business School, Chitkara University, Punjab, India

ABSTRACT

This chapter delves into the intersection of ethics and big data, with a primary focus on the ethical concerns arising from AI. The primary objective of this chapter is to highlight a novel approach that researchers might employ throughout the process of conducting a systematic literature review (SLR) to enhance efficiency and reduce costs associated with data synthesis and abstraction. Further, the conclusion emphasizes the need to navigate the intersection of ethics and big data, particularly concerning AI, presents a complex landscape of ethical concerns.

1. INTRODUCTION

We are witnessing a phenomenal increase in the number of technologies that are associated with artificial intelligence. According to Ray Kurzweil, it is postulated that by about 2029, computers with AI will reach a level of cognitive ability that is comparable to that of humans. If we project this forward to 2045, for example, we will have increased the sum of the ability of humans, living organisms, and computers by an amount of a billion. The concept of massive data insights and the use of AI has gained prominence in recent times and has evolved into a thriving area of study that has garnered significant

DOI: 10.4018/979-8-3693-1762-4.ch013

attention from both academics and professionals. The general availability of Big Data has increased alongside its expanding scope, particularly when the distribution of Big Data applications transitions to a perceptible mental approach (Needham, 2013; Sharma & Sood, 2022). As the number of individuals utilising several devices for connectivity increases, there is a corresponding rise in the interception and sharing of data. Artificial intelligence (AI) and the analysis of big data play a crucial role in the development and functioning of smart technology. Prominent examples of such systems include Google's engine for searching, Translate from Google, Amazon's Alexa virtual assistance, smartphones equipped with GPS tracking capabilities, medical and surgical robots, health and fitness software, energy-saving technologies, and numerous others. The increasing trend described has led to the development of novel research methodologies and protocols. However, it has also brought about significant shifts in societal norms and has generated debates concerning the ethical justifiability of research practices. The emergence of novel AI and massive data applications has sparked considerable interest towards ethical implications, particularly in relation to security as well as additional ethical considerations. Previous studies on AI have emphasized the importance of incorporating ethical considerations into the development and placement of AI technologies. These studies also highlight that AI systems can have significant societal impacts, and it is crucial to address ethical issues to ensure that these technologies are used responsibly and beneficially. However, AI ethics is an evolving field, and researchers are continuously adapting to address new challenges and concerns that emerge as AI technologies evolve. According to Boyd and Crawford (2012), there is a significant lack of understanding of the ethical issues that underlie the phenomenon of Big Data and AI. Given these circumstances, it is imperative to obtain a comprehensive comprehension of these issues and develop approaches to efficiently address them that encompass multiple stakeholders, especially the community at large. This approach is fundamental in order to ascertain that the advantages of these technologies surpass their drawbacks. These statements highlights the pressing necessity to engage in critical reflection, reconsideration, and reclamation of the authority associated with the expansive and ever-expanding field of 'Big Data' analysis (Dalton et al., 2016).This leads to the important question of how to balance benefits and downsides of such technologies. On the other hand, this subject has not been subjected to an adequate amount of investigation from either an academic or practical standpoint. The objective of this work is to perform a comprehensive review of existing literature in order to examine and analyse the difficulties associated with the convergence of artificial intelligence (AI) and Big Data values. Furthermore, in light of the evolving technologies and rapid globalisation, which are leading to the obsolescence of traditional paradigms, there is an imperative to comprehend the intricacies of Artificial Intelligence (AI) and Big Data. This chapter presents a series of investigative tasks for consideration.

- This paper aims to elucidate the implications of the converging of massive amounts of data and Artificial Intelligence (AI) in posing challenges to human beings.
- In order to gain comprehension regarding the ethical considerations that arise as a result of the use of AI and applications based on big data.

1.1 Understanding Big Data Analytics and AI

The recent advancements in Big Data, artificial intelligence (AI), and innovation based on data have resulted in significant social and across the industry benefits. Big data produces value in a variety of different disciplines, such as education and healthcare management, as well as the general public sector.

Yet the improper application of these technologies can result in the violation of data security and privacy laws, as well as ethical principles. The goals of this research place a high emphasis on the significance of protecting customers' personal information by requiring businesses to collect and make ethical use of customers' data, maintain dependable data security measures, and frequently analyses and upgrade their information management systems. The user's text is too brief to be rewritten in an academic manner. The concern around the unsettling nature of Big Data necessitates prompt consideration, particularly as society increasingly embraces the process of datafication. This phenomenon is characterised by the growing accessibility and efficiency of technology used to capture, gather, store, and analyse data. According to Strydom and Buckley (2020), Big Data is characterised by its dynamic nature, similar to the technologies that shape it. However, it is important to note that Big Data does not have a platform that is reliable. In recent years, the advancements in artificial intelligence (AI) capabilities, along with the use of Big Data, have garnered significant attention due to the potential rewards and associated threats that AI presents. The integration of artificial intelligence (AI) with the capabilities of big data is closely intertwined, as it enables the management of disparate data sets, facilitates connectivity, and facilitates the extraction of useful knowledge and predictive skills. Big data supplies the massive datasets needed by AI systems for training data and learning, i.e establishing baselines of normal behavior from which outliners can be identified. Big data processing refers to a distinct form of computation that distinguishes itself from traditional technology for processing based on three key factors: the sheer volume of data involved, the speed at which data is generated and sent, and the diverse range of data types encompassed. The process of utilising advanced techniques of analysis on extensive datasets is commonly referred to as "Big Database Analytics". This encompasses methodologies such as machine learning, forecasting, and information mining. The process of analysing large datasets can provide challenges due to the need for collecting and organising diverse data that follows distinct patterns or rules, sometimes referred to as a combination of data (Yenkar & Bartere, 2014). This study makes a scholarly contribution by investigating the legal and security risks associated with the convergence of big data, artificial intelligence, and security in the world of technology. It is anticipated that the implementation of policies that promote openness, ethics and transparency in digital technologies will have a favorable effect on many facets of citizens' lives.

1.2 AI: A Rutted Road to Revolution

Artificial Intelligence (AI)

What is the leading thing that springs to while you think of Artificial Intelligence, Is it related to robotics? The "Big Data"? Alternately, could it be the possibility of a shakeup in the industry? For a significant number of individuals, the initial perception of artificial intelligence (AI) is confined to its technological consumption and the potential ways in which it may alter our daily existence. Similar to how the introduction of personal computers initially sparked concerns about the potential displacement of human connections and utilisation, the recent advancements in artificial intelligence have generated apprehension among individuals around the possibility of being substituted. A civilization is capable of undergoing enormous shifts as a result of a revolution. Technologies that use AI (artificial intelligence) are widely recognised as crucial components within several businesses, encompassing sectors such as healthcare, manufacturing, finance, and retail. Nevertheless, the potential benefits of AI systems, such as enhancing productivity, cost reduction, and safety, have recently been accompanied by concerns regarding the possibility that these intricate systems may yield more ethical detriments than economic advantages (Pazzanese, 2020). According to Müller (2020), the impact of artificially intelligent (AI) and autonomous machines on the progress of humankind is substantial. Furthermore according to a study conducted by Hao and Qin (2020) and Sood et. al. (2022) a significant proportion of individuals belonging to the post-millennial generation hold the belief that they would engage in collaborative work with robotics and artificially intelligent systems (AI) within a timeframe of between eight and ten years. A speculation of these technologies is very complicated in context to determine the impact of AI. When people chat about AI, it's important to understand what its social effects are. When people think about AI, the ethics of it are generally ignored, which is a big problem for its general basis. People are suspicious of AI because of its broad theoretical implications, which have headed to the indication that self-learning machines will finally wipe out most of humanity.

AI and ethics converge in crucial ways. Ethical concerns surround AI's potential for bias, unfairness, and privacy violations. Ensuring AI progress and disposition is crucial to prevent harm. Transparency, fairness, and accountability are the major principles in sailing this complicated terrain. To strike a balance between AI innovation and well-being of society ethical guidelines and regulations are must. In the year 2019, Shohini Kundu, the author, acknowledged this problem and said, "Uniformity is indispensable to ethics and reliability"(Taricani et. al., 2020). We need to make choices based on more than just statistics. For hundreds of years, mutual trust, minimalizing harm, fairness, and equitability have been important parts of any system of reasoning that has survived. There are two peculiar measures for stimulating trust in a civilization and AI systems lack both robustness and accountability.

The subsequent section transitions to the literature review. Subsequently, the methodological part expounds upon the integration of artificial intelligence (AI) with other disciplines, encompassing an examination of both technological aspects and ethical implications. The aforementioned two parts analyse the consequences of these factors and establish the framework for future research endeavours.

2. LITERATURE REVIEW

2.1 Existing Approaches Towards Ethical AI

The previous part elucidated the significance of ethics within the realm of artificial intelligence. In this section, we will explore current methodologies and scholarly works that facilitate the examination and mitigation of these concerns. The present research examines many theoretical perspectives on the ethical and social implications of AI systems. Prior to delving into the realm of literature, it is imperative to provide a concise introduction to the fundamental concepts and terminology. The challenge of defining artificial intelligence (AI) is a complex endeavour, as the topic encompasses a multitude of concepts. According to Dignum (2019), artificial intelligence engages in the processing of information with the intention of achieving a specific objective. Although there exist alternative definitions that emphasise the outcomes of intelligent systems and their interpretation by humans (e.g. McCarthy et al., 1955), the aforementioned conceptualization of AI systems aligns with our objectives. The adoption of artificial intelligence (AI) in the banking and finance technology (fintech) sector is primarily motivated by its capacity to efficiently handle vast quantities of data and derive significant insights to inform decision-making processes (Daníelsson et al., 2022).Ethical AI is not solely regarded as a system that uses AI, but rather encompasses a comprehensive framework of principles, practices, and rules governing the development, deployment, and management of artificially intelligent systems.

2.2 Ethical Issues Arising From AI

The development and implementation of an ethics and human-centered artificial intelligence (AI) system should be undertaken with careful consideration of the values and impartial principles held by the society on which it impacts. Various proposals are being suggested in order to address the societal challenges posed by artificial intelligence. A new method called "Ethics by Design" has been suggested for AI study projects. Others make the case for teaching ethics. "Teaching AI professionals and students about ethics is an important part of the solution." It is imperative that those engaged in the study of machine learning, artificial intelligence, or data analytics receive comprehensive instruction and engage in discussions regarding the ethical, privacy, and security concerns associated with these fields. The publication by Edwards et al. (2019) presents an extensive examination of AI ethics standards, encompassing a range of principles pertaining to the ethical considerations of artificial intelligence. These principles include openness, fairness and equity, non-maleficence, accountability, confidentiality, generosity, autonomy and liberty, trust, environmental responsibility, worth, and togetherness. Several ethical difficulties related to artificial intelligence (AI) continue, primarily stemming from a longstanding tradition of scholarly discourse on the subject of ethics and AI in the literature (Stahl, 2021). Furthermore, it is worth noting that there is a significant emergence of this issue at an exceptional magnitude when viewed through the lens of policy, as highlighted by the (High-Level Advisory Council on AI, 2019). The subsequent enumeration presents an estimation of the many categories of ethical dilemmas that were identified through the examination of diverse case studies, accompanied by the valuable insights provided by specialists in this domain.

As long as there are possible benefits of AI in different fields, there are also many things that could go wrong. There are tough questions in most of these areas about how to find benefits and prices and what to do about them. A well-known example is the use of AI to make weapons that can fight themselves. There are many good reasons why replacing soldiers with robots might not be the best way to save lives. The rationales encompass a spectrum of considerations, spanning from pragmatic aspects such as the dependability of these systems, to political implications, such as the potential facilitation of initiating conflicts, and philosophical inquiries, such as the ethical implications of making human life-or-death decisions based on machine-generated input (Defence Innovation Board, 2019). According to Bynum (2008), the ideas of virtue ethics have been demonstrated to be applicable and significant in the realm of technological innovation since its inception. This may be observed in the contributions of Norbert Wiener (1954), a prominent figure in the development of digital technology.

2.3 Ethics by Design

Ethics is a scholarly topic of study that falls within the umbrella of philosophical thought, primarily concerned with the exploration and examination of ethical concerns. As "What is a good action?", "What is the value of a human life?", and "What is right or wrong?", "What is justice? "From the philosophical point of view Ethics is the branch of philosophy that deals with the issues pertaining to what is right and wrong, good and bad, and moral principles that guide human behavior and decision-making. It explores fundamental questions about how individuals and societies should act, what constitutes a morally virtuous life, and how to make ethical judgments. Ethics by design approach includes a set of best practices that focus on implementing ethical deliberations. These best practices include organizational aspects such

Figure 1. Different ethical issues

as the establishment of ethics board as well as advice for the actual process (Leidner and Plachouras, 2017). This targets to integrate "ethical decision routines in AI systems" (Hagendorff, 2020).

Design for values approach was applied to trustworthy AI and suggests spelling out the abstract idea of the ethical AI and spelling out the abstract idea to the field of ethical AI through a concrete set of values (Dignum, 2019). The value approach was discussed for the first time in the Human AI research project and introduced the design for values along with accountability, fairness, and transparency. The assumption is that processes of software and technology development are full of decisions that the "designers, developers and other stakeholders" have to make, "many of them of an ethical nature". This brings up the question that why would people adopt AI and Big Analytics. One of the key drivers of the adoption of AI is its ability to process vast amounts of data and extract valuable insights for decision-making purposes (Daníelsson et al., 2022).However, the use of AI and big data in the industry raises ethical and privacy concerns .This paper contributes to the literature by examining the ethical and privacy considerations linked with the intersection of big data, AI, and privacy in the digital finance industry. Further research can explore how the transition to new AI and data focused models can be implemented and the implementation of this update can be promoted to foster support among new users. Furthermore, future study might explore the potential influence of societal and cultural standards on the acceptance of AI and the utilisation of big data in other industries, including but not limited to banking and healthcare.

3. METHODOLOGY

In the course of this research, a process known as a systematic review was utilized to build a solid body of data that can be used to make suggestions to institutions, educators, and facilitators. The systematic review method is "a scientific process governed by a set of clear and demanding rules." Rules that are meant to show completeness, lack of bias, and openness and responsibility in technique and execution" (Dixon-Woods, 2011). This clarifies that the systematic literature review (SLR) is a method for analyzing and interpreting the outcomes of prior research projects that have looked into a specific research question, topic, or phenomenon. There are variant data sources, but here only Scopus have been used to get relevant information on ethics of AI and Big Data Analysis. The authors engaged in a variety of dialogues and conversations in order to find the pertinent digital sources. Subsequently, an examination was conducted on the designated digital repository with the objective of extracting pertinent information to effectively address the provided sources. The online resources were ultimately chosen based on author discussions and the evaluation of data sources, with Scopus being picked as the preferred option. These databases are globally recognized as the largest repositories of digital data, encompassing a vast collection of knowledge and studies pertaining to technology.

3.1 Research Objectives

The formulation of research objectives during the systematic evaluation of the literature phase is an essential component of a research study. The formation of research objectives indicates a comprehensive understanding of the subject matter as a whole and the specific research challenge at hand. A thorough investigation was undertaken to locate pertinent academic literature in order to enhance comprehension of the research issue. Ultimately, the objectives were ultimately formulated based on the research framework elucidated in the aforementioned scholarly references. The objectives of the study are as followed:

(i) The purpose of this discourse is to elucidate the manner in which the amalgamation of large amounts of data and the use of artificial intelligence (AI) poses an immense threat to the human race.
(ii) To learn about the moral issues that have come up because of AI and big data.

3.2 Search Strategy

The researchers conducted a comprehensive analysis of the research inquiries in order to identify the essential terms employed during the search procedure or investigation. The writers actively participated in a collaborative conversation in order to establish and finalise the specific search criteria. Additionally,

Table 1. Inclusion/exclusion criteria

No.	Inclusion Criteria
In1	Select the articles related to Big data and Ethics only
In2	Restrict studies to the time period of 2022 and 2023
In3	Consider articles of English language only
No.	**Exclusion Criteria**
Ex1	
Ex2	Exclude duplicate studies
Ex3	Ethics discussed in other Domains such as Humanities, Social Science etc.

they diligently extracted pertinent information from the selected repositories. Various types of string were manufactured, which ultimately made significant contributions to academic research.

("AI ethics" OR "Artificial Intelligence Ethics" OR "Ethics Related to AI") AND ("resistance" OR "barriers" OR "limitations" OR "challenges "OR "hurdles")

The search strings are constructed by utilizing the "AND" and "OR" operators to establish connections between several search phrases. The digital repositories that have been chosen possess a search mechanism that has been tailored to meet specific requirements. The execution of search strings is carried out through the utilization of personalized search mechanisms inside electronic data sources.

3.3 Inclusion/Exclusion Criteria

The inclusion/exclusion criteria are formulated to refine the search results and eliminate research that are insignificant, inaccessible, inapplicable, or of poor caliber. The criteria are devised by the authors, and subsequently ratified during the routine consensus conference. (Table 1)

3.4 Filtration Criteria

To filter the search string findings and remove irrelevant, not accessible, redundant and low quality, slack studies. The criteria are developed to filter the search string findings by the authors in a regular consensus meeting. In order to refine the search results and exclude unnecessary, inaccessible, redundant, and low-quality research, it is necessary to filter the keyword phrase findings. The criteria were formulated by the authors during a routine discussion meeting in order to refine the search string results.

3.5 Study Selection

The search query mentioned in Section 3.3 is employed to investigate the designated digital repositories. The search procedure was commenced in 2022 and concluded in 2023.In the initial phase, a total of 262 studies were retrieved from the Scopus database using the search string. The data was subsequently refined by applying a filter determined by the publishing stage, specifically limited to final publications, resulting in a total of 244 records. During the second phase of the evaluation process, the determination of whether or not to include for the 87 studies is conducted solely on the basis of the document type, specifically confined to articles. During the third phase, a temporal filter was applied, restricting the search results to the time period between 2022 and 2023. This filtering process resulted in the identification of 58 papers that were deemed relevant to the study. Finally, 27 primary studies are shortlisted using the subject area limited to Business Management, and Accounting, and Social Science.

4. REVIEW REPORTING

The data obtained from the chosen 27 main studies have been subjected to analysis and will be described in the following sections.

4.1 Articles Published During the Year

The distribution of primary research by year is depicted in Figure 2. The initial pertinent study was discovered in 2018, subsequent to which there has been a gradual however steady rise in the quantity of research publications. The execution of the SLR (Systematic Literature Review) string in the period from 2022 to 2023 has provided evidence that the field of AI and big-data ethical behaviour holds considerable importance and represents a cutting-edge research topic. Further investigation and exploration into the ethical implications of artificial intelligence (AI) and extensive data insights remains a significant area of research.

4.2 Articles Published in Different Subject Areas

The distribution of primary studies by study area is depicted in Fig.3. A list of subject areas related to AI ethics, along with general topics within each area were searched using variant keywords to search for significant documents in academic journals, books, and online resources. According to the results of the systematic literature review (SLR), it was found that a significant portion of the research undertaken falls within the domain of Social Sciences (35.4%). This is followed by research in the field of Computer Science (15.4%), the field of business administration (13.8%), and the fields of Humanities (10.8%).

4.3 Database From Scopus

The primary research that have been chosen are categorised into four main types: journal articles, conference papers (including workshops), book chapters, and magazine articles. This illustrates the utilisation of several articles derived from the Scopus database. Scopus is a widely used database in the realm of academic research, encompassing abstracts and citations. It provides access to a vast collection of scholarly literature, ensuring a comprehensive and diverse pool of sources for research. Scopus offers different citation analysis tools to track citations, h-index, and other bibliometric indicators. In addition, specific search terms were employed to retrieve relevant articles, conference proceedings, and other research materials from Scopus, with the aim of locating documents pertaining to the field of artificial intelligence morality.

Figure 2. Articles related to AI published during the year
Source: Scopus (2023)

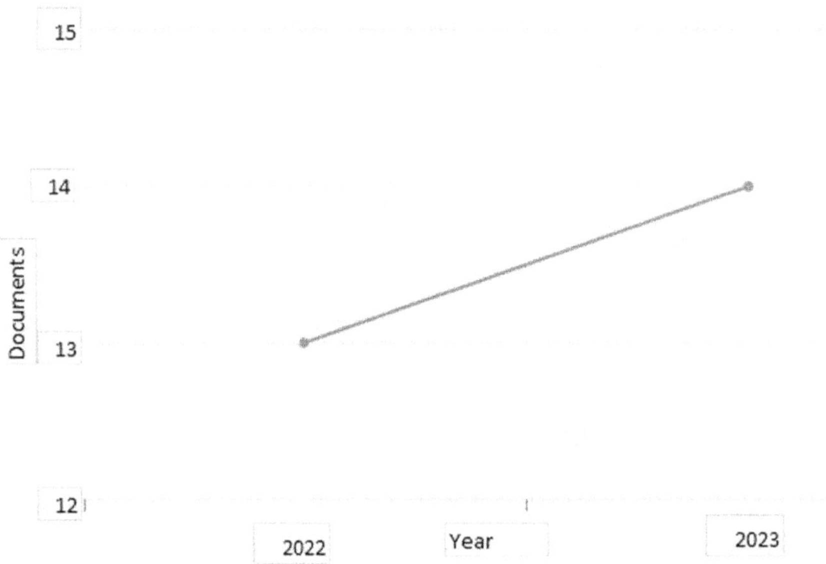

Figure 3. Articles published in different subject areas
Source: Scopus (2023)

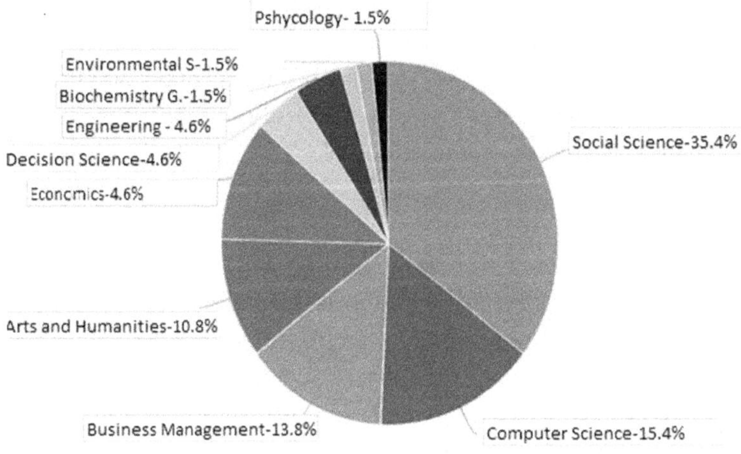

Figure 4. Articles published by different countries
Source: Scopus (2023)

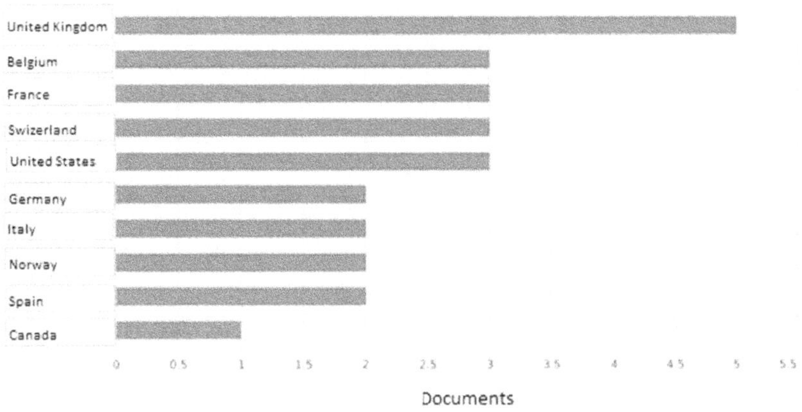

4.4 Articles Published by Different Countries

The selected primary studies of 27 documents among different countries delve into critical aspects of AI ethics, including unfairness, transparency, responsibility, and privacy. By using Scopus, we got an access on wealth of knowledge, analysis, and discourse on AI ethics, giving insights in this increasingly important and interdisciplinary field. Scopus is a comprehensive and multifunctional database that offers extensive coverage of research papers from a global perspective. By employing the country filter feature in Scopus, we conducted an examination of documents that have been authored by scholars and institutions affiliated with a certain country. Upon conducting a thorough analysis of the documents, it was determined that the United Kingdom emerged as the leading contributor of material pertaining to the ethical considerations of artificial intelligence (AI). Following the United Kingdom, Belgium, France, the United States, and Switzerland were ranked in descending order of significance, as illustrated in Figure 4.

5. RESULTS AND EVALUATION

The detail results to address RQ1 and RQ 2 are discussed in the following sections.

The systematic assessment of the 27 initial investigations reveals a range of diverse and complex factors that present challenges. The percentages of the problematic factors that have been found are presented in Figure 5. The subsequent information outlines the salient aspects of the difficulties that have garnered significant attention and citation within academic discourse. The conventional approach of centralised information development is frequently ineffective when used to crucial applications based on data (Mutlag et al., 2019). The purpose of the PRISMA Structure is to assist researchers enhance the communication of systematic assessments and meta-analytic Our attention has primarily been directed towards randomized trials; nevertheless, it is worth noting that PRISMA can also serve as a framework for publishing systematic analyses of various other forms of research, with a particular emphasis on analyses of solutions (Moher, et.al, 2010).

Figure 5. Frequency challenges related to AI ethics
Source: Scopus (2023)

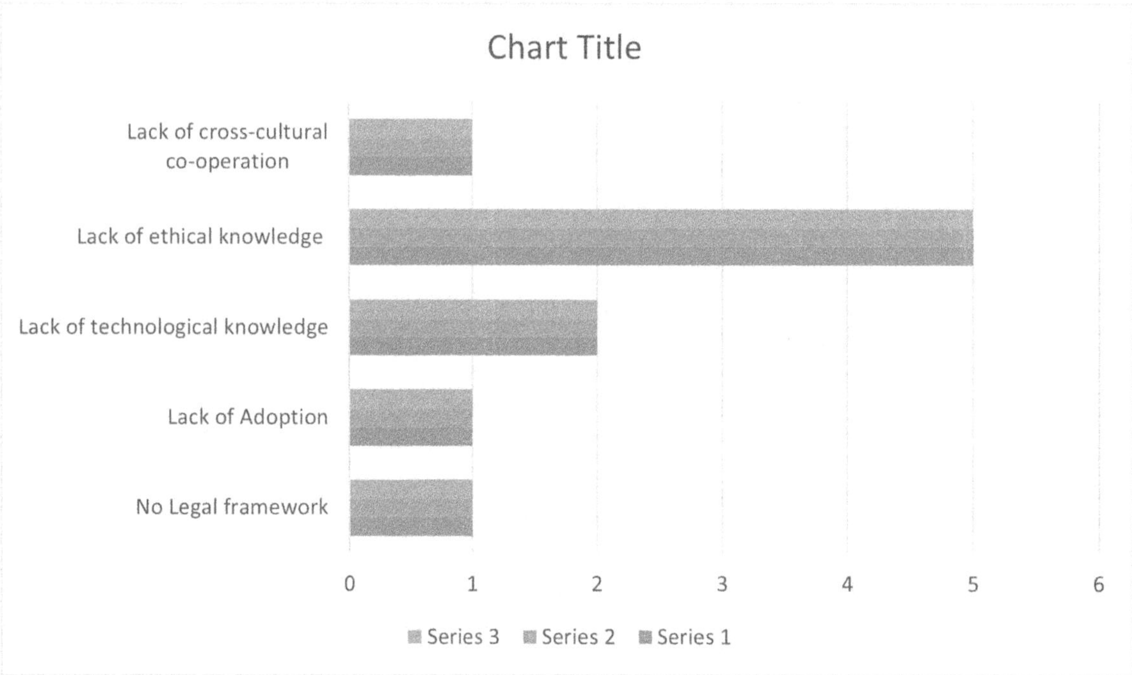

5.1 Lack of Moral Understanding

The absence of ethical knowledge is one of the main explanations why AI ethics in practice is not yet mature. Others contend that defining ethics in AI is impossible without a political approach, whereas AI system creation organisations believe that governmental agencies are ill-equipped to provide experts for this expanding field. Organisations that develop AI systems believe that governmental institutions are unable to provide experts for this emerging discipline. Management and technical personnel are unaware about the ethical and moral complexities posed by AI systems. AI ethics remain in their childhood, and there are not enough moral standards and models available to provide the AI industry with specific guidelines.

5.2 Lack of Technological Understanding

The policymakers have lack of technical knowledge due to which it becomes difficult for practicing AI ethics. Due to lack of technical understanding, a chasm has developed between the design of the system and ethical thought. Ethicists must possess the requisite abilities to comprehend technical knowledge within the context of their ethical framework.

Figure 6. PRISMA framework

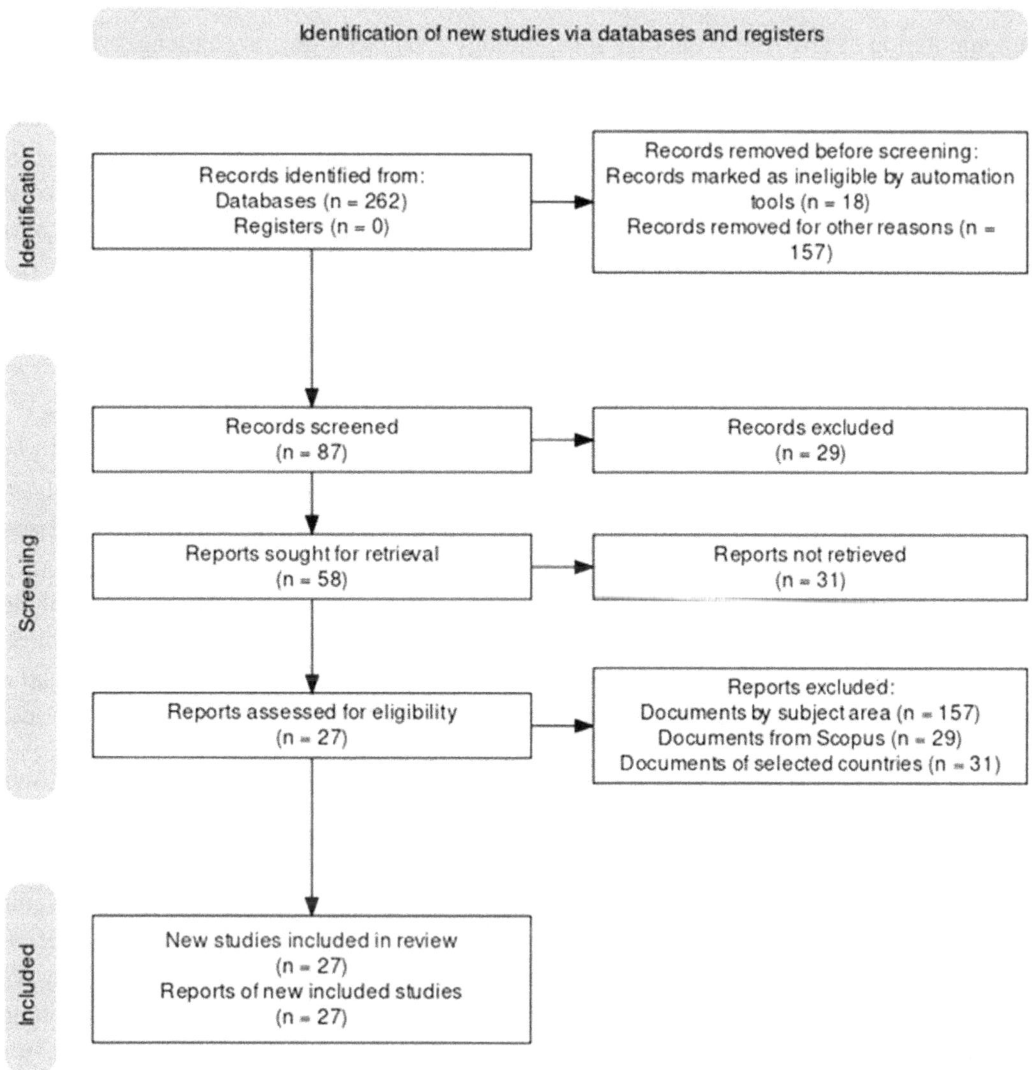

5.3 Lack of Cross-Cultural Operations

The absence of cross-cultural operations in AI's and other areas can provide a variety of obstacles and restrictions in their respective fields. Activities that include interaction, cooperation, or transactions between individuals or entities hailing from various cultural backgrounds are what we mean when we talk about cross-cultural operations. Cultural diversity manifests in distinct language usage, communication patterns, and non-verbal cues, hence engendering potential challenges such as misunderstandings, mistakes, and disrupted channels of communication. This can make it more difficult to collaborate effectively and make decisions.

5.4 Lack of Adoption

The implementation of artificially intelligent technology (AI) faces several challenges, including inadequate knowledge and skill, concerns over the protection of data and privacy, costs related to deployment, resistance to change, and ethical and legal considerations. The primary barrier to the mainstream use of artificial intelligence is to the growing significance of ethical considerations, which arise when organisations integrate AI into an expanding array of operational processes. The utilisation of artificial intelligence (AI) holds the capacity to provide a semblance of scientific credibility to inherent biases held by humans, frequently resulting in their magnification. Consequently, this phenomenon raises concerns about the reliability and suitability of AI for decision-making purposes.

5.5 No Legal Framework

The lack of a solid and real legal framework for AI presents the next level of difficulty for artificial intelligence and big data analysis. The existing paradigm for governing artificial intelligence revolves around the principles of human rights, humanist ethics, human dominance, and accountability. Consequently, there has been a notable shift towards prioritising the establishment and implementation of abstract regulations centred on constraining AI. These regulations are rooted in the safeguarding of rights for humans and the assertion of human dominance over AI. An instance of this methodology may be observed in the EU AI ACT, which was recently modified to accommodate the advancements in ChatGPT and GPT-4 innovations (Grady, 2023). However, it should be noted that the legislation does not adequately address the potential implications of AGI (artificial general intelligence) that may emerge in the future. According to the study conducted by Bubeck et al. (2023).

Analysis- The aforementioned problems outlined above offer a comprehensive examination of commonly and frequently highlighted factors that may serve as potential obstacles for the implementation of ethical practises in artificial intelligence. The predominant obstacle in the field of AI ethics is commonly recognized as a deficiency in ethical knowledge. In situations where individuals exhibit a deficiency in moral consciousness on a certain issue, it is plausible that consequential ethical lapses may transpire. Practitioners primarily prioritise creating software as their main responsibility, often overlooking ethical considerations due to little interest. The mitigation of ethical ambiguity in AI systems can only be achieved through the acquisition of ethical knowledge. The consistent consideration of ethical standards, codes, and regulations plays a crucial role in effectively overseeing the ethical principles inside artificial intelligence (AI) and autonomous systems. This discovery indicates that the domain of AI ethics concerns is very new and necessitates substantial research endeavours from several disciplines in order to reach a state of maturity. The importance of artificial intelligence (AI) technologies across several sectors necessitates urgent academic investigation to identify the pertinent obstacles that impede the integration of ethical considerations in AI. Furthermore, this study focuses exclusively on the detailed examination of problematic factors that occur with high frequency.

6. CONCLUSION AND FUTURE DIRECTIONS

In conclusion, research on the future of AI and Big data Analysis presents a dynamic landscape where technology, ethics, and societal implications intersect. As we continue to delve into AI's possibilities and challenges, it is crucial to maintain a multidisciplinary approach that balances technical advancements with the ethical considerations. When there is a lack of moral awareness on a particular problem, significant ethical errors might .The field of ethics in artificial intelligence (AI) has garnered considerable interest in recent years, prompting a need to comprehend the various obstacles and challenges that arise in the context of AI and big data analysis. The advent of Big Data has facilitated the collection and analysis of datasets across diverse settings, leading to the discovery of novel insights. This, coupled with advancements in Artificial Intelligence (AI) that rely on data, has enabled economic growth and various advantageous outcomes for individuals and society at large. This is because Big Data and the ability to collect and examine datasets from a wide range of settings and come up with new, surprising information go hand in hand. This rapidly growing phenomenon is expected to significantly impact numerous fields, including politics, law enforcement, economics, security, academia, medicine, and more. This study aims to examine two primary inquiries. Firstly, it seeks to explore the implications of the intersection of huge amounts of data and AI, or artificial intelligence, on human capabilities. Secondly, it aims to gain insights into the ethical considerations that arise from the utilisation of AI and the analysis of big data. Firstly, there is need to substantial emphases on governance, with intent to develop regulations and ethical norms for AI and 'actively participate' in the global governance of this technology'. Secondly, Collaboration across discipline lines is absolutely necessary in order to effectively solve these intricate problems. Researchers from a variety of disciplines, including computer science, philosophy, law, the social sciences, the business management, and public policy, need to collaborate in order to establish a complete knowledge of AI ethics and big data, as well as solutions that support the ethical and responsible use of technologies like these.

Moreover, the scope of this research can be expanded through the implementation of a survey targeting organisations that employ artificial intelligence (AI) technology. This survey would aim to explore the practical comprehension of AI ethics and ascertain the most effective strategies for addressing the complex variables associated with AI ethics, as well as establishing guidelines for ethical conduct. Furthermore, it is imperative to recognise that the ethical considerations surrounding artificial intelligence (AI) often overlap with the ethical concerns pertaining to technology at large, encompassing both virtual and traditional technologies. On the other hand, it does have a few quirks that one must be aware of and appropriately account for. The future of AI will be shaped not only by innovation but also by our ability to address ethical and societal concerns, ensuring that AI serves humanity's best interests and contributes positively to our rapidly evolving world. The primary goal is to contribute to the development of the future path of the massive data revolution, particularly in its intersection with advancements in AI, in a way that demonstrates a genuine consideration for fundamental ethical principles.

REFERENCES

AI, H. (2019). High-level expert group on artificial intelligence. *Ethics guidelines for trustworthy AI*, 6.

Board, D. I. (2019). AI Principles: Recommendations on the ethical use of artificial intelligence by the Department of Defense. *Supporting document. Defense Innovation Board*, *2*, 3.

Boyd, D., & Crawford, K. (2012). Critical questions for big data: Provocations for a cultural, technological, and scholarly phenomenon. *Information Communication and Society*, *15*(5), 662–679. doi:10.1080/1369118X.2012.678878

Bubeck, S., Chandrasekaran, V., Eldan, R., Gehrke, J., Horvitz, E., & Kamar, E. (2023). *Sparks of Artificial Intelligence: Early experiments with GPT-4*. Academic Press.

Bynum, T. (2008). *Computer and information ethics*. Academic Press.

Dalton, C. M., Taylor, L., & Thatcher, J. (2016). Critical data studies: A dialog on data and space. *Big Data & Society*, *3*(1), 2053951716648346. doi:10.1177/2053951716648346

Danielsson, J., Macrae, R., & Uthemann, A. (2022). Artificial intelligence and systemic risk. *Journal of Banking & Finance*, *140*, 106290. doi:10.1016/j.jbankfin.2021.106290

Dignum, V. (2019). *Responsible artificial intelligence: how to develop and use AI in a responsible way* (Vol. 2156). Springer. doi:10.1007/978-3-030-30371-6

Dixon-Woods, M., & Bosk, C. L. (2011). Defending rights or defending privileges? Rethinking the ethics of research in public service organizations. *Public Management Review*, *13*(2), 257–272. doi:10.1080/14719037.2010.532966

Duan, Y., Edwards, J. S., & Dwivedi, Y. K. (2019). Artificial intelligence for decision making in the era of Big Data–evolution, challenges and research agenda. *International Journal of Information Management*, *48*, 63–71. doi:10.1016/j.ijinfomgt.2019.01.021

Grady, P. (2023). *ChatGPT Amendment Shows the EU is Regulating by Outrage*. Available online at: https://datainnovation.org/2023/02/chatgpt-amendment-shows-the-eu-is-regulating-by-outrage/

Hao, Q., & Qin, L. (2020). The design of intelligent transportation video processing system in big data environment. *IEEE Access : Practical Innovations, Open Solutions*, *8*, 13769–13780. doi:10.1109/ACCESS.2020.2964314

Leidner, J. L., & Plachouras, V. (2017, April). Ethical by design: Ethics best practices for natural language processing. In *Proceedings of the First ACL Workshop on Ethics in Natural Language Processing* (pp. 30-40). 10.18653/v1/W17-1604

McCarthy, J., Minsky, M. L., Rochester, N., & Shannon, C. E. (2006). A proposal for the dartmouth summer research project on artificial intelligence, august 31, 1955. *AI Magazine*, *27*(4), 12–12.

Moher, D., Liberati, A., Tetzlaff, J., & Altman, D. G.Prisma Group. (2010). Preferred reporting items for systematic reviews and meta-analyses: The PRISMA statement. *International Journal of Surgery*, *8*(5), 336–341. doi:10.1016/j.ijsu.2010.02.007 PMID:20171303

Mutlag, A. A., Abd Ghani, M. K., Arunkumar, N., Mohammed, M. A., & Mohd, O. (2019, January). Enabling technologies for fog computing in healthcare IoT systems. *Future Generation Computer Systems*, *90*, 62–78. doi:10.1016/j.future.2018.07.049

Needham, J. (2013). *Disruptive possibilities: How big data changes everything*. O'Reilly Media, Inc.

Pazzanese, C. (2020). Ethical concerns mount as AI takes bigger decision-making role in more industries. *The Harvard Gazette, 26.*

Sharma, V., & Sood, D. (2022). The role of artificial intelligence in the insurance industry of India. In *Big data analytics in the insurance market* (pp. 287–297). Emerald Publishing Limited. doi:10.1108/978-1-80262-637-720221017

Sood, K., Dhanaraj, R. K., Balusamy, B., Grima, S., & Uma Maheshwari, R. (Eds.). (2022). *Big Data: A game changer for insurance industry*. Emerald Publishing Limited. doi:10.1108/9781802626056

Stahl, B. C., & Stahl, B. C. (2021). Ethical issues of AI. *Artificial Intelligence for a better future: An ecosystem perspective on the ethics of AI and emerging digital technologies*, 35-53.

Strydom, M. J., & Buckley, S. B. (2020). Big Data Intelligence and Perspectives in Darwinian Disruption. In *AI and Big Data's Potential for Disruptive Innovation* (pp. 1–43). IGI Global. doi:10.4018/978-1-5225-9687-5.ch001

Taricani, E., Saris, N., & Park, P. A. (n.d.). *Beyond technology: The ethics of artificial intelligence*. Academic Press.

Vincent, C. M. (2020). Ethics of Artificial Intelligence and Robotics. *The Stanford Encyclopedia of Philosophy*. Retrieved January 15, 2021 from https://plato.stanford.edu/archives/win2020/entries/ethics-ai/

Yenkar, V., & Bartere, M. (2014). Review on data mining with big data. *International Journal of Computer Science and Mobile Computing*, *3*(4), 97–102.

Chapter 14
Comprehending Algorithmic Bias and Strategies for Fostering Trust in Artificial Intelligence

U. Sidhi Menon
https://orcid.org/0009-0007-2338-8246
Christ University, India

Theresa Siby
Christ University, India

Natchimuthu Natchimuthu
https://orcid.org/0000-0002-2461-7801
Christ University, India

ABSTRACT

Fairness is threatened by algorithm bias, systematic and unfair disparities in machine learning results. Amazon's AI-driven hiring tool favoured men. AI promised data-driven, impartial decision-making, but it has revealed sector-wide prejudice, perpetuating systematic imbalances. The algorithm's bias is data and design. Biassed historical data and feature selection and pre-processing can bias algorithms. Development is harmed by human biases. Algorithm prejudice impacts money, education, employment, and crime. Diverse and representative data collection, understanding complicated "black box" algorithms, and legal and ethical considerations are needed to address this bias. Despite these issues, algorithm bias elimination techniques are emerging. This chapter uses secondary data to study algorithm bias. Algorithm bias is defined, its origins, its prevalence in data, examples, and issues are discussed. The chapter also tackles bias reduction and elimination to make AI a more reliable and impartial decision-maker.

DOI: 10.4018/979-8-3693-1762-4.ch014

INTRODUCTION

The Dartmouth Summer Research Project on Artificial Intelligence, held in 1956, is largely seen as the seminal event that established artificial intelligence as a distinct subject.

The project had a duration of around six to eight weeks and mostly consisted of an extensive brainstorming session. Originally, there were eleven scientists and mathematicians who intended to attend the event. Although not all of them showed up, more than ten additional individuals came for brief periods of time.

During the early 1950s, the topic of "thinking machines" was referred to by other titles, including cybernetics, automata theory, and sophisticated information processing. The multitude of names indicates the diverse range of intellectual perspectives.

In 1955, John McCarthy, a youthful Assistant Professor of Mathematics at Dartmouth College, made the decision to assemble a collective with the purpose of elucidating and advancing concepts related to artificial intelligence. He selected the moniker 'Artificial Intelligence' for the emerging discipline. He selected the name partly due to its neutrality, in order to steer clear of a restricted emphasis on automata theory and to avoid the strong emphasis on analogue feedback in cybernetics. Additionally, he wanted to avoid the possibility of having to recognise Norbert Wiener as an authoritative figure or engage in arguments with him (Moor, 2006).

Today, machine learning (ML) and artificial intelligence (AI) are extensively utilised across various sectors of our economy to make critical decisions that have wide-ranging consequences. Examples include employers using ML algorithms to evaluate job applications, financial institutions using ML tools to assess individual creditworthiness for loan approvals, retailers employing recommendation algorithms for product suggestions, doctors relying on algorithms for medical decision support, and some courts in the United States utilizing algorithms like COMPAS (Correctional Offender Management Profiling for Alternative Sanctions) for predicting recidivism (Van Dijck, 2022). Initially, ML algorithms were seen as a way to reduce historical bias and discrimination, with the promise of objective, data-driven decision-making (Fu et al., 2020).

For inclusive growth, can we envision a fair and equitable world where access to quality healthcare, nutritious food, and basic human necessities is available to everyone, regardless of age, gender, or social class? The question arises whether data-driven technologies like artificial intelligence and data science can achieve this goal or if existing biases that affect real-world outcomes will seep into the digital realm as well. This highlights the significant potential risk of biased AI. The journey to managing and mitigating this risk starts with comprehending how bias can creep into algorithms and why detecting it can be challenging. While AI holds the promise of creating a better and more equitable world, if left unregulated, it could also perpetuate historical inequalities. Fortunately, businesses can take steps to minimize this risk and use AI systems and decision-making software with confidence (Best & Rao, 2021).

Before delving deeper into this topic, let's establish the definitions of Algorithm, Algorithm Bias, Machine Learning, and Artificial Intelligence. Fundamentally, an algorithm can be defined as a collection of instructions or principles employed in computational or problem-solving tasks, typically carried out by a computing device (Köchling & Wehner, 2020). Within the field of computation, algorithms are transformed into software applications that have the capability to analyse incoming data using predetermined rules and produce corresponding output. Algorithms are integral to decision-making and advisory procedures across various domains of society (Silva & Kenney, 1960). Algorithmic bias is the term used to describe specific characteristics of an algorithm that result in unfair or subjective outcomes, displaying a preference for one group or entity over another (LibertiesEU, 2021).

The basic goal of machine learning, a subfield encompassed by artificial intelligence and computer science, is to replicate human learning mechanisms through the utilisation of data and algorithms. The accuracy of the aforementioned phenomenon improves progressively with time and is a fundamental element within the expanding field of data science. The algorithms in question have undergone training to provide classifications, predictions, and extract valuable insights inside data mining projects through the application of statistical approaches. These insights are critical for decision-making across a range of applications and businesses, and they should ideally have an impact on key performance metrics. The need for data scientists is anticipated to rise with the continued proliferation of big data because these professionals are crucial for finding pertinent business issues and the data required to answer them (IBM, n.d.).

The area of artificial intelligence (AI), which has only been for around 60 years, comprises many different fields, theories, and methods, including computer science, statistical analysis, probability theory, and computational neuroscience. Its main goal is to mimic cognitive functions of humans. AI growth, which began to emerge around the time of World War II and has closely matched technological advancements, has allowed computers to undertake increasingly complicated activities that were previously the sole preserve of humans. (History of Artificial Intelligence - Artificial Intelligence - www.coe.int, n.d.).

REAL CASE REPORTS OF ALGORITHM BIASES

Case reports serve as poignant reminders of the tangible consequences of algorithmic bias on individuals and communities. This inquiry into case reports seeks to elucidate the particular occurrences that exemplify the practical consequences and moral deliberations associated with prejudiced algorithms. Let's examine some real-life instances where the algorithm exhibited bias in the various fields which it was applied.

According to the Harvard Business Review, it was claimed in 1988 by the UK Commission for Racial Equality that a computer programme utilised by a medical school in the United Kingdom exhibited bias towards female and non-European candidates. The computer algorithm employed to choose candidates for interview shown a bias towards women and those with non-European names. The software application was designed to simulate the decision-making process employed by human admissions officers, achieving a level of accuracy ranging from 90 to 95 percent. According to Manyika (2022), the educational establishment in question exhibited a higher enrolment rate of non-European students compared to other medical institutions in London. The hiring and resume shortlisting algorithm developed by Amazon's software engineers in 2014 exhibited a prejudice against female candidates, resulting in their exclusion from the selection process. In 2015, it was shown that the system exhibited bias against women in technical occupations. As a result of concerns pertaining to bias and equity, Amazon recruiters made the decision to abstain from employing the programme for the purpose of assessing potential candidates (Omowole, 2022).

In addition, a study conducted by McKinsey revealed that the proportion of female CEOs in the United States is at 27%. However, it is noteworthy that a mere 11% of the images retrieved from an image search for "CEOs" depict women in this leadership role (Silberg & Manyika, 2019). The report by McKinsey also includes an examination of racial disparities in online advertising targeting, as investigated by Latanya Sweeney. Sweeney's research revealed that searches for the term "arrest" were more likely to generate advertisements featuring names commonly associated with African Americans, as opposed to names associated with individuals of white ethnicity. According to Sweeney (2013), it was postulated that visitors might have clicked on various variations more frequently for different queries, resulting in the algorithm displaying them more frequently. This occurred despite the equal presentation of several advertisement language variations, both with and without the term "arrest". The paper additionally emphasises that, based on research conducted by Jon Kleinberg, algorithms have the potential to aid in mitigating racial disparities within the criminal justice system. Another study revealed that candidates who have been historically underserved have significant advantages when computerised financial underwriting approaches are employed (Van Dijck, 2022). The COMPAS (Correctional Offender Management Profiling for Alternative Sanctions) system was utilised in Broward County, Florida, for the purpose of predicting the likelihood of reoffending. According to Mattu (2023), ProPublica's Julia Angwin and other researchers have demonstrated that COMPAS (Correctional Offender Management Profiling for Alternative Sanctions) exhibited a disproportionate tendency to misclassify African-American defendants as "high-risk" compared to white defendants, with the former group being affected at a rate nearly twice as high.

Based on a recent study conducted by PwC, it was determined by the US Department of Commerce that facial recognition artificial intelligence (AI) exhibits a tendency to inaccurately identify individuals belonging to racial and ethnic minority groups (Best & Rao, 2022). According to a separate investigation conducted by Georgia Tech, it was demonstrated that artificial intelligence (AI)-driven autonomous vehicles had increased difficulty in detecting individuals with darker skin tones (Cuthbertson, 2019). In the domain of natural language processing (NLP), which pertains to the use of artificial intelligence (AI) to facilitate computers in comprehending and deciphering human language, the presence of bias against individuals based on their ethnicity, gender, or disability has been acknowledged. The researchers also emphasise a study undertaken by Krishna (2020) from the University of Melbourne, which examines how algorithms can potentially perpetuate gender prejudices against women. In order to illustrate the potential of AI models to propagate existing societal biases on a large scale, researchers devised an experimental recruitment algorithm that replicated the gender biases observed in human recruiters. A study conducted at Stanford University unveiled notable racial disparities within automated speech recognition systems, wherein voice assistants exhibited a misidentification rate of 35% for keywords uttered by Black users, compared to a misidentification rate of 19% for terms said by White users.

The World Economic Forum's report highlights a study conducted on a risk prediction algorithm in the healthcare domain, which is utilised by an estimated 200 million individuals in the United States. The analysis uncovered a notable bias in favour of white patients as opposed to black patients. Furthermore, the report elucidates the apprehensions expressed by researchers in the domain of cancer detection, since current algorithms exhibit limitations in accurately discerning skin cancer in persons with higher levels of melanin pigmentation (Omowole, 2022). In a comparable vein, a recent investigation pertaining to models for mortgage loan approval has revealed that the efficacy of these prognostic systems is diminished when used to minority cohorts, mostly owing to the paucity of comprehensive credit history data accessible for such groups, particularly those belonging to low-income brackets. The presence of inconsistent credit history data has a role in the discrepancies observed in home loan approval rates among majority and minority populations. According to a study conducted by Scott Nelson of the University of Chicago and Laura Blattner of Stanford University, it was shown that individuals belonging to low-income and minority groups had credit history data that is less thorough. This finding provides valuable insights into the disparities observed in mortgage approval outcomes (Andrews, 2021). Regrettably, the underlying source of this bias has predominantly evaded detection and clarification as a result of insufficiently comprehensive data analysis. The healthcare risk prediction system, when implemented on a population of around 200 million Americans, revealed a disparity in the treatment received by white patients compared to black patients, with white patients being afforded more favourable therapy. The algorithm's development problem can be attributed to insufficient testing across a range of racial groups prior to its adoption. According to a study conducted by Carnegie Mellon University in 2015, there was a significant presence of gender bias observed in the advertisements displayed on Google and Facebook platforms. It was discovered that in the same year, Google inadvertently mislabelled photographs of black individuals as "Gorillas." In 2016, Microsoft made the decision to deactivate Tay, an AI chatbot, due to its utilisation of racial insults. The argument contends that ProPublica's research of its predictive model for future criminal behaviour in the United States exhibited unjust targeting towards specific racial groupings. In the same year, it was determined that a machine learning algorithm employed for the evaluation of participants in a global beauty competition exhibited bias in favour of individuals possessing darker skin tones. A multitude of individuals from various regions across the globe submitted photographs to Beauty.AI, with the assumption that an advanced algorithm devoid of human biases would impartially determine their aesthetic appeal, thereby providing an accurate definition of human beauty. The computer autonomously acquired the ability to discern human aesthetics by analysing a vast collection of photos from numerous beauty pageants, thereby identifying patterns associated with winning contestants. In contrast, artificially intelligent algorithms shown a rapid decline in performance, as the computer made determinations of winners only based on the criterion of skin tone (Sen, 2017).

Types of Algorithm Bias

As we explore the realm of algorithmic decision-making, it becomes clear that bias can appear in different ways. To successfully address and minimise these concerns, it is necessary to clearly identify and understand the many forms of algorithmic bias.

Selection Bias: This bias originates when the dataset utilised to train an algorithm lacking representativeness of the entire population or exhibits a skewed distribution towards a specific group. As a result, the algorithm may have difficulties in generating precise forecasts for demographic groupings that are not adequately represented. (Mehrabi et al., 2021)

Labelling Bias: Labelling bias arises when errors or inconsistencies in the labelling of training data introduce bias into the algorithm. For instance, if data annotators apply labels influenced by their own biases, the algorithm may learn and perpetuate these biases.

Historical Bias: Historical bias stems from systemic discrimination within historical data, which serves as the basis for training algorithms. If past decisions or practices were discriminatory, the algorithm may unintentionally assimilate and perpetuate these biases in its predictions. (Mehrabi et al., 2021)

Amplification Bias: Amplification bias refers to a situation in which an algorithm intensifies or magnifies pre-existing biases that are inherent in the training data. An illustration of this phenomenon can be observed when an algorithm for generating sentencing recommendations is trained on a dataset that exhibits prejudice. In such cases, the system may inadvertently propose lengthier punishments for certain demographic groups, thus inadvertently perpetuating the existing bias. (Nitika Sharma, 2023).

Figure 1. Types of algorithm bias

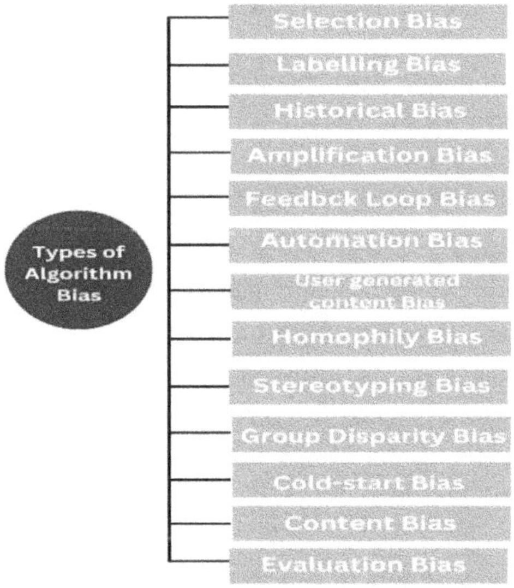

Feedback Loop Bias: Feedback loop bias materializes when an algorithm's recommendations or decisions influence real-world outcomes, subsequently shaping the training data. This feedback loop can intensify and reinforce existing biases, creating a harmful cycle.

Automation Bias: Automation bias pertains to the inclination of individuals to place more trust in algorithmic recommendations than their own judgment, even when the algorithm exhibits bias. This can result in biased decisions based on algorithmic suggestions.

User-Generated Content Bias: In platforms where users generate content, such as social media or online reviews, algorithmic systems may inadvertently promote or suppress specific content based on user behavior. This can lead to biased content recommendations.

Homophily Bias: Homophily bias arises when an algorithm relies on the principle that individuals with similar characteristics tend to share similar preferences. This can create filter bubbles, where users are exposed only to content and information aligning with their existing beliefs and opinions (Annie Brown, 2020).

Stereotyping Bias: Algorithms may perpetuate stereotypes related to race, gender, age, or other attributes when making predictions or recommendations. Stereotyping bias can lead to unfair and inaccurate outcomes.

Group Disparity Bias: This bias involves variations in algorithmic outcomes among different demographic groups. Group disparity bias can manifest as unequal access to resources, opportunities, or services.

Cold-Start Bias: The phenomenon known as cold-start bias is observed in recommendation systems, wherein algorithms encounter difficulties in generating precise recommendations for novel users or objects that possess insufficient data. This bias has the potential to lead to less than ideal initial user experiences.

Content Bias: Content bias materializes when an algorithm selects or prioritizes content based on pre-existing biases. For instance, in a news recommendation system, content with a particular political leaning may be favoured.

Evaluation Bias: This bias arises when the evaluation metrics used to assess algorithm performance themselves exhibit bias. Biased evaluation metrics can lead algorithms to optimize for incorrect objectives.

It is crucial to understand that these forms of bias are not mutually exclusive, and algorithms can exhibit multiple types of bias simultaneously. The detection and mitigation of these biases are essential to ensure fair and equitable algorithmic outcomes (Annie Brown, 2020).

Reasons of Algorithm Bias

An essential measure in promoting openness and accountability in artificial intelligence systems is to investigate the fundamental factors contributing to algorithm bias. Let us analyse the primary factors behind this trend. The occurrence of Algorithm Bias can be attributed to several significant reasons:

Input Bias, Training Bias, and Programming Bias: Input Bias arises when the data used as inputs to algorithms is non-representative, lacks information, carries historical bias, or is otherwise flawed. Algorithms are influenced by the nature and quality of the data they analyse, which may cause biases in the output data to persist. Training Bias is linked to the data used to train algorithms, where data misclassification or inappropriate assessment can lead to biased outcomes. Algorithms may learn from correlations instead of causation, potentially overlooking relevant factors. Programming Bias refers to biases ingrained in an algorithm's design, including subjective rules programmed into it. Learning algorithms can also develop bias from their interactions with human users and assimilation of new data.

De-Biasing Data Challenges: Removing sensitive attributes, such as race, from datasets as a means of achieving fairness is often insufficient because correlated attributes can serve as proxies for the sensitive ones. For example, even without race data, ZIP codes may indicate a person's race based on

Figure 2. Reasons of algorithm bias

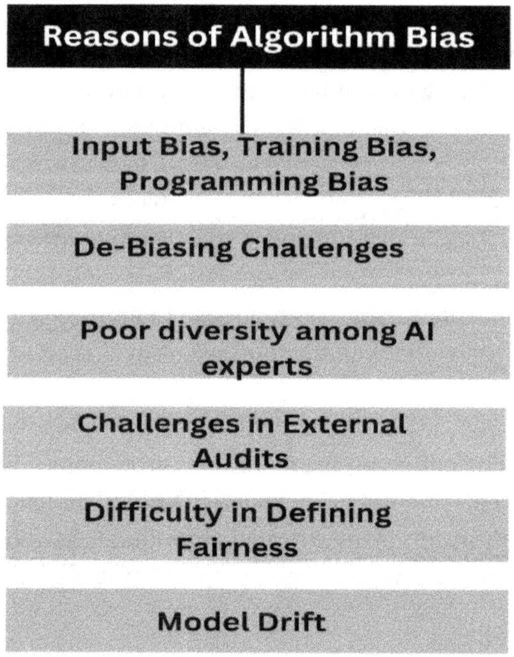

residential patterns. Research has shown that removing sensitive columns doesn't prevent systematic discrimination. Some experts suggest retaining sensitive columns and using them as levers to monitor and correct bias during model training (Kelley, 2023).

Poor Diversity Among AI Experts: Bias in AI systems is a result of a dearth of diversity among AI professionals (AI Bias Is Personal for Me. It Should Be for You, Too., n.d.). A more diverse team can offer a wider range of perspectives and identify biases that a homogenous team might overlook. For instance, Joy Buolamwini, a Ghanaian-American computer scientist, uncovered biases in facial recognition tools that performed poorly on darker skin tones. Joy Buolamwini, a graduate researcher at the MIT Media Lab and founder of the Algorithmic Justice League, remains quickly combating biases in decision-making software, specifically in facial recognition technology. Her fascination with this field originated during her undergraduate studies when she noticed that social robots encountered challenges in recognising her face in comparison to those with lighter complexion. Consistently seeing this phenomenon in many situations, she came to recognise the presence of unconscious bias in technology. The issue arises from the utilisation of benchmark datasets within the facial recognition community, which may exhibit a deficiency in diversity, resulting in a distorted portrayal of advancements. Buolamwini underscores the necessity of scrutinising the representativeness of benchmarks and underscores the significance of gathering varied data.

The ramifications of partial algorithms transcend beyond facial recognition, impacting domains such as automation, critical decision-making (insurance, loans, recidivism risk), and even admissions determinations. Buolamwini emphasises the potential for continuing past disparities if algorithms rely on obsolete data. She promotes the idea of making algorithm outcomes transparent and testing for biases, regardless of the fact that the underlying algorithms are still kept private. Buolamwini recognises the difficulties in being transparent because of business sensitivity, but highlights the responsibility of companies in dealing with biases. Reflecting on her personal journey as a black woman from Mississippi who achieved the prestigious titles of Rhodes Scholar and Fulbright Fellow, Buolamwini contemplates the potential consequences of algorithmic decision-making on persons from varied backgrounds. She recognises that she has gained advantages and chances due to specific privileges, emphasising the significance of taking identity perspectives into account in algorithmic decision-making(Tucker, 2018). A more diverse team might have identified and addressed this issue earlier.

Challenges in External Audits: External audits of AI algorithms are valuable for detecting biases, especially in high-stakes applications. However, privacy regulations like General Data Protection Regulation and California Consumer Privacy Act make it difficult to access both the model and the training data. Companies are constrained by privacy regulations, limiting external assessments of their algorithms.

Difficulty in Defining Fairness: Fairness is a complex concept that requires a clear definition before it can be implemented in algorithms. With over 30 different mathematical fairness definitions, stakeholders need to reach a consensus on which definition to use, complicating the task for technologists and data scientists.

Model Drift: Some algorithms are designed to continuously learn from data, making them vulnerable to becoming biased over time. An example is Microsoft's chatbot "Tay," which initially started innocently but quickly became biased and offensive based on the conversations it learned from. Continuous learning algorithms can evolve in unexpected and undesirable ways. These factors collectively contribute to the persistence of bias in AI systems, making it a complex and challenging issue to address (7 Reasons for Bias in AI and What to Do about It - insideBIGDATA, 2022).

Figure 3. Sectors affected by algorithm bias

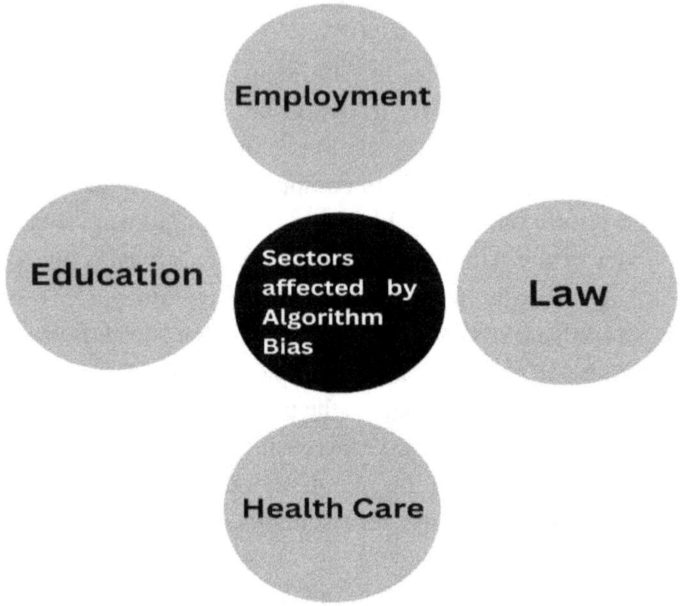

Sectors Affected by Algorithm Bias

The ramifications of algorithmic bias reverberate across diverse industries, presenting obstacles and ethical dilemmas in domains that primarily depend on data-centric decision-making. Understanding the impact of prejudice in many sectors is crucial for promoting impartiality and equality.

Education

No one should be denied access to education because of their age, caste, creed, race, sexual orientation, and ethnicity or on any other grounds. Recent examples, however, demonstrate that the algorithm is learning the current database and performing upon the training to choose students, and because of a discrepancy between the outcomes an algorithm is meant to predict and what it really predicts, bias occurs. Additionally, biases have been documented in which students have been treated unfairly based on their grades and class or school.

The algorithm developed by Ofqual made projections regarding the expected academic performance of students in a certain educational institution, rather than providing an accurate assessment of individual students' overall achievements over the academic year. The algorithm utilised prior school performance as a predictor, resulting in a higher probability of grade reduction for high-achieving students in underperforming schools. Ofqual's grading algorithm raises concerns regarding the ethical appropriateness of assigning grades to pupils based on the quality of their schools rather than utilising more personalised measures of achievement. Furthermore, the algorithm exhibited bias towards students enrolled in private schools by assigning greater significance to teacher evaluations in educational institutions characterised by smaller class sizes. According to Smith (2020),

Employment

In this field, algorithm bias has been thoroughly documented. There are many opportunities for the algorithm to learn, forecast, and choose candidates based on the organization's previous recruiting practices since organizations utilize various software programmes, engines, and now AI to shortlist resumes and complete the hiring process. The underrepresentation of women in Amazon's training data is likely what led to the bias in the algorithm used to filter resumes that was discovered.

This application evaluated candidates by looking for trends in the resumes that were submitted to the business over a ten-year period. Amazon's dataset and hiring practices, however, certainly reflected the fact that men predominate in the tech sector. Therefore, the computer recognised gender-related traits in the dataset even though gender did not include as a variable in its model. (Omowole, 2022c) The hiring algorithm used by Amazon is an excellent illustration of how bias in a dataset (such as the underrepresentation of women in STEM areas and in positions of leadership in technology companies) can distort decisions against underrepresented groups.

Healthcare

Nobody would dispute the existence of algorithm bias in the healthcare sector. In reality, an industry that ought to be honest and fair is not. Everyone should have access to medical facilities around the world, regardless of who they are. But regrettably, new research has shown that algorithms have the potential to reinforce current racial inequalities in healthcare, which emphasises the significance of carefully considering the labels used during algorithm development. The algorithm for numerous therapies and testing does not identify patients of colour. Specifically, compared to white patients with the same number of chronic illnesses, black patients were less likely to be selected by the algorithm as candidates for potentially helpful care programmes. (Wiens et al., 2020)

Despite all the drawbacks, there is yet hope for an effective algorithm programme to improve the industry. Reinforcement learning (RL), for instance, can help with medical decision-making by integrating patient data gathered from electronic health record (EHR) systems. These data can be used by RL, a sort of machine learning, to create therapy guidelines (B. D. Smith et al., 2023).

Law

Currently, algorithm bias in the legal industry is a significant issue. The potential for bias in these algorithms has grown significantly as legal institutions have depended on the past records, personal bias and now more on artificial intelligence and machine learning algorithms to assist in decision making processes. Unfair outcomes, which disproportionately affect marginalised people, can and have been caused by biases in training data, errors in design, or historical injustices. To respect the ideals of justice and equality before the law, it is crucial to ensure fairness, accountability, and openness in the creation and use of legal algorithms. One of the most important steps in moving the legal industry towards a more just and equitable system for everybody is to address algorithmic bias. People of colour are wrongly identified by face recognition AI more frequently than people with white skin, for example, according to a US Department of Commerce research. This finding raises concerns about the likelihood that law enforcement could unfairly detain people of colour by utilizing face recognition technology. In fact, false positives from facial recognition software have already resulted in erroneous detentions.

METHODS FOR DETECTING ALGORITHM BIAS

Identifying algorithmic bias is a crucial measure to guarantee fair and impartial results in artificial intelligence. This investigation into the different approaches to detecting bias is to provide insight into the methods used to analyse and correct apparent discriminatory practises.

Government Policies

In the present context, algorithm bias in government policies is a major issue. The possibility of bias entering these systems is a major concern as governments depend more on historical algorithms and artificial intelligence to make judgements regarding resource distribution, law enforcement, and public services. These algorithms produce unfair results that disproportionately affect marginalised people because they have biases built into them from the data they are trained on. Governments must place a high priority on openness, equity, and ongoing audits of these algorithms to make sure they do not support or exacerbate already-existing inequities. In the increasingly data-driven governance environment of today, addressing algorithm bias in government policy is crucial to upholding the principles of equity and justice. It was hard to say that the Medicaid algorithm in Arkansas and the unemployment algorithm in Michigan were biased because there was no indication that the decisions were more likely to be incorrect for men than for women or for Black people than for White ones. Instead, these algorithms have inadequate decision-making capabilities and poor algorithmic design. However, the citizens who lost their jobs and need access to healthcare were disproportionately affected by these inaccurate estimates. The challenges victims had when attempting to challenge these algorithmic conclusions serve as a reminder of the necessity for technical due process, or the right to meaningfully engage with and challenge algorithmic judgements. This analysis aims to pinpoint potential biases, errors, and disparities within the dataset. Researchers can employ statistical methods to gauge whether particular groups are underrepresented or overrepresented in the data. Techniques such as data preprocessing, which includes reweighting or resampling, can help in rectifying these imbalances and mitigating bias.

Fairness Metrics: The development of fairness metrics offers a means to quantitatively assess bias in algorithmic outcomes. These metrics serve to evaluate discrepancies in outcomes among different demographic groups. Common fairness metrics include disparate impact, equal opportunity, and statistical parity. By applying these metrics, researchers can objectively measure the extent of bias within algorithmic predictions.

Auditing and Review: Auditing algorithmic systems for bias involves conducting in-depth examinations of the system's decision-making processes. This may entail a review of the algorithm's source code, parameters, and training data. This auditing process proves particularly valuable when the decision-making process is intricate and not easily interpretable.

Testing with Synthetic Data: The use of synthetic data aids in assessing an algorithm's behaviour in a controlled setting. By creating synthetic datasets that mimic the characteristics of real-world data, researchers can systematically introduce bias and gauge how the algorithm responds. This approach can pinpoint vulnerabilities and areas requiring bias mitigation.

User Feedback and Case Studies: Actively soliciting feedback from users who engage with algorithmic systems can be a valuable tool in bias detection. Users may report instances where they perceive the system's decisions as being unfair or biased. Furthermore, conducting case studies on specific algorithmic decisions and their real-world implications can shed light on potential sources of bias.

Challenges in Detecting Algorithm Bias

Although there is an increasing recognition of algorithmic bias, the process of identifying and reducing these biases is not simple." It is crucial to navigate the complex terrain of obstacles in identifying algorithm bias in order to strengthen the basis of fair and impartial artificial intelligence.

Opaque Algorithms: Many algorithmic systems, particularly those rooted in machine learning and deep learning, are inherently opaque. Their decision-making processes are complex and challenging to interpret. This opacity presents a significant challenge in the identification and comprehension of bias sources within the system.

Dynamic Data: Data used for training algorithms can evolve over time, reflecting shifts in society and changing norms. This dynamic nature of data poses challenges in maintaining fairness in algorithmic decisions (Nitika Sharma, 2023).

Intersectional Bias: People often belong to multiple demographic groups, and bias can intersect in intricate ways. Detecting intersectional bias, where a person's multiple identity facets contribute to discrimination, proves challenging.

Legal and Ethical Concerns: The legal and ethical landscape concerning algorithm bias is continually evolving. Ensuring that bias detection methods adhere to legal requirements and ethical standards is a complex task.

Bias Amplification: Algorithms may inadvertently magnify existing biases present in the training data. Detecting this form of bias amplification necessitates the use of advanced techniques and comprehensive data analysis.

Ethical Considerations in Algorithmic Bias

Transparency and Accountability: Transparency in algorithmic decision-making holds utmost importance. Developers must be held accountable for the outcomes generated by their algorithms, and mechanisms for reviewing and challenging algorithmic decisions should be firmly established.

Fairness and Equity: Ensuring that algorithms operate with fairness and equity aligns with moral imperatives. The development and utilization of algorithms should adhere to principles of fairness and social justice.

Informed Consent: Users engaging with algorithmic systems must be well-informed about how their data is utilized and the potential biases present in the system. Maintaining ethical standards necessitates obtaining informed consent.

Bias Mitigation: Ethical considerations underscore the need for proactive steps to mitigate bias within algorithms, even before it is identified. Proactive efforts to reduce bias prove essential.

Diverse Development Teams: Ensuring the presence of diverse teams of developers and data scientists can aid in mitigating bias during algorithm development. Diverse perspectives contribute to a more comprehensive understanding of potential sources of bias (Nitika Sharma, 2023).

Challenges in Mitigating Algorithm Bias

Addressing algorithm bias requires more than just identifying it; it involves dealing with a variety of obstacles, including interpretability issues and ethical implications. This investigation of the difficul-

Figure 4. Ethical consideration in algorithm bias

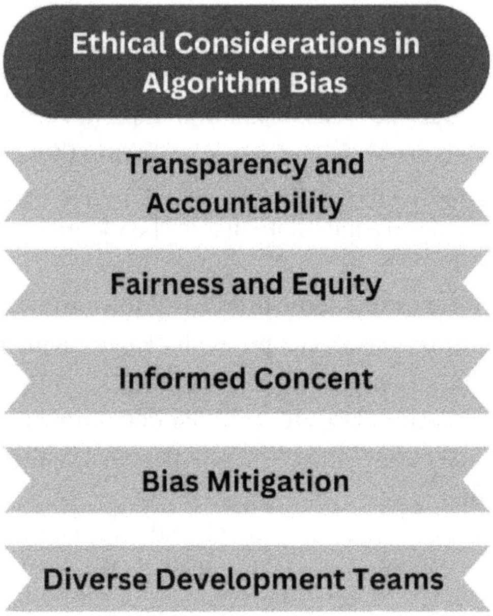

ties of reducing algorithmic bias seeks to provide insight into the complexities and obstacles involved in promoting fairness.

At the **organizational level**, there are several challenges related to team composition and engagement with third-party developers and algorithm vendors:

The issue of limited diversity within teams: Tech companies often lack diversity, primarily consisting of male, affluent, and white employees. The problem extends beyond the pipeline issue, encompassing biases in recruitment, hiring, promotions, harassment, and outdated workplace policies. The homogeneity of technology workplaces contributes to biased algorithm development.

Insufficient understanding of social science and domain-specific knowledge: Algorithm development teams typically comprise individuals with STEM backgrounds, such as data scientists, computer scientists, and engineers. These individuals approach problems with a mathematical perspective, which differs significantly from the social science and philosophical approach required to address societal issues. Integrating social science knowledge into algorithm development is crucial for tackling bias, but it's often overlooked, particularly in smaller companies and startups. (Eubanks, 2018)

"Unknown Unknowns": The presence of bias may not be readily apparent during the initial development of an algorithm. Teams may lack awareness of the subsequent effects of their data and decisions, particularly in the absence of social science knowledge. The task of retrospectively identifying and isolating these "unknown unknowns" might pose significant challenges, hence resulting in unanticipated biases inside algorithms. (Hao, K., 2020).

Focus on Technical Bias and Solutions: Organizations often focus on technical bias in data and solutions that attempt to "de-bias" data as a band-aid approach. This narrow focus on technical aspects may not address more subtle forms of bias and societal considerations.

At an **industry-wide level**, the rapidly evolving nature of AI presents challenges:

Industry pace and Market Priorities: Technology businesses operate within a dynamic industry characterised by rapid product innovation, wherein market competition hinges on the timely introduction of novel offerings. The incorporation of ethical considerations and the mitigation of bias necessitate significant investments of time and resources, frequently resulting in reactive actions such as product recalls rather than proactive efforts to solve fundamental concerns. (Leetaru, K., 2019)

Lack of Regulations and Guidance: The AI industry is now characterised by a lack of comprehensive legal frameworks, with emerging rules often being imprecise and high-level, missing practical and implementable tools. The establishment of basic values is impeded by geopolitical divisions, hence impeding international consensus on norms and impeding progress. (Pichai, S., 2020)

Persistence of Black Box Algorithms: Numerous artificial intelligence systems exhibit a "black box" characteristic, rendering their decision-making procedures impervious to external examination. There is a growing desire from both regulators and consumers for the utilisation of "white box" models that offer greater explain ability. However, it is important to acknowledge that there exists a trade-off between the level of explain ability and the predictive accuracy of these models. (Hulstaert, L., 2019)

Constraints in Machine Learning: Machine learning (ML) systems frequently employ binary categories that tend to oversimplify intricate facets of identity, such as gender and disability, so resulting in the elimination of nuanced identities. (Disability, Bias, and AI; Report, 2023)

IP Laws Impact Algorithm Transparency: Intellectual property (IP) laws protect algorithm inputs, making it challenging to understand how algorithms were developed and how they make decisions, particularly for organizations using existing algorithmic systems (Nkonde, 2021).

On a broader **societal level**, historical inequities, education gaps, and differing definitions of fairness contribute to bias:

Historical Disparities & Data: Biased algorithms are deeply rooted in historical inequalities and power disparities. Acquiring data that accurately represents all identities has significant challenges, as it is sometimes accompanied by privacy problems and may occasionally overlook certain dimensions such as the gender spectrum and disability.

Poor Diversity in STEM: The poor of diversity in STEM fields reinforces power imbalances of decision-making in Artificial Intelligence Systems.

The utilisation of obsolete educational approaches: The neglect of ethics, social science, and design thinking education from the curricula of data and computer scientists imposes constraints on their capacity to effectively mitigate any negative consequences associated with AI systems.

Legal Restrictions: The present legal frameworks that require the implementation of algorithms that are blind to race and gender may unintentionally sustain prevailing disparities and impede the identification of discriminatory practises..

Definition of "Fairness": The concept of fairness varies across disciplines, and there is no universally accepted definition. Developers struggle to express fairness in mathematical terms, and there are trade offs between different definitions, making it challenging for businesses to adopt a specific definition (Silberg & Manyika, 2019).

These challenges collectively contribute to the persistence of bias in AI systems at multiple levels of society and industry.

STRIVING FOR TRUSTWORTHY ARTIFICIAL INTELLIGENCE: STEPS TO REDUCE AI BIAS

AI has reached a level of sophistication where it plays a crucial role in making important decisions. To ensure trustworthiness and reduce bias and its associated risks, the following measures can be implemented:

Identify Unique Vulnerabilities: Different industries, such as banks, retailers, and utilities, face distinct risks related to potential AI bias. Understanding where bias can infiltrate data sets and algorithms, as well as its potential impact, is essential. Prioritize mitigation efforts based on these specific vulnerabilities to minimize financial, operational, and reputational risks.

Control Data: The adequacy of conventional data controls in identifying issues connected to AI bias may be limited. Particular emphasis should be placed on historical data and data obtained from external sources. It is advisable to use caution when dealing with "proxies" or correlations that may be biased inside data sets. The implementation of strategies aimed at generating "synthetic data" has the potential to mitigate bias by addressing gaps in the available information.

Govern AI at AI Speed: Artificial intelligence (AI) systems are frequently designed to operate continuously and leverage data from several areas inside an organisation. The concept of governance ought to be implemented in a continuous and comprehensive manner, including the entire organisation. This entails the establishment of clearly comprehensible frameworks, toolkits, as well as standardised definitions and controls. This technique guarantees that both artificial intelligence experts and clients can conform to regulations and detect issues prior to their escalation.

Diversify Your Team: The recognition of bias is contingent upon one's perspective, and the inclusion of diverse teams can serve to mitigate the potential oversight of bias. An inclusive team composition should encompass individuals with expertise in data science, business management, and diverse professional backgrounds, including but not limited to legal, accounting, sociological, and ethical domains. This amalgamation of perspectives enables a comprehensive examination of bias risks and the development of effective mitigation strategies.

Validate independently and continuously: Similar to managing any other major risk, establish an independent line of defense to continuously analyze data and algorithms for fairness. This can be an internal team or a trusted third party with a proven methodology. Technology tools like Bias Analyzer can automate this process and provide insights into mitigation actions.

Stay Informed: Given the fast-paced nature of AI research, business leaders should stay updated on the latest developments. Organizations like the AI Now Institute, the Partnership on AI, and the Alan Turing Institute's Fairness, Transparency, Privacy group offer valuable resources for learning more about AI ethics and fairness.

Establish Responsible Processes: In order to ensure the proper implementation of artificial intelligence (AI), it is crucial to establish resilient systems designed to address and minimise bias. Utilise a diverse range of technical tools and operational practises, including the incorporation of internal "red teams" or the usage of third-party audits. Prominent technological companies, such as Google AI and IBM, provide recommended standards and frameworks for achieving this purpose.

Engage in Fact-Based Conversations: Utilise sophisticated methodologies to examine and identify inherent biases inside machine systems, hence facilitating evidence-driven dialogues pertaining to potential biases originating from human influences. Algorithms have the capability to operate in conjunction with human decision-makers, and the utilisation of "explainability techniques" can aid in identifying the reasoning behind decisions. This facilitates a more comprehensive comprehension of disparities and promotes enhancements in processes driven by humans. (Best & Rao, 2021; Manyika, 2022).

Leverage Human-Machine Collaboration: Implement "human-in-the-loop" systems that allow humans to double-check AI recommendations or choose from options provided by the AI. Transparency about the AI's confidence in its recommendations aids human decision-making.

Take a Multidisciplinary Approach: There is a need to allocate additional resources towards the expansion of diversity within the realm of bias research. Furthermore, it is imperative to prioritise privacy concerns when facilitating the progress of collaborative efforts. The incorporation of transparency in design decisions and the integration of ethical considerations into computer science curricula have the potential to foster the development of artificial intelligence (AI) in a more responsible manner.

Invest in Diversifying the AI Field: A more heterogeneous AI community has the potential to enhance its ability to anticipate, evaluate, and mitigate prejudice. This underscores the need for investments in education and opportunities, exemplified by the initiatives undertaken by AI4ALL, a nonprofit organisation dedicated to cultivating a diverse and inclusive pool of AI professionals in marginalised groups. Artificial intelligence (AI) exhibits significant potential for both commercial enterprises and the broader community (Jensen, 2020). However, the establishment of trust in these systems is contingent upon their capacity to generate impartial outcomes. Through collaborative efforts, individuals can effectively address the issue of bias in artificial intelligence (AI), so enabling the realisation of its potential advantages while simultaneously confronting the societal dilemmas associated with human prejudices. (Best & Rao, 2021; Manyika, 2022).

CONCLUSION

In conclusion, algorithm bias is a significant and complex issue that affects various sectors, including education, employment, healthcare, law, and government policies. The use of algorithms and artificial intelligence has the potential to bring about more efficient and fair decision-making processes, but it also carries the risk of perpetuating and even exacerbating existing biases in society. Detecting and mitigating algorithm bias is crucial for ensuring fairness and equity in automated systems. Various methods can be employed to detect algorithm bias, including data analysis and pre-processing, the use of fairness metrics, auditing and review, testing with synthetic data, and gathering user feedback. These approaches help in identifying bias within algorithms and understanding its extent and impact. However, there are challenges in the process, including the opacity of some algorithms, the dynamic nature of data, intersectional bias, legal and ethical concerns, and the potential for bias amplification. Ethical considerations play a vital role in addressing algorithm bias. Transparency, accountability, fairness, and equity are core principles that should guide the development and use of algorithms. Informed consent is essential to ensure that users are aware of how their data is being used and the potential biases present in the system. Bias mitigation efforts should be proactive, and having diverse development teams can help identify and address potential sources of bias.

Addressing algorithm bias is an ongoing endeavour that requires collaboration among technology companies, researchers, policymakers, and the public. It is a complex issue that requires continuous monitoring and adaptation as algorithms evolve and data changes. The ultimate goal is to harness the benefits of artificial intelligence and algorithms while ensuring that they contribute to a more equitable and just society. As we move forward, it is essential to recognize that algorithm bias is not an insurmountable problem. With the right strategies, tools, and a commitment to ethical principles, we can work towards algorithms that make decisions that are fair, unbiased, and aligned with our societal values. The journey to a more equitable use of artificial intelligence and machine learning in various sectors of our economy is ongoing, but it is a journey worth undertaking for the betterment of society as a whole.

In conclusion, the issue of algorithmic bias is a pressing concern as we continue to integrate artificial intelligence into various aspects of our lives. Understanding and addressing this bias is pivotal for building trustworthy AI systems. By acknowledging that AI can perpetuate societal biases, we take the first step in mitigating its adverse impacts. To foster trust in artificial intelligence, it is imperative that developers, researchers, and policymakers prioritize transparency, fairness, and accountability. This means actively working to detect and eliminate bias in AI systems through methods like data analysis, fairness metrics, and user feedback. Additionally, proactive efforts to diversify development teams can help in identifying and rectifying sources of bias. As AI technologies evolve, we must also keep pace with ongoing monitoring and adaptation to ensure that they align with our societal values. Ethical principles, such as informed consent, must guide AI development to guarantee that users are aware of potential biases and data usage. In this dynamic landscape, we have the opportunity to harness the benefits of AI while ensuring fairness and equity. Through ethical principles, collaborative efforts, and a commitment to transparency, we can navigate the complexities of algorithmic bias and steer AI towards a more trustworthy and equitable future.

REFERENCES

AI bias is personal for me. It should be for you, too. (n.d.). PwC. https://www.pwc.com/us/en/tech-effect/ai-analytics/artificial-intelligence-bias.html

Andrews, E. L. A. (2021, August 6). *How flawed data aggravates inequality in credit.* Stanford HAI. https://hai.stanford.edu/news/how-flawed-data-aggravates-inequality-credit

Annie Brown. (2020, February 7). Biased Algorithms Learn From Biased Data: 3 Kinds Biases Found In AI Datasets. *Forbes.* Retrieved October 12, 2023, from https://www.forbes.com/sites/cognitive-world/2020/02/07/biased-algorithms/?sh=37f7add076fc

Are robots sexist? UN report shows gender bias in talking digital tech. (2019, May 22). UN News. https://news.un.org/en/story/2019/05/1038691

Awan, A. A. (2023, July 17). *What is Algorithmic Bias?* https://www.datacamp.com/blog/what-is-algorithmic-bias

Best & Rao. (2021). *Understanding algorithmic bias and how to build trust in AI.* PwC. Retrieved October 8, 2023, from https://www.pwc.com/us/en/tech-effect/ai-analytics/algorithmic-bias-and-trust-in-ai.html

Cuthbertson, A. (2019, March 6). Self-driving cars more likely to drive into black people, study claims. *The Independent.* https://www.independent.co.uk/tech/self-driving-car-crash-racial-bias-black-people-study-a8810031.html#:~:text=Researchers%20at%20the%20Georgia%20Institute,into%20them%2C%20the%20authors%20note

Definition of algorithm. (2023, September 28). *Merriam-Webster Dictionary*. https://www.merriam-webster.com/dictionary/algorithm

Dilmegani, C. (2023, September 11). *Bias in AI: What it is, Types, Examples & 6 Ways to Fix it in 2023*. AIMultiple. https://research.aimultiple.com/ai-bias/

Disability, Bias, and AI Report. (2023, September 27). AI Now Institute. Retrieved October 8, 2023, from https://ainowinstitute.org/publication/disabilitybiasai-2019

Eubanks. (2018, January). *Automating Inequality: How High-Tech Tools Profile, Police, and Punish the Poor*. St. Martin's Press, Inc. https://doi.org/ doi:10.5555/3208509

Fu, R., Huang, Y., & Singh, P. V. (2020). AI and Algorithmic Bias: Source, Detection, Mitigation and Implications. SSRN *Electronic Journal*. doi:10.2139/ssrn.3681517

Gillis, A. S. (2023, July 31). *Algorithm*. https://www.techtarget.com/whatis/definition/algorithm

Hao, K. (2020, April 2). *This is how AI bias really happens-and why it's so hard to fix*. Retrieved from https://www.technologyreview.com/s/612876/ this-is-how-ai-bias-really-happensand-why-itsso-hard-to-fix/?utm_source=newsletters&utm_medium=email&utm_campaign=the_algorithm.unpaid. engagement

History of Artificial Intelligence. (n.d.). Artificial Intelligence. https://www.coe.int/en/web/artificial-intelligence/history-of-ai

Hulstaert, L. (2019, March 14). *Machine learning interpretability techniques*. Retrieved from https://towardsdatascience.com/machine-learninginterpretability-techniques-662c723454f3

IBM. (n.d.). *What is Machine Learning?* Retrieved October 8, 2023, from https://www.ibm.com/topics/machine-learning

insideBIGDATA. (2022, February 107 *Reasons For Bias In AI and What To Do About It*. https://insidebigdata.com/2022/02/09/7-reasons-for-bias-in-ai-and-what-to-do-about-it/

Jensen, B. (2020, September 1). *AI4ALL: Diversifying the Future of Artificial Intelligence*. https://hai.stanford.edu/news/ai4all-diversifying-future-artificial-intelligence

Kelley, S. (2023, March 6). Removing Demographic Data Can Make AI Discrimination Worse. *Harvard Business Review*. https://hbr.org/2023/03/removing-demographic-data-can-make-ai-discrimination-worse

Köchling, A., & Wehner, M. C. (2020, November 1). *Discriminated by an algorithm: a systematic review of discrimination and fairness by algorithmic decision-making in the context of HR recruitment and HR development*. Business Research. doi:10.1007/s40685-020-00134-w

Krishna, T. H. (2020, December 2). *Entry barriers for women are amplified by AI in recruitment algorithms, study finds*. Newsroom. https://www.unimelb.edu.au/newsroom/news/2020/december/entry-barriers-for-women-are-amplified-by-ai-in-recruitment-algorithms,-study-finds#:~:text=Research%20showed%20even%20basic%20algorithms,resum%C3%A9s%2C%20according%20to%20new%20research

Leetaru, K. (2019, January 21). *Why Is AI And Machine Learning So Biased? The Answer Is Simple Economics*. Retrieved from https://www.forbes. com/sites/kalevleetaru/2019/01/20/why-is-ai-andmachine-learning-so-biased-the-answer-is-simpleeconomics/#51eb3979588c

Liberties, E. U. (2021, May 18). *Algorithmic Bias: Why and How Do Computers Make Unfair Decisions?* Retrieved October 8, 2023, from https://www.liberties.eu/en/stories/algorithmic-bias-17052021/43528

Manyika, J. (2022, November 17). What do we do about the biases in AI? *Harvard Business Review*. https://hbr.org/2019/10/what-do-we-do-about-the-biases-in-ai

Mehrabi, N., Morstatter, F., Saxena, N., Lerman, K., & Galstyan, A. (2021, July 13). A Survey on Bias and Fairness in Machine Learning. *ACM Computing Surveys*, *54*(6), 1–35. doi:10.1145/3457607

Moor, J. (n.d.). The dartmouth college artificial intelligence conference: The next fifty years. *AI Magazine*, *27*(4).

Nitika Sharma. (2023, September 8). *Understanding Algorithmic Bias: Types, Causes and Case Studies*. Analytics Vidhya. Retrieved October 12, 2023, from https://www.analyticsvidhya.com/blog/2023/09/understanding-algorithmic-bias/

Nkonde, M. (2021, September 13). Is AI Bias a Corporate Social Responsibility Issue? *Harvard Business Review*. https://hbr.org/2019/11/is-ai-bias-a-corporate-social-responsibility-issue

Omowole, A. (2022, November 8). *Research shows AI is often biased. Here's how to make algorithms work for all of us*. World Economic Forum. https://www.weforum.org/agenda/2021/07/ai-machine-learning-bias-discrimination/

Pichai, S. (2020, January 20). *Why Google thinks we need to regulate AI*. Retrieved from https://www.ft.com/ content/3467659a-386d-11ea-ac3c-f68c10993b04

Sen, A. (2017, August 9). *When artificial intelligence goes wrong*. Mint. https://www.livemint.com/Technology/VXCMw0Vfilaw0aIlnD1v2O/When-artificial-intelligence-goes-wrong.html

Silberg, J., & Manyika, J. (2019, June 6). *Tackling bias in artificial intelligence (and in humans)*. McKinsey & Company. https://www.mckinsey.com/featured-insights/artificial-intelligence/tackling-bias-in-artificial-intelligence-and-in-humans

Silva & Kenney. (1960). Algorithms, Platforms, and Ethnic Bias: An Integrative Essay. *Phylon, 55*(1-2), 9–37. https://www.jstor.org/stable/26545017

Smith, B. D., Khojandi, A., & Vasudevan, R. K. (2023). Bias in reinforcement Learning: A review in healthcare applications. *ACM Computing Surveys*, *56*(2), 1–17. doi:10.1145/3609502

Smith, H. (2020). Algorithmic bias: Should students pay the price? *AI & Society*, *35*(4), 1077–1078. doi:10.100700146-020-01054-3 PMID:32952313

SweeneyL. (2013). Discrimination in online ad delivery. Social Science Research Network. https://doi.org/ doi:10.2139/ssrn.2208240

Tucker, I. (2018, October 11). "A white mask worked better": why algorithms are not colour blind. *The Guardian*. https://www.theguardian.com/technology/2017/may/28/joy-buolamwini-when-algorithms-are-racist-facial-recognition-bias

Unni, M. *V., S*, R., Kar, R., Bh, R., V, V., & Johnson, J. M. (2023, March 2). Effect of VR Technological Development in the Age of AI on Business Human Resource Management. *2023 Second International Conference on Electronics and Renewable Systems (ICEARS)*. 10.1109/ICEARS56392.2023.10085258

Unni. (2020, April). Does Digital and Social Media Marketing Play a Major Role in Consumer Behaviour? *International Journal of Research in Engineering, Science and Management, 3*(4), 272–278. https://www.ijresm.com/Vol.3_2020/Vol3_Iss4_April20/IJRESM_V3_I4_63.pdf

Van Dijck, G. (2022, June 10). Predicting Recidivism Risk Meets AI Act. *European Journal on Criminal Policy and Research, 28*(3), 407–423. Advance online publication. doi:10.100710610-022-09516-8

White & Case LLP. (2017, January 23). *Algorithms and bias: What lenders need to know*. JD Supra. Retrieved October 8, 2023, from https://www.jdsupra.com/legalnews/algorithms-and-bias-what-lenders-need-67308/

Wiens, J., Price, W. N. II, & Sjoding, M. W. (2020). Diagnosing bias in data-driven algorithms for healthcare. *Nature Medicine, 26*(1), 25–26. doi:10.103841591-019-0726-6 PMID:31932798

KEY TERMS AND DEFINITIONS

Algorithm: An algorithm refers to a systematic procedure employed for the purpose of executing a computation or resolving a problem. Algorithms, whether implemented in hardware or software, operate as a systematic series of instructions that execute preset operations in a sequential manner (Gillis, 2023).

Algorithm Bias: When a computer system makes repeated, systematic mistakes that lead to unjust results, such as favouring one random set of users over another, this is referred to as algorithmic bias. It's a common worry nowadays as applications for artificial intelligence (AI) and machine learning (ML) permeate more and more of our daily lives (Awan, 2023).

Artificial Intelligence: Artificial intelligence (AI) refers to the replication of human intelligence in computer systems and machines. This is accomplished through developing algorithms, software, and hardware that enable these systems to perform tasks often associated with human intelligence. The aforementioned activities encompass problem-solving, acquisition of knowledge, comprehension of natural language, recognition of patterns, and decision-making informed by data and acquired knowledge (Unni et al., 2023).

Bias: Bias is an unfair inclination or prejudice in favour of or against something, leading to unbalanced outcomes in various contexts.

Decision Making: Decision making entails the act of choosing a specific option or path of action from a range of available alternatives (Unni, 2020).

Fairness: The concept of fairness varies across disciplines, and there is no universally accepted definition. Developers struggle to express fairness in mathematical terms, and there are trade-offs between different definitions, making it challenging for businesses to adopt a specific definition (Silberg & Manyika, 2019).

Machine Learning: Machine learning, a specialised domain within the realm of artificial intelligence (AI) and computer science, is dedicated to the utilisation of data and algorithms in order to replicate the learning process observed in humans. Through this iterative approach, the system's precision is gradually enhanced (IBM, n.d.).

Chapter 15

A Critical Data Ethics Analysis of Algorithmic Bias and the Mining/Scraping of Heirs' Property Records

Robin Throne
https://orcid.org/0000-0002-3015-9587
University of the Cumberlands, USA

ABSTRACT

The data and research ethics surrounding artificial intelligence (AI), machine learning (ML), and data mining/scraping (DMS) have been widely discussed within scholarship and among regulatory bodies. Concurrently, the scholarship has continued to examine land rights within the Gullah Geechee community for heirs' property land rights and land dispossession along the Gullah Geechee Cultural Heritage Corridor (GGCHC). This chapter presents the results of a critical data ethics analysis for the risks of data brokerage, algorithmic bias, the use of DMS by dominant groups external to the GGCHC, and the ensuing privacy implications, discrimination, and ongoing land dispossession of heirs' property owners. Findings indicate a gap in documented research for heirs' property records, yet Gullah Geechee algorithmic bias was evident. Further research is needed to understand better the data privacy protections needed for heirs' property records and ongoing scrutiny of local versus federal policy for data privacy protections specific to heirs' property records.

INTRODUCTION

The data ethics and privacy challenges surrounding artificial intelligence (AI), machine learning (ML), and data mining/scraping (DMS) have been well documented within scholarship and among regulatory bodies (Dobbs & Gaither, 2023; Throne, 2022; Strobel & Shokri, 2022). Recently, social media companies have been levied unprecedented fines and other penalties for these invasions and data extractions, and the U.S. Congress is considering proliferation protections for the use of AI (Throne, 2022).

DOI: 10.4018/979-8-3693-1762-4.ch015

A subset of the scholarship has focused explicitly on the destructive nature of algorithmic bias, digital discrimination, and threats to data privacy (Strobel & Shokri, 2022). Opportunities exist for the design of "robust algorithms that are also accurate and privacy preserving" (Strobel & Shokri, 2022, p. 49), yet those algorithms designed for data extraction used for predatory or nefarious purposes may not consider fairness, equity, or even data privacy.

For example, Jackson et al. (2019) stressed the need for data science to consider the long history of social injustices against vulnerable populations:

Another lesson learned is that where data on vulnerable populations exists, partnering with data scientists derived from those vulnerable populations can help to disentangle an algorithm's inferential ability from a manifesting of implicit bias in data collection. Data science must include vulnerable populations in the research design, analysis and inference of data findings in order to make interpretations that are valuable and meaningful to those populations. Whether focused on social science, biomedical applications or preventing the harvesting of large scale genomic data from vulnerable populations with no clear reciprocal benefit to them, the inclusion of these diverse population and perspectives can improve data science. (p. 7)

Specifically, LaPointe and Yale (2022) noted the unique challenges of underrepresented minority property owners who may be displaced by tax delinquency.

While scientists, researchers, and human protections professionals may understand and comply with the need to consider these ethical aspects of DMS for research purposes established in policy, others who desire to use these technologies for financial gain may desire the continued lack of policy or even consider data access over data privacy. However, as Simshaw (2022) reported, the limited digitization of heirs' property records may have impeded AI, ML, and DMS use. This data ethics analysis explored the existing literature for any reported use of DMS among heirs' property records.

Specifically, the social injustices, biases, discrimination, and land dispossession experienced by heirs' property (tenancy in common) owners along the Gullah Geechee Cultural Heritage Corridor (GGCHC) have been well documented. In prior work, the chapter author, with others (Throne, 2020; Versey & Throne, 2021), used critical inquiry and intersectionality to address the heirs' property challenges along the GGCHC and, specifically, the contemporary challenges of the multiple heirs to property whereby common land originated when the original owner died intestate, and the property was passed down generation to generation outside of probate[1]. Heirs' property exists across states, regions, and indigenous populations, including Appalachia, Native Americans, Hispanic residents of Texas, and rural African Americans (Bailey & Thomson, 2022; Simshaw, 2022). Frankly, for the GGCHC heirs' property owners, the financial interests of often-White property developers and land speculators have been reported as predatory and unscrupulous in methods used to acquire these often Black-owned generational properties along the GGCHC (Bailey & Thomson, 2022).

Further, as Bailey and Thompson (2022) summarized, "A more pernicious version of this occurs when land speculators and real estate developers find or create cleavages within families and use unscrupulous but legal means to gain ownership of lands that may have been in the family for generations" (p. 2). Therefore, the constructs of algorithmic bias and DMS were considered within these contexts for this analysis. Legal vulnerabilities for this population of heirs' property owners have included tax sales and predatory partition actions as legal mechanisms for GGCHC land dispossession (Light, 2022). As Tucker (2023) noted, "It is useful to think about how and whether privacy regulation reduces, doesn't affect or augments the potential for algorithmic discrimination" (p. 16). Thus, this chapter aimed to examine the use of algorithmic bias and the use of DMS with GGCHC digitized heirs' property records.

BACKGROUND

Jones and Pippin (2019) noted that "advances in the development and availability of digitized property data is changing the way people look at property" (p. 3), yet also stressed the historical challenges with land title in digital format as "gathering information about land titles required a trip to the 'deed room' in a local county courthouse as well as time-consuming and laborious title search experience" (p. 3). The authors described the potential of digitization of heirs' property records at the local level and concluded:

In short, a great deal of data exists that could tell us more about heirs' property. While every parcel, family, and community will have different and distinct stories to tell, the availability of digitized property data is increasing in a way that has great potential to deepen our understanding of the heirs' property phenomenon more generally. Learning more about the land and its ownership is possible at scales that may not have been feasible until now. (p. 8)

While Parks (2022) confirmed the opportunities existent within DMS, risks are also significant for personally identifiable information and said, "Scraping of such data in bulk can harm individual privacy, undermine democracy, and potentially even physically endanger us" (p. 945). Further, while data scraping of publicly available information provides opportunities for scholars, researchers, advertisers, and others, it can also be used for malicious or nefarious purposes. For uses such as heirs' property records, Park (2022) noted the legal vulnerabilities of publicly available information that includes personally identifiable information and reported, "In many cases, an individual never knows that their personal information has been made public, making it impossible for them to consent to its publication. Some personal information is made public through lawful government public records" (p. 923). While such risks exist, current privacy rules fail to adequately protect individual's personally identifiable information from DMS (Park, 2022). "Today's privacy statutes do not do enough to address this issue, allowing businesses to scrape and repurpose our personal information with near impunity," Park (2022, p. 945) concluded.

Similarly, Winters-Michaud et al. (2023) confirmed that "most analysis tends to be done either for a single county or performed at the county level rather than at the parcel level" (p. 6). The authors highlighted the vulnerability of heirs' property in rural and agricultural settings and urban heirs' property that can also be at risk for land dispossession (Winters-Michaud et al., 2023). The authors created an algorithm to identify heirs' property parcels. They noted the chance for error due to delay in recording owner names after the original owner was deceased or in other human errors in record transactions and the respective continued refinement needed for the algorithm. Such algorithms created by scholars and researchers raise the question of whether land developers and speculators can use them for financial gain and whether algorithmic bias may be reproduced in such uses.

For example, Strobel and Shokri (2022) noted that adversaries can affect a minor manipulation to cross the model's decision boundary and violate algorithmic fairness. The authors further demonstrated how adversarial manipulation can disrupt trustworthy ML. For data privacy implications, they concluded,

To obtain a precise definition for data privacy, we need to differentiate between general patterns that apply to the entire population, which we would want to reveal, and the patterns that apply to specific data of individual users, which we want to keep private. Hence, learning anything about individual data records beyond general patterns should be considered a privacy violation. (p. 45)

Therefore, fairness, equity, and explainability are essential elements of algorithmic fairness and may reduce algorithmic bias (Jackson, 2019; Strobel & Shokri, 2022). In contrast, Simshaw (2022) stressed that another form of AI bias lies in the fact that algorithms rely on digitized data, and due to the lack of digitization for heirs' property records may arise as a form of legal injustice for an already marginalized population. "Without marketable title to the land in the form of records, members of these communities struggle to use their land as collateral for securing loans and accessing credit, and even sometimes struggle to prevent their land from being taken by the government" (Simshaw, 2022, p. 194).

In the early 2000s, literature reviews as systematic analyses exponentially increased concurrently with the rise of evidence-based research (Kraus et al., 2022). As recommended for systematic reviews, the chapter author's, among others, previous research was used to formulate the keywords and key phrases used for literature sourcing, which included data ethics, data policy, data privacy, data mining and/or scraping, Gullah Geechee, heirs' property, algorithmic bias, and heirs' property records, as well as filtered since 2019. While data mining and data scraping are often used synonymously across disciplines, the distinction between mining and scraping typically refers to data mining as gathering large datasets for analysis. In contrast, data scraping involves the process of gathering large datasets. Due to the overlap in meaning, these data extraction terms are used concomitantly as DMS for this chapter's purposes. For example, Kraus et al. (2022) noted that critical analysis has allowed scholars to subjectively utilize their expertise and experience to critically evaluate scholarly literature critically, even using non-systematic means for source collection. This type of analysis can be used as foundational research to explore the underexplored areas of literature maintained in the academic database corpora (Kraus et al., 2022).

Scholarly critical reviews also allow the engagement of systematically sourced literature using academic database filtering and other bibliometric methodologies to advance theory and practice or initial recommendations for future research to advance both theory and practice. Due to the advances in ubiquitous AI, ML, and DMS post-pandemic technologies and the respective expanding scholarship, only current sources since 2022 were included in the final selection. Sources between 2019 and 2021 were retained solely for background. While returns for algorithmic bias and DMS were more robust, only two sources were included in the final sample for GGCHC algorithmic bias and four for DMS of heirs' property records, specifically. Therefore, the apparent paucity of scholarship and apparent gaps are addressed in the research recommendations that follow the source analysis.

Bargeman (2023) reported on the racial wealth gap that persists for heirs' property owners within the GGCHC. The author cited the data challenges with heirs' property records and quoted from a *Regional Matters* report from 2021, whereby Carpenter and Waddell noted the challenges in the identification of heirs' property,

Heirs' property is notoriously challenging to track and quantify, as evidenced by the studies conducted thus far... Data that reports heirs' property is incomplete, and only the laborious efforts of retrieving and parsing county-level records and connecting with landowners directly would reveal its true scope.

Further, Bargemen (2023) called out the risks of heirs' property owners from predatory bad actors who seek to acquire the property for financial gain through fraudulent schemes. For example, the author offered, "When victims of natural disasters who own heirs' property are denied assistance and cannot afford repairs, their homes are at risk of condemnation and demolition, thereby further increasing the wealth gap" (Bargemen, 2023, p. 347) and may fall outside the federal ownership rules for disaster assistance due to unclear property titles. In turn, further land dispossession may occur if the owners fall prey to predatory schemes in turbulent times of distress (Bargeman, 2023; Jones & Pippen, 2019).

Data Mining/Scraping of Property Records

Data scraping involves techniques whereby a computer program extracts data from human-readable outputs, often synonymously with web scraping (Cambridge, 2023). Whereas data mining has been defined as a form of knowledge discovery in data through the process of discovering patterns, trends, and other information in large datasets, often using AI and ML (IBM, 2023). For the purposes of this chapter, these processes have been aggregated as DMS. Due to the proliferation of data centers and warehousing, these methods have accelerated use over the past two decades. Association rules, neural networks, decision trees, K-nearest neighbor, competitor analysis, parsing, vertical aggregation, and text pattern matching are all forms of DMS (Cambridge, 2023; IBM, 2023). Web scraping can also simulate human web searching using bots for data crawling, yet data crawling may access data from any source, not solely web pages (Reviglio, 2022). For public property records, data crawling is used to access publicly available data despite the use of privacy regimes to protect individual privacy (Reviglio, 2022). Several DMS applications and services have been developed specifically for real estate.

Many authors and researchers have noted the outmoded electronic applications that exist for accessing publicly available property records. Cattanach and Greenberg (2021) specifically noted the threats to property owners' privacy with the cybersecurity challenges of these systems. "When coupled with the greatly enhanced technical ability of third parties to retrieve and analyze (legally or otherwise) troves of data and extract the commercial value from it, the tension between unfettered public access to information held by government and the legitimate concerns about privacy and confidentiality has never been more acute" (Cattanach & Greenberg, 2021, p. 1). In addition to property record databases, data brokers routinely scrape public records datasets, including motor vehicle and driver licensing databases, criminal records, employment background reports, voting registration, and other publicly available state and local licensing databases (Cattanach & Greenberg, 2021; Reviglio, 2022). While DMS continues, governmental regulations to protect individual privacy have not kept up with the technological advances and pose risks to data privacy (Cattanach & Greenberg, 2021).

Similarly, Sherman (2021) highlighted the growing risks to data privacy from these forms of data brokerage, specifically "the practice of buying, aggregating, selling, licensing, and otherwise sharing individuals' data" that largely remains unregulated (p. 2). This is further complicated by the nebulous and non-regulated definition of "data broker" and data sharing opportunities (p. 2). For example, in a study of 10 of the largest data brokers, datasets of 6.5 billion individual property records were commercially available from U.S.-based LexisNexis, advertising that these datasets also identified relatives, associates, and neighbors" (Sherman, 2021). In conclusion, Sherman (2021) stressed the need for governmental regulation to protect individual data privacy in the digital era,

There are virtually no controls on the data brokerage industry (data broker firms specifically) and on the practice of data brokerage itself (the broader buying, licensing, and sharing of data that underpins these companies' operation). Americans also do not have federal privacy rights to gain insight into the data brokerage ecosystem's surveillance of them, nor do they have federal rights to demand that incorrect data is corrected... All these harms—to Americans' civil rights, to U.S. national security, and to U.S. democracy writ large—will only persist without further regulation. (p. 12)

Likewise, Reviglio (2022) called out data brokerage as a risk to not only individual data privacy but also national security, primarily due to the complex and oblique relationships between big tech and data brokers. The author noted the stealth and third-party exchanges[2] (i.e., ISP providers, online platforms and databases, social media, insurance/financial products and platforms, targeted advertisers, smartphone application extraction, and other data brokers) often utilized by data brokers, which raises threats to individual data privacy and requires examination of the "systemic, economic and political role of the data broker industry" (Reviglio, 2022, p. 22). However, it is essential to note that governmental entities also gather and share datasets (Cattanach & Greenberg, 2021). Thus, these varied and complex challenges continue for regulatory changes that need to be addressed for AI, ML, and DMS, and likely a need to distinguish data brokerage from other aspects of DMS (Cattanach & Greenberg, 2021; Reviglio, 2022).

GULLAH GEECHEE CULTURAL HERITAGE CORRIDOR AND ALGORITHMIC BIAS

While the scholarship surrounding GGCHC and algorithmic biases has been relatively limited, no sources were returned for algorithmic bias and heirs' property. However, research into algorithmic racial bias continues exponentially and deserves continued attention for use with specific populations. For example, Kozlowski, with others (2022a, 2022b), reported, "There is no such thing as algorithmic neutrality. The automatic inference of authors' race based on their features in bibliographic databases is itself an algorithmic process that needs to be scrutinized, as it could implicitly encode bias, with major impact in the over and under representation of racial groups" (p. 2, 2022b). Further, the authors called for continued examination of racial bias in data science, using qualitative and quantitative methods as equitable algorithmic approaches to account for minorities and other marginalized groups remain limited (Kozlowski et al., 2022b).

In other work, Kozlowski et al. (2022a) noted that inequalities in science have long been problematic, and too many minoritized authors limit their work to the scientific disciplines and research that align with their specific "gendered and racialized social identities" (p. 6). DMS requires algorithmic approaches yet creates a dilemma for scientists when "using biased instruments to study bias only replicates the very inequities they hope to address" (Kozlowski et al., 2022a, p. 13). Kozlowski et al. (2022b) concluded:

Ultimately, scientometrics researchers utilizing race data are responsible for preserving the integrity of their inferences by situating their interpretations within the broader socio-historical context of the people, places, and publications under investigation. In this way, they can avoid preserving unequal systems of race stratification and instead contribute to the rigorous examination of race and science intersections toward a better understanding of the science of science as a discipline. Once again, we quote Zuberi: "The racialization of data is an artifact of both the struggles to preserve and to destroy racial stratification." (p. 13)

Likewise, Tucker (2023) noted the challenges inherent to measuring and calibrating privacy harms, which may explain "the shift in the policy debate towards questions of algorithmic bias or discrimination" (p. 16). Like Kozlowski et al. (2022b), Tucker (2023) noted the complexities of achieving algorithmic fairness and whether privacy protections foster algorithmic discrimination. The author considered the policy implications and noted,

Privacy regulation might demand that firms reduce the amount of personally identifiable information available - this might hinder firms and the government's ability to audit their algorithms and identify instances of bias. Privacy regulation could also restrict the use of data by algorithms, which give rise to algorithmic discrimination. Since the direction of the interaction between privacy regulation and algorithmic bias is unclear, this makes it an important area for empirical research. (p. 16)

The dangers of policymaking for algorithmic decision-making may lie in the very nature of policy based on political ideologies or the financial interests of those policymakers (Phillips-Brown, 2023). For example, Benabdallah et al. (2022) used arts-based research of Gullah land to examine AI bias through an ongoing artwork of *Speculative Landscapes*. While not directly related to DMS use with heirs' property records, the authors offer stark and visually apparent GGCHC land dispossession and algorithmic bias in AI-based imaging of property:

However, as the property values went up, real estate capitalists found loopholes that allowed for the slow erosion of the communally held Gullah land. Today, where there were once Gullah farms, churches, schools, and graveyards, there are vacation resorts and golf courses. Maps are often tied to colonialist ideas of ownership, boundaries, and territories. They also create documented versions of a reality that are assumed to be true. With this mapping in mind, Chariell created a "deepfake" version of an area where a historical grave site was being encroached upon by a vacation resort. Using segmentation, they removed all of the buildings, speedboats, and signs of disturbance around the cemetery site as a restorative speculative gesture. (p. 93)

In closing, the authors questioned this visual dichotomy of reality in the use of AI with satellite images to illustrate contrasts of "truth" and "alternate reality" through speculative images to illustrate GGCHC land dispossession (Benabdallah et al., 2022). Further, the authors noted that through art, AI bias may become visually evident to "materialize complex relationships with machine learning; ground these relationships in the present and the personal; and point to generative ways of engaging with biased systems around us" (Benabdallah et al., 2022, p. 85).

DATA MINING/SCRAPING HEIRS' PROPERTY RECORDS

A total of four gleaned sources provided tangential alignment with DMS of heirs' property records, and three were returned for analysis from 2022. Two sources met the criteria for analysis of these constructs from 2023. Regrettably, no research sources were located specifically on heirs' property record digital extraction using DMS. Thus, the limitations of research into data privacy and predatory practices with digital heirs' property records were evident. However, it is essential to note that grey literature was not used for this analysis, and only peer-reviewed sources were considered. The organizational and agency reports from those working fervently to reverse heirs' property land dispossession may bolster future research into these challenges[3].

First, while quite tangentially related to the focus of the analysis, Monroe-White (2022) outlined the need for specific data science frameworks to be established so that the complexities of the Gullah-Geechee diaspora are considered by design. By educating data science students in the diasporic experiences of GGCHC descendants, fairness and equity in data science may ultimately provide "mitigation of data harms by having a more diverse and inclusive data science workforce capable of identifying and challenging biased datasets and algorithms pre-deployment" (Monroe-White, 2022, p. 5).

Next, Simmons-Jenkins et al. (2022) reported on the cycle of environmental and social injustice impacts on Gullah/Geechee Heritage Sites. The researchers reported on the inconsistencies and increased risk to African American cemeteries within the GGCHC as digitized records of cemeteries did not align with physical maps and states' physical inventories of historic cemeteries (Simmons-Jenkins et al., 2022). The report noted the U.S. Senate Bill 3667, introduced in February 2022, which would direct the Department of the Interior to establish the United States African-American Burial Grounds Preservation Program (2022) within the National Park Service to remedy the inconsistencies and provide resources for location verification of these historic sites (Simmons-Jenkins, 2022). However, as of the research publication in June 2022, the researchers reported, "the Gullah/Geechee descendants of those buried have not received a written report or any acknowledgment of tests confirming no additional burials existed within the private owner's footprint" (Simmons-Jenkins et al., 2022, p. 5).

From a policy perspective, protections for these historic cemeteries are critically tangential to the data protection policy needed for the GGCHC heirs' property owners as "access to land, in life and in death, mirrors the level of injustice prevalent in society" (Simmons-Jenkins et al., 2022, p. 5) and particular to this target population that has experienced land dispossession (Light, 2022). Simmons-Jenkins et al. (2022) concluded for segments of the GGCHC analyzed, which reinforced prior research findings that noted the ongoing GGCHC land dispossession:

Based on the social caste system of that era, the burial lands to which Black people had access 50-150 years ago most likely carried less value than other property comparatively. As land speculators prioritized developing undervalued property over higher-priced land, burial grounds that served as the sacred resting places of Black people became the most cost-effective option. (p. 5)

Finally, while no evidence was returned for the use of DMS outside of the GGCHC, long-term heirs' property researchers Bailey and Thomson (2022), Thomson and Bailey (2023), and Dobbs and Gaither (2023)[4] utilized the CoreLogic database of county property tax records to continue to determine accurate estimations of the amount of heirs' property in the U.S.[5] Bailey and Thomson (2022) concluded that decades of land dispossession may be well documented in individual county courthouses across states and provide a record of the actors involved in the heirs' property dispossession. This claim may provide a grain or even partial pathway of evidence as to the need for examination of whether DMS or other ubiquitous technologies are/or have been used by these actors to locate and target vulnerable heirs' property.

As Thomson and Bailey (2023) reported, "...private companies such as CoreLogic have collected data from over 3,000 counties and county-equivalent jurisdictions in the United States and made these data available for purchase. Increased availability of such data over the past five years has made large-scale heirs' property research possible." While the data privacy implications of these trends in other disciplines have been well documented, the implications specific to GGCHC heirs' property remain unknown. The authors proposed revenue solutions for the resulting wealth inequities in addition to other social injustices and harms along the GGCHC and other states that stress the egregious nature of these losses and the need for reparations:

Legalized theft of Gullah/Geechee lands has occurred within living memory—over the course of the past 72 years. Dispossession of the Gullah/Geechee led directly to the accumulation of wealth by those who used or benefitted from the use of unscrupulous means to gain control of the land. These actions should be recognized, publicly acknowledged, and explicitly included as justification for legislation associated with generating revenue for reparations. Using the tax system to tap into a small part of that wealth to fund reparations for a definable population represents one possible mechanism to address past injustices. (Bailey & Thomson, 2022, p. 18)

In a first study to expand the scope of heirs' property research across all U.S. states, Dobbs and Gaither (2023) outlined the various electronic means used to assess the scope and location of heirs' property across U.S. states. The researchers used a commercial parcel dataset product and noted the complexities and nuances in identifying the data fields to accurately target heirs' property records. In addition to identifying 444,172 heirs' parcels across U.S. states (not including territories), including the four lower Atlantic GGCHC states. They noted, "The patterns that we saw in the distribution of these data across the country are consistent with places of historical marginalization and with prior efforts locating heirs' parcels for specific places" (Dobbs & Gaither, p. 19). Results from this study and the Bailey and Thomson (2022) and Thomson and Bailey (2023) may also provide maps to areas of the U.S. for local policy analysis of heirs' property data privacy and guide further research of algorithmic bias in the use of AI, ML, and DMS.

FUTURE RESEARCH DIRECTIONS

The findings from this initial critical review highlighted the gap in policy and data privacy protection challenges for GGCHC heirs' property records as digitization has become the norm, making these records subject to AI, ML, and DMS. The record has been clear as to how and why heirs' property records have been subject to historical and ongoing predatory or nefarious land dispossession and are not in sync with property rights, property vulnerabilities, and owner data privacy protection. However, documented empirical research for these inferences remains limited. Further research is needed to better understand the policies needed to ensure data privacy protection for this target population and heirs' property records, as well as continued examination for transparency into how and why these records result in legal actions. Thus, case study research is recommended as more than anecdotal evidence may inform the specific property challenges within state and local policy contexts.

For example, Bailey and Thomson (2022) explained that the origins of these records remain at a county level, and digitization/aggregation has risen over the past five years, which may foster data extraction on a larger scale. Further, Kraus et al. (2022) recommended that "expanding [the] number of databases, journals, periodicals, automated approaches, and semi-automated procedures that use text mining and machine learning can offer researchers the ability to source new, relevant research and forecast the citations of influential studies. This enables them to determine further relevant articles" (p. 2585). Examining these technologies for literature sourcing may extract a larger sample size of empirical research surrounding the DMS of heirs' property records and algorithmic bias and discrimination of GGCHC populations. Therefore, rigorous corpus analysis as continued research is needed to fully gain a sense of the scope and depth of the research into these issues for heirs' property records, and if so, recommendations and solutions may be needed to inform policy for data privacy protections as the scope of the problem remains unknown amid the rise of ubiquitous technologies and data digitization/ aggregation.

In addition, the evident gap in the research for the contemporary use of DMS of heirs' property records offers an opportunity for researchers to consider this specific niche within data extraction amid the growth of data digitization/aggregation. Kraus et al. (2022) outlined steps to rigorous corpus collection and analysis that may be worthy of implementation for those considering further work. Monroe-White (2022) recommended better data science education,

As providers, curators and instructors of data best practices and systems, we are responsible for preparing members of the data science workforce to intelligently contend with the socio-technical complexities of their work, create liberatory data science pedagogy and curricula (Castillo-Montoya, Abreu and Abad, 2019; Johnson and Elliott, 2020) and advocate for the use of data to empower Black, Indigenous and marginalized people of color. (p. 5)

From a policy perspective, Simmons-Jenkins et al. (2022) recommended policy actions such as the need for local government regulation to exercise stewardship over burial grounds as "culturally sensitive, environmentally vulnerable sites" (p. 6). Further, they also recommended that federal policy changes be considered,

The legacy of environmental and social injustice impacts to Gullah/Geechee burial sites can be mitigated. . . . More broadly, federal legislation must provide redress for the discriminatory practices of the past and the environmentally unjust policies of the present that perpetuate Black land devaluation in life that leads to desecration in death. (pp. 6-7)

Bargeman (2023) highlighted the need for multi-faceted efforts from multiple federal agencies to tackle the complex and pervasive problem of heirs' property to remedy the persistent racial wealth gap. Further, the author recommended federal policy use as "the pre-existing federal mandates of FEMA, USDA, and the CRA provide authority and incentive for both public and private actors to get involved" (p. 348). Finally, Reviglio (2022) called for more research into the under-regulation of data brokerages, and Dobbs and Gaither (2023) noted that their expanded model can be used to identify heirs' property parcels across U.S. states to expand the scholarship of heirs' property, which may also provide a foundation for future research.

CONCLUSION

As the use of AI, ML, and DMS for mining heirs' property records remains largely unknown, the critical review was not without findings. Algorithmic bias is evident among the GGCHC, which may make it susceptible to continued heirs' property land dispossession as publicly identifiable information has been noted as specifically susceptible to DMS. However, an apparent paucity of scholarship and apparent gaps were found and subsequently addressed in the research recommendations. The under-regulation of data brokerages further complicates policy implications as heirs' property records reside at a county level, and local data privacy policy may primarily affect these records rather than the federal data privacy policy and regulations underway amid this climate with federal governmental agencies. In addition to the continued research necessary for the ongoing examination of the use of ubiquitous technologies to gain access to heirs' property and whether this access and use lead to legal consequences for heirs' property owners, policy considerations at local, state, and federal levels require ongoing scrutiny.

REFERENCES

African-American Burial Grounds Preservation Act. S. 3667, 117th Cong. (2022). https://www.congress.gov/bill/117th-congress/senate-bill/3667

Bailey, C., & Thomson, R. (2022). Heirs property, critical race theory, and reparations. *Rural Sociology*, *87*(4), 1219–1243. doi:10.1111/ruso.12455

Bargeman, K. B. (2023). The heirs' property dilemma: How more robust federal policies can help narrow the racial wealth gap. North Carolina Banking Institute, 27(1), 320-348.

Benabdallah, G., Alexander, A., Ghosh, S., Glogovac-Smith, C., Jacoby, L., Lustig, C., ... Rosner, D. (2022, June). Slanted speculations: Material encounters with algorithmic bias. In *Designing Interactive Systems Conference* (pp. 85-99). 10.1145/3532106.3533449

Cambridge. (2023). Data scraping. In *The Cambridge dictionary*. Cambridge University Press.

Carpenter, S., & Waddell, S. R. (2021). *Whose land is it? Heirs' property and its role in generational land retention. Regional Matters*. Federal Reserve Bank of Richmond.

Cattanach, R. E., & Greenberg, J. M. (2021). Critical update needed! Addressing cybersecurity challenges of public records in the Digital Age. *The Urban Lawyer*, *51*(3), 427–456.

Dobbs, G. R., & Gaither, C. J. (2023). *How much heirs' property is there? Using LightBox Data to estimate heirs' property extent in the US*. Southern Rural Development Center. https://scholarsjunction.msstate.edu/srdctopics-heirsproperty/1/

Egger, R., Kroner, M., & Stöckl, A. (2022). Web scraping: Collecting and retrieving data from the web. In *Applied data science in tourism: Interdisciplinary approaches, methodologies, and applications* (pp. 67–82). Springer. doi:10.1007/978-3-030-88389-8_5

IBM. (2023). *What is data mining?* https://www.ibm.com/topics/data-mining

Jackson, L., Kuhlman, C., Jackson, F., & Fox, P. K. (2019). Including vulnerable populations in the assessment of data from vulnerable populations. *Frontiers in Big Data*, *2*, 19. doi:10.3389/fdata.2019.00019 PMID:33693342

Jones, S., & Pippin, J. S. (2019). Learning about the land: What can tax appraisal data tell us about heirs' properties? *SRS-244*, 3-8.

Kozlowski, D., Larivière, V., Sugimoto, C. R., & Monroe-White, T. (2022a). Intersectional inequalities in science. *Proceedings of the National Academy of Sciences of the United States of America*, *119*(2), e2113067119. doi:10.1073/pnas.2113067119 PMID:34983876

Kozlowski, D., Murray, D. S., Bell, A., Hulsey, W., Larivière, V., Monroe-White, T., & Sugimoto, C. R. (2022b). Avoiding bias when inferring race using name-based approaches. *PLoS One*, *17*(3), e0264270. doi:10.1371/journal.pone.0264270 PMID:35231059

Kraus, S., Breier, M., Lim, W. M., Dabić, M., Kumar, S., Kanbach, D., Mukherjee, D., Corvello, V., Piñeiro-Chousa, J., Liguori, E., Palacios-Marqués, D., Schiavone, F., Ferraris, A., Fernandes, C., & Ferreira, J. J. (2022). Literature reviews as independent studies: Guidelines for academic practice. *Review of Managerial Science*, *16*(8), 2577–2595G. doi:10.100711846-022-00588-8

LaPointC.YaleS. O. M. (2022). Property tax sales, private capital, and gentrification in the US. doi:10.2139/ssrn.4219360

Light, T. H. (2022). Frankenstein's monster: Constructing a legal regime to regulate race and place. *Southern Cultures*, *28*(3), 74–89. doi:10.1353cu.2022.0027

Monroe White, T. (2022). Emancipating data science for Black and Indigenous students via liberatory datasets and curricula. *IASSIST Quarterly*, *46*(4). Advance online publication. doi:10.29173/iq1007

Parks, A. M. (2021). Unfair collection: Reclaiming control of publicly available personal information from data scrapers. *Michigan Law Review*, *120*, 913.

Phillips-Brown, M. (2023). *Algorithmic neutrality.* arXiv preprint, 2303.05103.

Reviglio, U. (2022). The untamed and discreet role of data brokers in surveillance capitalism: A transnational and interdisciplinary overview. *Internet Policy Review*, *11*(3), 1–27. doi:10.14763/2022.3.1670

Sherman, J. (2021). Data brokers and sensitive data on US individuals. *Duke University Sanford Cyber Policy Program, 9*.

Simmons-Jenkins, G., Miller, S. E., & Murray, E. J. (2022, June). One feeds the other: The cycle of environmental and social injustice impacts on Gullah/Geechee Heritage Sites. *Spark: Elevating Scholarship on Social Issues.*

Simshaw, D. (2022). Access to AI justice: Avoiding an inequitable two-tiered system of legal services. *Yale Journal of Law & Technology*, *24*, 150–226.

Strobel, M., & Shokri, R. (2022). Data privacy and trustworthy machine learning. *IEEE Security and Privacy*, *20*(5), 44–49. doi:10.1109/MSEC.2022.3178187

Thomson, R., & Bailey, C. (2023). *Identifying heirs' property: Extent and value across the South and Appalachia.* Southern Rural Development Center. https://scholarsjunction.msstate.edu/srdctopics-heirsproperty/2/

Throne, R. (2020). Dispossession of land cultures: Women and property tenure among Lowcountry heirs in the Gullah Geechee Corridor. In *Multidisciplinary issues surrounding African Diasporas* (pp. 152–174). IGI Global. doi:10.4018/978-1-5225-5079-2.ch007

Throne, R. (2021). Land as agency: A critical autoethnography of Scandinavian acquisition of dispossessed land in the Iowa Territory. In *Indigenous research of land, self, and spirit* (pp. 118–131). IGI Global.

ThroneR. (2022). Adverse trends in data ethics: The AI Bill of Rights and human subjects protections. *Information Policy & Ethics eJournal.* doi:10.2139/ssrn.4279922

Tucker, C. (2023). The economics of privacy: An agenda. In *The economics of privacy*. University of Chicago Press.

Versey, H. S., & Throne, R. (2021). A critical review of Gullah Geechee midlife women and heirs' property challenges along the Gullah Geechee Cultural Heritage Corridor. In *Examining international land use policies, changes, and conflicts* (pp. 46–64). IGI Global. doi:10.4018/978-1-7998-4372-6.ch003

Winters-Michaud, C., Burnett, W., Callahan, S., Keller, A., Williams, M., & Harakat, S. (2023). Land-use patterns on heirs' property in the American South. *Applied Economic Perspectives and Policy*, •••, 1–15. doi:10.1002/aepp.13354

ADDITIONAL READING

Egger, R., Kroner, M., & Stöckl, A. (2022). Web scraping. In *Applied Data Science in Tourism* (pp. 67–82). Springer. doi:10.1007/978-3-030-88389-8_5

Gaither, C. J., Carpenter, A., McCurty, T. L., & Toering, S. (2019). *Heirs' property and land fractionation: Fostering stable ownership to prevent land loss and abandonment* (Vol. SRS-244). United States Department of Agriculture Forest Service.

Kumaresan, U., & Ramanujam, K. (2022). A framework for automated scraping of structured data records from the deep web using semantic labeling: Semantic scraper. *International Journal of Information Retrieval Research, 12*(1), 1–18. doi:10.4018/IJIRR.290830

Mazilu, M. C. (2022). Web scraping and ethics in automated data collection. In *Education, Research and Business Technologies* (pp. 285–294). Springer. doi:10.1007/978-981-16-8866-9_24

Rahman, R. U., Wadhwa, D., Bali, A., & Tomar, D. S. (2020). The emerging threats of web scraping to web applications security and their defense mechanism. In *Encyclopedia of criminal activities and the deep web* (pp. 788–809). IGI Global. doi:10.4018/978-1-5225-9715-5.ch053

Stein, S., & Carpenter, A. (2022). Heir's property in an urban context (No. 2427-2022-041). American Economic Association.

Throne, R. (2022). Free the data but not the human subjects identifiers. SSRN *Electronic Journal.* doi:10.2139/ssrn.4304179

Xiao, G. (2022). Data misappropriation: A trade secret cause of action for data scraping and a new paradigm for database protection. *The Columbia Science and Technology Law Review, 24*(1), 125–172. doi:10.52214tlr.v24i1.10456

KEY TERMS AND DEFINITIONS

Algorithmic Bias: Algorithmic bias is the discrimination caused by algorithmic decision-making that occurs when one group is unfairly or arbitrarily disadvantaged over another (Kim & Cho, 2022).

Algorithmic Discrimination: Algorithmic discrimination occurs amidst algorithmic bias when algorithmic decision-making allows one group to be unfairly or arbitrarily disadvantaged over another (Kim & Cho, 2022).

Algorithmic Fairness: Algorithmic fairness is the intentional examination of models for fair and equitable algorithms to reduce bias. "To construct a fair algorithmic model, "three criteria need to be considered: fairness, expressiveness, and utility. Fairness can be evaluated by the three measures introduced by how the model is treated fairly without bias between groups. Expressiveness is how the value after applying the method of processing data expresses the information of the original data. It can be evaluated by the performance obtained from the various classifiers. Utility is an evaluation of tasks that the AI model must perform" (Kim & Cho, 2022, p. 2).

Algorithmic Neutrality: Neutrality is the unconditional absence of bias, and numerous scientists have noted the impossibility and illusion of algorithmic neutrality (Kozlowski et al., 2022b; Phillips-Brown, 2023). Algorithmic neutrality cannot exist when the training data is human data, as bias is intrinsically human, and human bias cannot be eliminated, only reduced.

Data Brokers: Data brokers are typically companies or organizations that handle the exchange of much of the internet data content including internet companies, advertisers, retailers, trade associations, ad-tech groups, data analytics firms, and credit agencies (Reviglio, 2022).

Data Mining/Scraping: While data mining and data scraping are often used synonymously across disciplines, the distinction between mining versus scraping typically refers to data mining as the gathering of large datasets for analysis, while data scraping involves the process of gathering large datasets. Due to the overlap in meaning, for the purposes of this chapter, the terms are concomitant.

ENDNOTES

1 For detailed analyses of heirs' property fractional interest and protections, see Gaither et al.'s (2019) technical report from the USDA National Forestry Service: *Heirs property and land fractionation: Fostering stable ownership to prevent land loss and abandonment.*

2 First-party DMS is often gained via consent and/or user activity.

3 See Center for Heirs Property Preservation, among other organizations.

4 The Southern Rural Development Center released a related brief (Heirs' Property Issue Brief 23-1) of a synthesis of these two studies to assess the scope and magnitude of heirs' property: https://srdc.msstate.edu/sites/default/files/2023-06/HP-brief-Bailey-Dobbs-Gaither-Thomson-6.2023-final.pdf.

5 Bailey and Thomson (2022) have made available an index to the research-based variables operationalized for this calculation at https://onlinelibrary.wiley.com/action/downloadSupplement?doi=10.1111%2Fruso.12455&file=ruso12455-sup-0001-Supinfo.docx.

Compilation of References

Abaido, G. M. (2020). Cyberbullying on social media platforms among university students in the United Arab Emirates. *International Journal of Adolescence and Youth*, *25*(1), 407–420. doi:10.1080/02673843.2019.1669059

Abu-Shanab, E., & Al-Jamal, N. (2015). Exploring the Gender Digital Divide in Jordan. *Gender, Technology and Development*, *19*(1), 91–113. doi:10.1177/0971852414563201

Adiratna, H., & Wulansari, A. (2021). Factors Influencing Purchase Intention of Elancing Using UTAUT Model: A Case Study of Mahajasa. *Malaysian Journal of Social Sciences and Humanities*, *6*(9), 563–564. doi:10.47405/mjssh.v6i9.1056

African-American Burial Grounds Preservation Act. S. 3667, 117th Cong. (2022). https://www.congress.gov/bill/117th-congress/senate-bill/3667

Afzal, A., Khan, S., Daud, S., Ahmad, Z., & Butt, A. (2023). Addressing the Digital Divide: Access and Use of Technology in Education. *Journal of Social Sciences Review*, *3*(2), 883–895. doi:10.54183/jssr.v3i2.326

Agarwal, R., & Weill, P. (2012). The benefits of combining data with empathy. *MIT Sloan Management Review*, *54*(1), 35.

Agrawal, A., Gans, J. S., & Goldfarb, A. (2019). Exploring the impact of artificial intelligence: Prediction versus judgment. *Information Economics and Policy*, *47*, 1–6. doi:10.1016/j.infoecopol.2019.05.001

Aguilar, S. J. (2020). Guidelines and tools for promoting digital equity. *Information and Learning Science*, *121*(5/6), 285–299. doi:10.1108/ILS-04-2020-0084

AI bias is personal for me. It should be for you, too. (n.d.). PwC. https://www.pwc.com/us/en/tech-effect/ai-analytics/artificial-intelligence-bias.html

AI, H. (2019). High-level expert group on artificial intelligence. *Ethics guidelines for trustworthy AI*, 6.

Aissaoui, N. (2021). The digital divide: a literature review and some directions for future research in light of COVID-19. *Global Knowledge, Memory and Communication, 71*(8/9), 686–708. https://doi.org/https://doi.org/10.1108/GKMC-06-2020-0075

Aitken, M., Toreini, E., Carmichael, P., Coopamootoo, K., Elliott, K., & van Moorsel, A. (2020). Establishing a social licence for Financial Technology: Reflections on the role of the private sector in pursuing ethical data practices. *Big Data & Society*, *7*(1). doi:10.1177/2053951720908892

Aizenberg, E., & Van Den Hoven, J. (2020). Designing for human rights in AI. *Big Data & Society*, *7*(2), 2053951720949566. doi:10.1177/2053951720949566

Akhtar, P., Frynas, J. G., Mellahi, K., & Ullah, S. (2019). Big data-savvy teams' skills, big data-driven actions and business performance. *British Journal of Management*, *30*(2), 252–271. doi:10.1111/1467-8551.12333

Al-Adwan, A. S., Kokash, H., Adwan, A. A., Alhorani, A., & Yaseen, H. (2020). Building customer loyalty in online shopping: The role of online trust, online satisfaction and electronic word of mouth. *International Journal of Electronic Marketing and Retailing*, *11*(3), 278–306. doi:10.1504/IJEMR.2020.108132

Alalwan, A. (2018). Investigating the impact of social media advertising features on customer purchase intention. *International Journal of Information Management*, *42*, 65–77. doi:10.1016/j.ijinfomgt.2018.06.001

Alalwan, A. A., Rana, N. P., Dwivedi, Y. K., & Algharabat, R. (2017). Social media in marketing: A review and analysis of the existing literature. *Telematics and Informatics*, *34*(7), 1177–1190. doi:10.1016/j.tele.2017.05.008

Allison, G. (2017). The thucydides trap. *Foreign Policy*, *9*(6), 73–80. Advance online publication. doi:10.7551/mitpress/9780262028998.003.0006

Ameen, N., Tarhini, A., Reppel, A., & Anand, A. (2021). Customer experiences in the age of artificial intelligence. *Computers in Human Behavior*, *114*, 106548. doi:10.1016/j.chb.2020.106548 PMID:32905175

Analytica, O. (2023). How are global businesses managing today's political risks? 2023 survey report. *WTW*, 1-34.

Anand, A., Sagar, S. R., & Kumar, C. (2023). How India Is Able to Control Inflation During the Russia-Ukraine War. In Cases on the Resurgence of Emerging Businesses (pp. 237-251). IGI Global. doi:10.4018/978-1-6684-8488-3.ch017

Andrews, E. L. A. (2021, August 6). *How flawed data aggravates inequality in credit*. Stanford HAI. https://hai.stanford.edu/news/how-flawed-data-aggravates-inequality-credit

Andriole, S. J. (2017). Five myths about digital transformation. *MIT Sloan Management Review*, *58*(3), 20–22.

Annie Brown. (2020, February 7). Biased Algorithms Learn From Biased Data: 3 Kinds Biases Found In AI Datasets. *Forbes*. Retrieved October 12, 2023, from https://www.forbes.com/sites/cognitiveworld/2020/02/07/biased-algorithms/?sh=37f7add076fc

Antonio, A., & Tuffley, D. (2014). The Gender Digital Divide in Developing Countries. *Future Internet*, *6*(4), 673–687. doi:10.3390/fi6040673

Antonopoulos, A. M. (2014). *Mastering Bitcoin: Unlocking Digital Cryptocurrencies*. O'Reilly Media, Inc.

Antonova, N. (2014). Psychological Effectiveness of Interactive Advertising in Russia. *Journal of Creative Communications*, *10*(3), 303–311. Advance online publication. doi:10.1177/0973258615614426

Antwi, S. (2021). "I just like this e-Retailer": Understanding online consumers repurchase intention from relationship quality perspective. *Journal of Retailing and Consumer Services*, *61*, 102568. doi:10.1016/j.jretconser.2021.102568

Araújo, C. S., Magno, G., Meira, W., Almeida, V., Hartung, P., & Doneda, D. (2017). Characterizing videos, audience and advertising in YouTube channels for kids. In *Lecture Notes in Computer Science (Including Subseries Lecture Notes in Artificial Intelligence and Lecture Notes in Bioinformatics)*. Elsevier. doi:10.1007/978-3-319-67217-5_21

Are robots sexist? UN report shows gender bias in talking digital tech. (2019, May 22). UN News. https://news.un.org/en/story/2019/05/1038691

Arora, M., Prakash, A., Dixit, S., Mittal, A., & Singh, S. (2022). A critical review of HR analytics: visualization and bibliometric analysis approach. *Information Discovery and Delivery*.

Arora, P., & Narula, S. (2018). Linkages between service quality, customer satisfaction and customer loyalty: A literature review. *Journal of Marketing Management*, *17*(4), 30.

Ashfaq, M., Yun, J., Yu, S., & Loureiro, S. M. C. (2020). I, Chatbot: Modeling the determinants of users' satisfaction and continuance intention of AI-powered service agents. *Telematics and Informatics*, *54*, 101473. doi:10.1016/j.tele.2020.101473

Ashford, N. A. (2000). An innovation-based strategy for a sustainable environment. In *Innovation-oriented environmental regulation: theoretical approaches and empirical analysis* (pp. 67–107). Physica-Verlag HD. doi:10.1007/978-3-662-12069-9_5

Askari, G., Gordji, M. E., Shabani, S., & Filipe, J. A. (2020). *Game theory and trade tensions between advanced economies*. Academic Press.

Asrani, C. (2020). *Bridging the Digital Divide in India : Barriers to Adoption and Usage*. Issue June.

Atik, S. O., & Ipbuker, C. (2021). Integrating convolutional neural network and multiresolution segmentation for land cover and land use mapping using satellite imagery. *Applied Sciences (Basel, Switzerland)*, *11*(12), 5551. doi:10.3390/app11125551

Autocar. (2019, October 10). *New Kia experience centre inaugurated in New Delhi.* Autocar India. https://www.autocarindia.com/car-news/new-kia-experience-centre-inaugurated-in-new-delhi-414447

AutoS. (2023). https://briteskoda.com/experience-center/

Awan, A. A. (2023, July 17). *What is Algorithmic Bias?* https://www.datacamp.com/blog/what-is-algorithmic-bias

Aydın, A., & Bensghir, T. K. (2019). Digital Data Sovereignty: Towards a Conceptual Framework. In *2019 1st International Informatics and Software Engineering Conference (UBMYK)* (pp. 1-6). IEEE. 10.1109/UBMYK48245.2019.8965469

Bacher, N., & Manowicz, A. A. (2020). *Digital auto customer journey-An analysis of the impact of digitalization on the new car sales process and structure.* www.ijsrm.com

Bagheri, M. (2020). Disruptive technologies or Big-Bang disruption: A research gap in marketing studies. *Proceedings of IC Mark Tech, 2019*, 229–241.

Bailey, C., & Thomson, R. (2022). Heirs property, critical race theory, and reparations. *Rural Sociology*, *87*(4), 1219–1243. doi:10.1111/ruso.12455

Bakke, R., & Barland, G. (2021). Governance Challenges in Decentralized Marketplaces: A Regulatory Perspective. *Journal of Business Regulation*, *14*(3), 289–308.

Bakke, R., & Barland, G. (2022). Blockchain and Decentralized Marketplaces: A Comprehensive Review. *Journal of Financial Technology*, *1*(1), 45–58.

Balakrishnan, J., & Manickavasagam, J. (2016). User Disposition and Attitude towards Advertisements Placed in Facebook, LinkedIn, Twitter and YouTube. *J. Electron. Commer. Organ.*. doi:10.4018/JECO.2016070102

Ballard, S., Chappell, K. M., & Kennedy, K. (2019). Judgment call the game: Using value sensitive design and design fiction to surface ethical concerns related to technology. In *Proceedings of the 2019 on Designing Interactive Systems Conference* (pp. 421-433). 10.1145/3322276.3323697

Balog, K. (2020). The concept and competitiveness of agile organization in the fourth industrial revolution's drift. *Strategic Management*, *25*(3), 14–27. doi:10.5937/StraMan2003014B

Bandara, D. M. D. (2020). Impact of Social Media Advertising on Consumer Buying Behavior: With Special Reference to Fast Fashion Industry. In *The Conference Proceedings of 11th International Conference on Business & Information ICBI.* University of Kelaniya.

Bandi, S., & Kothari, A. (2022). Artificial Intelligence: An Asset for the Financial Sector. *Impact of Artificial Intelligence on Organizational Transformation*, 2.

Bansal, N. (2020). *3 trends driving the auto industry's shift to dealer digitization*. https://www.thinkwithgoogle.com/intl/en-apac/consumer-insights/consumer-journey/3-trends-driving-the-auto-industrys-shift-to-dealer-digitization/

Baramidze, T. (2018). *The Effect of Influencer Marketing on Customer Behaviour. The Case of YouTube Influencers in Makeup Industry*. Vytautas Magnus University.

Bargeman, K. B. (2023). The heirs' property dilemma: How more robust federal policies can help narrow the racial wealth gap. North Carolina Banking Institute, 27(1), 320-348.

Barlow, C. (2020). *Social Media Marketing 2020: A Guide to Brand Building Using Instagram, YouTube, Facebook, Twitter, and Snapchat, Including Specific Advice on Personal Branding for Beginners*. Independently Published.

Barry, T. M. (1987). The development of the hierarchy of effects: An historical perspective. *Curr. Issues Res. Advert.*, *10*, 251–295. doi:10.1080/01633392.1987.10504921

Barton, D., & Court, D. (2012). Making advanced analytics work for you: A practical guide to capitalizing on big data. *Harvard Business Review*, *90*(10), 79–83. PMID:23074867

Bauer, H. (2018). *The Digital Customer Journey in the Automobile Industry - A Quick-Check for the Retail Environment*. Seinajoki University of Applied Sciences. Retrieved from https://urn.fi/URN:NBN:fi:amk-201805219194

Baum, D., Spann, M., Fuller, J., & Thürridl, C. (2019). The impact of social media campaigns on the success of new product introductions. *Journal of Retailing and Consumer Services*, *50*, 289–297. doi:10.1016/j.jretconser.2018.07.003

BBC. (2021, March 31). *Huawei's business damaged by US sanctions despite success at home*. Retrieved September 21, 2023, from https://www.bbc.com/news/technology-56590001

Behara, G. K., & Khandrika, T. (2020). Blockchain as a disruptive technology: Architecture, business scenarios, and future trends. In *AI and Big Data's Potential for Disruptive Innovation* (pp. 130–173). IGI Global. doi:10.4018/978-1-5225-9687-5.ch006

Belk, R. W. (2013). Extended self in a digital world. *The Journal of Consumer Research*, *40*(3), 477–500. doi:10.1086/671052

Bellamy, R. K., Dey, K., Hind, M., Hoffman, S. C., Houde, S., Kannan, K., Lohia, P., Mehta, S., Mojsilovic, A., Nagar, S., Ramamurthy, K. N., Richards, J., Saha, D., Sattigeri, P., Singh, M., Varshney, K. R., & Zhang, Y. (2019). Think your artificial intelligence software is fair? Think again. *IEEE Software*, *36*(4), 76–80. doi:10.1109/MS.2019.2908514

Bellman, K., Landauer, C., Dutt, N., Esterle, L., Herkersdorf, A., Jantsch, A., ... Tammemäe, K. (2020). Self-aware cyber-physical systems. *ACM Transactions on Cyber-Physical Systems, 4*(4), 1-26.

Benabdallah, G., Alexander, A., Ghosh, S., Glogovac-Smith, C., Jacoby, L., Lustig, C., ... Rosner, D. (2022, June). Slanted speculations: Material encounters with algorithmic bias. In *Designing Interactive Systems Conference* (pp. 85-99). 10.1145/3532106.3533449

Benabdelouahed, R., & Dakouan, D. (2020). The Use of Artificial Intelligence in Social Media: Opportunities and Perspectives. *Expert Journal of Marketing*, *8*(1), 82–87.

Benkoël, D. (2020). *What consumers really think about Trusted Digital IDs*. Retrieved from https://dis-blog.thalesgroup.com/mobile/2020/02/11/qa-what-consumers-really-think-about-trusted-digital-ids/

Bentham, J., & Mill, J. S. (2004). Utilitarianism and other essays. Academic Press.

Bentzen, M. M. (2016). The Principle of Double Effect Applied to Ethical Dilemmas of Social Robots. In Robophilosophy/TRANSOR (pp. 268-279). Academic Press.

Bergin, P., Choi, W. J., & Pyun, J. (2023). *Catching Up by 'Deglobalizing': Capital Account Policy and Economic Growth.* doi:10.3386/w30944

Berman, P. (2018). *Digital Identity As a Basic Human Right.* Retrieved from https://impakter.com/digital-identity-basic-human-right/

Best & Rao. (2021). *Understanding algorithmic bias and how to build trust in AI.* PwC. Retrieved October 8, 2023, from https://www.pwc.com/us/en/tech-effect/ai-analytics/algorithmic-bias-and-trust-in-ai.html

Best Practices for Digital Identity Verification. (2022). Retrieved from https://integrity.aristotle.com/2022/05/best-practices-for-digital-identity-verification/

Bharadiya, J. P. (2023). A Comparative Study of Business Intelligence and Artificial Intelligence with Big Data Analytics. *American Journal of Artificial Intelligence*, *7*(1), 24.

Bigne, E., Andreu, L., Hernandez, B., & Ruiz, C. (2018). The impact of social media and offline influences on consumer behaviour, An analysis of the low-cost airline industry. *Current Issues in Tourism*, *21*(9), 1014–1032. doi:10.1080/13683500.2015.1126236

Bijmolt, T. H., Broekhuis, M., De Leeuw, S., Hirche, C., Rooderkerk, R. P., Sousa, R., & Zhu, S. X. (2021). Challenges at the marketing–operations interface in omni-channel retail environments. *Journal of Business Research*, *122*, 864–874. doi:10.1016/j.jbusres.2019.11.034

Bitner, M. J., Brown, S. W., & Meuter, M. L. (2000). Technology infusion in service encounters. *Journal of the Academy of Marketing Science*, *28*(1), 138–149. doi:10.1177/0092070300281013

Bleize, D. N. M., & Antheunis, M. L. (2019). Factors influencing purchase intent in virtual worlds: A review of the literature. *Journal of Marketing Communications*, *25*(4), 403–420. doi:10.1080/13527266.2016.1278028

Blockchain in Digital Identity. (2023). Retrieved from https://consensys.net/blockchain-use-cases/digital-identity/

Board, D. I. (2019). AI Principles: Recommendations on the ethical use of artificial intelligence by the Department of Defense. *Supporting document. Defense Innovation Board*, *2*, 3.

Boateng, H., & Okoe, A. F. (2015). Consumers' attitude towards social media advertising and their behavioral response: The moderating role of corporate reputation. *Journal of Research in Interactive Marketing*, *9*(4), 299–312. doi:10.1108/JRIM-01-2015-0012

Boddington, P., & Boddington, P. (2017). Does AI Raise Any Distinctive Ethical Questions? *Towards a code of ethics for artificial intelligence*, 27-37.

Bogers, M., Chesbrough, H., & Moedas, C. (2018). Open innovation: Research, practices, and policies. *California Management Review*, *60*(2), 5–16. doi:10.1177/0008125617745086

Böhme, R., Christin, N., Edelman, B., & Moore, T. (2015). Bitcoin: Economics, Technology, and Governance. *The Journal of Economic Perspectives*, *29*(2), 213–238. doi:10.1257/jep.29.2.213

Bolton, R. N., Parasuraman, A., Hoefnagels, A., Migchels, N., Kabadayi, S., Gruber, T., Loureiro, Y. K., & Solnet, D. (2013). Understanding Generation Y and their use of social media: A review and research agenda. *Journal of Service Management*, *24*(3), 245–267. doi:10.1108/09564231311326987

Botelho, F. H. F. (2021). Accessibility to digital technology: Virtual barriers, real opportunities. *The Official Journal of RESNA, 33*, 27–34. https://doi.org/https://doi.org/10.1080/10400435.2021.1945705

Bouncken, R. B., Reuschl, A. J., & Ratzmann, M. (2020). The future of the sharing economy: A comprehensive review and synthesis. *Technological Forecasting and Social Change*, *150*, 119791.

Boyd, D., & Crawford, K. (2012). Critical questions for big data: Provocations for a cultural, technological, and scholarly phenomenon. *Information Communication and Society*, *15*(5), 662–679. doi:10.1080/1369118X.2012.678878

Brams, S. J. (1993). Theory of moves. *American Scientist*, *81*(6), 562–570.

Braw, E. (2023). How to "Friendshore". American Enterprise Institute for Public Policy Research.

Breuss, F. (2022). Who wins from an FTA induced revival of world trade? *Journal of Policy Modeling*, *44*(3), 653–674. doi:10.1016/j.jpolmod.2022.05.003

Brock, J. K.-U., & von Wangenheim, F. (2019). Demystifying AI: What digital transformation leaders can teach you about realistic artificial intelligence. *California Management Review*, *61*(4), 110–134. doi:10.1177/1536504219865226

Brooks-Patton, B., & Noor, S. (2023, April). Block Place: A Novel Blockchain-based Physical Marketplace System. In *South east Con 2023, 927-934.* IEEE. doi:10.1109/SoutheastCon51012.2023.10115212

Brusseau, J. (2023). AI human impact: Toward a model for ethical investing in AI-intensive companies. *Journal of Sustainable Finance & Investment*, *13*(2), 1030–1057. doi:10.1080/20430795.2021.1874212

Brynjolfsson, E., & Collis, A. (2019). How should we measure the digital economy? *Harvard Business Review*, *97*(6), 140–148.

Bubeck, S., Chandrasekaran, V., Eldan, R., Gehrke, J., Horvitz, E., & Kamar, E. (2023). *Sparks of Artificial Intelligence: Early experiments with GPT-4*. Academic Press.

Buck, C., & Reith, R. (2020). Privacy on the road? Evaluating German consumers' intention to use connected cars. *International Journal of Automotive Technology and Management*, *20*(3), 297–318. doi:10.1504/IJATM.2020.110408

Buckley, R. P., Zetzsche, D. A., Arner, D. W., & Tang, B. W. (2021). Regulating artificial intelligence in finance: Putting the human in the loop. *The Sydney Law Review*, *43*(1), 43–81.

Budd, J. M. (1988). A bibliometric analysis of higher education literature. *Research in Higher Education*, *28*(2), 180–190. doi:10.1007/BF00992890

Bulsara, H. P., & Vaghela, P. S. (2022). Millennials Online Purchase Intention Towards Consumer Electronics: Empirical Evidence from India. *Indian Journal of Marketing*, *52*(2), 53–70. doi:10.17010/ijom/2022/v52/i2/168154

Buterin, V. (2014). *A Next-Generation Smart Contract and Decentralized Application Platform*. Ethereum White Paper. Retrieved from https://ethereum.org/whitepaper/

Bynum, T. (2008). *Computer and information ethics*. Academic Press.

Cabrera, J., Loyola, M. S., Magaña, I., & Rojas, R. (2023, June). Ethical dilemmas, mental health, artificial intelligence, and llm-based chatbots. In *International Work-Conference on Bioinformatics and Biomedical Engineering* (pp. 313–326). Springer Nature Switzerland. doi:10.1007/978-3-031-34960-7_22

Calp, M. H. (2020). The role of artificial intelligence within the scope of digital transformation in enterprises. In *Advanced MIS and digital transformation for increased creativity and innovation in business* (pp. 122–146). IGI Global. doi:10.4018/978-1-5225-9550-2.ch006

Cambridge. (2023). Data scraping. In *The Cambridge dictionary*. Cambridge University Press.

Campbell, C., Sands, S., Ferraro, C., Tsao, H. Y. J., & Mavrommatis, A. (2020). From data to action: How marketers can leverage AI. *Business Horizons*, *63*(2), 227–243. doi:10.1016/j.bushor.2019.12.002

Can, C. M. (2022). Temporal Theory and US-China Relations. *Journal of Strategic Security*, *15*(2), 1–16. doi:10.5038/1944-0472.15.2.1985

CaoL. (2020). AI in finance: A review. Available at SSRN 3647625.

Carpenter, S., & Waddell, S. R. (2021). *Whose land is it? Heirs' property and its role in generational land retention. Regional Matters*. Federal Reserve Bank of Richmond.

Casey, M. J., & Vigna, P. (2018). *The Truth Machine: The Blockchain and the Future of Everything*. St. Martin's Press.

Cath, C. (2018). Governing artificial intelligence: Ethical, legal and technical opportunities and challenges. *Philosophical Transactions. Series A, Mathematical, Physical, and Engineering Sciences*, *376*(2133), 20180080. doi:10.1098/rsta.2018.0080 PMID:30322996

Cattanach, R. E., & Greenberg, J. M. (2021). Critical update needed! Addressing cybersecurity challenges of public records in the Digital Age. *The Urban Lawyer*, *51*(3), 427–456.

Chabowski, B. R., Samiee, S., & Hult, G. T. M. (2013). A bibliometric analysis of the global branding literature and a research agenda. *Journal of International Business Studies*, *44*(6), 622–634. doi:10.1057/jibs.2013.20

Chandra, A. (2022). *Bridging the Digital Gender Divide*. doi:10.4018/978-1-7998-8594-8.ch002

Chatila, R., Renaudo, E., Andries, M., Chavez-Garcia, R. O., Luce-Vayrac, P., Gottstein, R., Alami, R., Clodic, A., Devin, S., Girard, B., & Khamassi, M. (2018). Toward self-aware robots. *Frontiers in Robotics and AI*, *5*, 88. doi:10.3389/frobt.2018.00088 PMID:33500967

Chatterjee, S., Chaudhuri, R., Vrontis, D., Thrassou, A., & Ghosh, S. K. (2021). Adoption of artificial intelligence-integrated CRM systems in agile organizations in India. *Technological Forecasting and Social Change*, *168*, 120783. doi:10.1016/j.techfore.2021.120783

Chaudhary, P., & Sharma, K. K. (2021). Effects of Covid-19 on De-globalization. *Globalization, Deglobalization, and New Paradigms in Business*, 133–153. doi:10.1007/978-3-030-81584-4_8

Chauhan, C. (2020, December 30). *Smaller dealerships are here. Did COVID expedite them?* ET Auto. https://auto.economictimes.indiatimes.com/news/aftermarket/smaller-dealerships-are-here-did-covid-expedite-them/80019721

Chen, Q., Lu, Y., Gong, Y., & Xiong, J. (2023). Can AI chatbots help retain customers? Impact of AI service quality on customer loyalty. *Internet Research*.

Chen, D., & Swan, M. (2021). Scalability Concerns in Decentralized Marketplaces: A Critical Analysis. *International Journal of Blockchain and Distributed Ledger Technology*, *4*(1), 23–38.

Chen, J., & Wang, L. (2020). Decentralization and Efficiency: Exploring New Avenues in Business Models. *Technological Forecasting and Social Change*, *176*, 120890.

Chen, W.-K., Ling, C.-J., & Chen, C.-W. (2022). What affects users to click social media ads and purchase intention? The roles of advertising value, emotional appeal and credibility. *Asia Pacific Journal of Marketing and Logistics*. Advance online publication. doi:10.1108/APJML-01-2022-0084

Chesbrough, H. (2020). The Role of Open Innovation in Disruptive Business Models. *California Management Review*, *62*(3), 5–23.

China Org. (2023). *The Belt and Road Initiative.* BRI. Retrieved September 20, 2023, from http://belt.china.org.cn/

Chong, A. Y.-L. (2013). Predicting m-commerce adoption determinants: A neural network approach. *Expert Systems with Applications*, *40*(2), 523–530. doi:10.1016/j.eswa.2012.07.068

Chouldechova, A. (2017). Fair prediction with disparate impact: A study of bias in recidivism prediction instruments. *Big Data*, *5*(2), 153–163. doi:10.1089/big.2016.0047 PMID:28632438

Chris, A. (2006). *The long tail: Why the future of business is selling less of more*. Hyperion.

Christensen, C. M. (1997). *The Innovator's Dilemma: When New Technologies Cause Great Firms to Fail.* Harvard Business Review Press.

Chung, K., Thaichon, P., & Quach, S. (2022). Types of artificial intelligence (AI) in marketing management. In *Artificial intelligence for marketing management* (pp. 29–40). Routledge. doi:10.4324/9781003280392-4

Chung, M., Ko, E., Joung, H., & Kim, S. J. (2020). Chatbot e-service and customer satisfaction regarding luxury brands. *Journal of Business Research*, *117*, 587–595. doi:10.1016/j.jbusres.2018.10.004

Chungviwatanant, T., Prasongsukam, K., & Chungviwatanant, S. (2016). A study of factors that affect consumer's attitude toward a "skippable in-stream ad" on YouTube. *Au Gsb E J.*, *9*, 83–96.

Clauberg, R. (2020). Challenges of digitalization and artificial intelligence for modern economies, societies and management. *RUDN Journal of Economics*, *28*(3), 556–567. doi:10.22363/2313-2329-2020-28-3-556-567

Cockburn, I. M., Henderson, R., & Stern, S. (2018). The impact of artificial intelligence on innovation: An exploratory analysis. In *The economics of artificial intelligence: An agenda* (pp. 115–146). University of Chicago Press.

Colbert, A., Yee, N., & George, G. (2016). From the editors: The digital workforce and the workplace of the future. Academy of Management Journal, 59(3), 731–739.

Colladon, A. F. (2020). Forecasting election results by studying brand importance in online news. *International Journal of Forecasting*, *36*(2), 414–427. doi:10.1016/j.ijforecast.2019.05.013

Conway, P. (1996). *Preservation in the digital world.* Council on Library and Information Resources.

Cordova-Buiza, F., Urteaga-Arias, P. E., & Coral-Morante, J. A. (2022). Relationship between Social Networks and Customer Acquisition in the Field of IT Solutions. *IBIMA Business Review.*, *2022*, 631332. Advance online publication. doi:10.5171/2022.631332

Correa, H. (2001). Game theory as an instrument for the analysis of international relations. *Ritsumeikan Annual Review of International Studies*, *14*(2), 187–208.

Cortellazzo, L., Bruni, E., & Zampieri, R. (2019). The role of leadership in a digitalized world: A review. *Frontiers in Psychology*, *10*, 1938. doi:10.3389/fpsyg.2019.01938 PMID:31507494

Crosby, M., Pattanayak, P., Verma, S., & Kalyanaraman, V. (2016). Blockchain technology: Beyond bitcoin. *Applied Innovation*, *2*, 6–10.

Culnan, M. J., McHugh, P. J., & Zubilaga, J. I. (2010). How large U.S. companies can use Twitter and other social media to gain business value. *MIS Quarterly Executive, 9*(4), 243–259.

Cummins, N., & Schuller, B. W. (2020). Five crucial challenges in digital health. *Frontiers in Digital Health*, *2*, 536203. doi:10.3389/fdgth.2020.536203 PMID:34713029

Cuthbertson, A. (2019, March 6). Self-driving cars more likely to drive into black people, study claims. *The Independent.* https://www.independent.co.uk/tech/self-driving-car-crash-racial-bias-black-people-study-a8810031.html#:~:text=Researchers%20at%20the%20Georgia%20Institute,into%20them%2C%20the%20authors%20note

Cuzzolin, F., Morelli, A., Cirstea, B., & Sahakian, B. J. (2020). Knowing me, knowing you: Theory of mind in AI. *Psychological Medicine*, *50*(7), 1057–1061. doi:10.1017/S0033291720000835 PMID:32375908

Dabrock, P., Tretter, M., Braun, M., & Hummel, P. (2021). Data sovereignty: A review. *Big Data & Society*, *8*(1). Advance online publication. doi:10.1177/2053951720982012

Dagnon, S. (2018). *Using Chatbots for Social Media Marketing*. Available at: https://mavsocial.com/chatbots-social-media-marketing/

Dahiya, M. (2017). A tool of conversation: Chatbot. *International Journal on Computer Science and Engineering*, *5*(5), 158–161.

Dai, H., Haried, P., & Salam, A. F. (2011). Antecedents of online service quality, commitment and loyalty. *Journal of Computer Information Systems*, *52*(2), 1–11.

Dalton, C. M., Taylor, L., & Thatcher, J. (2016). Critical data studies: A dialog on data and space. *Big Data & Society*, *3*(1), 2053951716648346. doi:10.1177/2053951716648346

Danielsson, J., Macrae, R., & Uthemann, A. (2022). Artificial intelligence and systemic risk. *Journal of Banking & Finance*, *140*, 106290. doi:10.1016/j.jbankfin.2021.106290

Danks, D., & London, A. J. (2017). Algorithmic Bias in Autonomous Systems. In Ijcai (Vol. 17, No. 2017, pp. 4691-4697). doi:10.24963/ijcai.2017/654

Dash, M., Sahu, R., & Pandey, A. (2018). Social media marketing impact on the purchase intention of millennials. *International Journal of Business Information Systems*, *28*(2), 147. doi:10.1504/IJBIS.2018.10012924

Das, K., Patel, J., Sharma, A., & Shukla, Y. (2023). Creativity in marketing: Examining the intellectual structure using scientometric analysis and topic modeling. *Journal of Business Research*, *154*, 113384. doi:10.1016/j.jbusres.2022.113384

Davenport, T. H. (2019). Can we solve AI's "trust problem"? *MIT Sloan Management Review*, *60*(2), 1–5.

Davis, F. D., Bagozzi, R. P., & Warshaw, P. R. (1989). User acceptance of computer technology: A comparison of two theoretical models. *Management Science*, *35*(8), 982–1003. doi:10.1287/mnsc.35.8.982

Dawar, N., & Bendle, N. (2018). Marketing in the age of Alexa. *Harvard Business Review*, *96*(3), 80–86.

De Moura, N., Chatila, R., Evans, K., Chauvier, S., & Dogan, E. (2020). *Ethical decision making for autonomous vehicles. In 2020 IEEE intelligent vehicles symposium (iv)*. IEEE.

Deepa, M. M., Sajan, S., & Gupta, T. (2023). The intersection of technology and society: An analysis of the digital divide. *European Chemical Bulletin*, *12*(12), 1106–1116. doi:10.48047/ecb/2023.12.si12.09

Definition of algorithm. (2023, September 28). *Merriam-Webster Dictionary*. https://www.merriam-webster.com/dictionary/algorithm

Dehghani, M., Niaki, M. K., Ramezani, I., & Sali, R. (2016). Evaluating the influence of YouTube advertising for attraction of young customers. *Computers in Human Behavior*, *59*, 165–172. doi:10.1016/j.chb.2016.01.037

De, R., Pandey, N., & Pal, A. (2020). Impact of digital surge during Covid-19 pandemic: A viewpoint on research and practice. *International Journal of Information Management*, *55*, 102171. Advance online publication. doi:10.1016/j.ijinfomgt.2020.102171 PMID:32836633

Derek Chun, W. S., Siu, H. Y., Wai, M. C., & Chi, Y. L. (2022). Transformation Conference paper Bridging the Gap Between Digital Divide and Educational Equity by Engaging Parental Digital Citizenship and Literacy at Post-Covid-19 Age in the Hong Kong Context. *The Post-Pandemic Landscape of Education and Beyond: Innovation and Transformation*, 165–182.

DeviS.KotianS.KumavatM.PatelD. (2022). Digital Identity Management System Using Blockchain. SSRN. doi:10.2139/ssrn.4127356

Dewar, R., & Xie, Y. (2021). Disruptive Innovation and Firm Performance: A Review. *Technological Forecasting and Social Change*, *167*, 120667.

Diakopoulos, N. (2016). Accountability in algorithmic decision making. *Communications of the ACM*, *59*(2), 56–62. doi:10.1145/2844110

Diebold, G., & Castro, D. (2023). Digital Equity 2.0: How to Close the Data Divide. Center for Data Collection.

Digital Identity Roadmap Guide. (2018). Retrieved from https://tinyurl.com/4fe4kwb9

Dignum, V. (2022). *Responsible Artificial Intelligence—from Principles to Practice.* arXiv preprint arXiv:2205.10785.

Dignum, V. (2019). *Responsible artificial intelligence: how to develop and use AI in a responsible way* (Vol. 2156). Springer. doi:10.1007/978-3-030-30371-6

Dilmegani, C. (2023, September 11). *Bias in AI: What it is, Types, Examples & 6 Ways to Fix it in 2023.* AIMultiple. https://research.aimultiple.com/ai-bias/

Disability, Bias, and AI Report. (2023, September 27). AI Now Institute. Retrieved October 8, 2023, from https://ainowinstitute.org/publication/disabilitybiasai-2019

Dixon-Woods, M., & Bosk, C. L. (2011). Defending rights or defending privileges? Rethinking the ethics of research in public service organizations. *Public Management Review*, *13*(2), 257–272. doi:10.1080/14719037.2010.532966

Dobbs, G. R., & Gaither, C. J. (2023). *How much heirs' property is there? Using LightBox Data to estimate heirs' property extent in the US.* Southern Rural Development Center. https://scholarsjunction.msstate.edu/srdctopics-heirsproperty/1/

Domeyer, A., McCarthy, M., Pfeiffer, S., & Schert, G. (2020). *How governments can deliver on the promise of digital ID.* Retrieved from https://www.mckinsey.com/industries/public-sector/our-insights/how-governments-can-deliver-on-the-promise-of-digital-id

Donthu, N., Kumar, S., Mukherjee, D., Pandey, N., & Lim, W. M. (2021). How to conduct a bibliometric analysis: An overview and guidelines. *Journal of Business Research*, *133*, 285–296. doi:10.1016/j.jbusres.2021.04.070

Dornberger, R., Inglese, T., Korkut, S., & Zhong, V. J. (2018). Digitalization: Yesterday, today and tomorrow. *Business Information Systems and Technology 4.0: New Trends in the Age of Digital Change*, 1-11.

Dove, E. S. (2018). The EU general data protection regulation: Implications for international scientific research in the digital era. *The Journal of Law, Medicine & Ethics*, *46*(4), 1013–1030. doi:10.1177/1073110518822003

Dremel, C., Herterich, M., Wulf, J., Waizmann, J.-C., & Brenner, W. (2017). How AUDI AG established big data analytics in its digital transformation. *MIS Quarterly Executive*, *16*(2), 81–100.

Duan, Y., Edwards, J. S., & Dwivedi, Y. K. (2019). Artificial intelligence for decision making in the era of Big Data–evolution, challenges and research agenda. *International Journal of Information Management, 48*, 63–71. doi:10.1016/j.ijinfomgt.2019.01.021

Duffett, R. G. (2015). Effect of Gen Y's affective attitudes towards facebook marketing communications in South Africa. *The Electronic Journal on Information Systems in Developing Countries, 68*(1), 1–27. doi:10.1002/j.1681-4835.2015.tb00488.x

Duffett, R. G. (2015). Facebook advertising's influence on intention-to-purchase and purchase amongst Millennials. *Internet Research, 25*(4), 498–526. doi:10.1108/IntR-01-2014-0020

Duffett, R. G., Edu, T., & Negricea, I. C. (2019). YouTube marketing communication demographic and usage variables influence on Gen Y's cognitive attitudes in South Africa and Romania. *The Electronic Journal on Information Systems in Developing Countries, 85*(5), 1–13. doi:10.1002/isd2.12094

Duffett, R. G., Edu, T., Negricea, I. C., & Zaharia, R. M. (2020). Modelling the effect of YouTube as an advertising medium on converting intention-to-purchase into purchase. *Transformations in Business & Economics, 19*, 112–132.

Duffett, R. G., Petroşanu, D. M., Negricea, I. C., & Edu, T. (2019). Effect of YouTube marketing communication on converting brand liking into preference among Millennials regarding brands in general and sustainable offers in particular: Evidence from South Africa and Romania. *Sustainability (Basel), 11*(3), 1–24. doi:10.3390u11030604

Dunford, M., & Qi, B. (2020). Global reset: COVID-19, systemic rivalry and the global order. *Research in Globalization, 2*, 100021. doi:10.1016/j.resglo.2020.100021

Dwivedi, Y. K., Hughes, L., Ismagilova, E., Aarts, G., Coombs, C., Crick, T., Duan, Y., Dwivedi, R., Edwards, J., Eirug, A., Galanos, V., Ilavarasan, P. V., Janssen, M., Jones, P., Kar, A. K., Kizgin, H., Kronemann, B., Lal, B., Lucini, B., ... Williams, M. D. (2021). Artificial Intelligence (AI): Multidisciplinary perspectives on emerging challenges, opportunities, and agenda for research, practice and policy. *International Journal of Information Management, 57*, 101994. doi:10.1016/j.ijinfomgt.2019.08.002

Economic Times. (2023, September 14). *India's solar imports from China down nearly 80% by $2 billion in H1 2023: Ember.* Retrieved October 1, 2023, from https://economictimes.indiatimes.com/industry/renewables/indias-solar-imports-from-china-down-nearly-80-by-2-billion-in-h1-2023-ember/articleshow/103667346.cms

Egger, R., Kroner, M., & Stöckl, A. (2022). Web scraping: Collecting and retrieving data from the web. In *Applied data science in tourism: Interdisciplinary approaches, methodologies, and applications* (pp. 67–82). Springer. doi:10.1007/978-3-030-88389-8_5

Eichengreen, B. (2023). Globalization: Uncoupled or unhinged? *Journal of Policy Modeling, 45*(4), 685–692. Advance online publication. doi:10.1016/j.jpolmod.2023.02.008

El Sawy, O. A., Kraemmergaard, P., Amsinck, H., & Lerbech Vinther, A. (2016). How LEGO built the foundations and enterprise capabilities for digital leadership. *MIS Quarterly Executive, 15*(2), 141–166.

Ellitan, L., & Richard, A. (2022). The influence of online shopping experience, customer satisfaction and adjusted satisfaction on online repurchase intention to Tokopedia consumers in Surabaya. *Budapest International Research and Critics Institute-Journal (BIRCI-Journal), 5*(2), 16504-16516.

Erkan, I., & Evans, C. (2016). The influence of eWOM in social media on consumers' purchase intentions: An extended approach to information adoption. *Computers in Human Behavior, 61*, 47–55. doi:10.1016/j.chb.2016.03.003

Eskicioglu, A. M., Town, J., & Delp, E. J. (2003). Security of digital entertainment content from creation to consumption. *Signal Processing Image Communication*, *18*(4), 237–262. doi:10.1016/S0923-5965(02)00143-1

Esteller-Cucala, M., Fernandez, V., & Villuendas, D. (2020). Towards data-driven culture in a Spanish automobile manufacturer: A case study. *Journal of Industrial Engineering and Management*, *13*(2), 228–245. doi:10.3926/jiem.3042

EU. (2019, January 31). *EU-Japan trade agreement enters into force*. Retrieved September 16, 2023, from https://ec.europa.eu/commission/presscorner/detail/en/IP_19_785

Eubanks. (2018, January). *Automating Inequality: How High-Tech Tools Profile, Police, and Punish the Poor.* St. Martin's Press, Inc. https://doi.org/ doi:10.5555/3208509

Euchner, J., & Ganguly, A. (2020). Disruptive Business Models and the Digital Transformation: A Review. *Journal of Business Models*, *8*(3), 44–61.

Evans, M. (2019). *Build A 5-star customer experience with artificial intelligence*. https://www.forbes.com/sites/allbusiness/2019/02/17/customer-experience-artificialintelligence/#1a30ebd415bd

Evanschitzky, H., Wangenheim, F. V., & Woisetschläger, D. M. (2011). Service & solution innovation: Overview and research agenda. *Industrial Marketing Management*, *40*(5), 657–660. doi:10.1016/j.indmarman.2011.06.004

Faguet, J. P. (2012). Decentralization and Governance. *World Development*, *41*, 67–74.

Fang, Y. H. (2019). An app a day keeps a customer connected: Explicating loyalty to brands and branded applications through the lens of affordance and service-dominant logic. *Information & Management*, *56*(3), 377–391. doi:10.1016/j.im.2018.07.011

Fan, X., Chai, Z., Deng, N., & Dong, X. (2020). Adoption of augmented reality in online retailing and consumers' product attitude: A cognitive perspective. *Journal of Retailing and Consumer Services*, *53*, 101986. doi:10.1016/j.jretconser.2019.101986

Farrugia, C., Grima, S., & Sood, K. (2022). The General Data Protection Regulation (GDPR) for risk mitigation in the insurance industry. In Big Data: A game changer for insurance industry (pp. 265-302). Emerald Publishing Limited.

Farrukh, M., Raza, A., Meng, F., & Wu, Y. (2022). CMS at 13: A retrospective of the journey. *Chinese Management Studies*, *16*(1), 119–139. doi:10.1108/CMS-07-2020-0291

Far, S. B., & Rad, A. I. (2022). Applying digital twins in metaverse: User interface, security, and privacy challenges. *Journal of Metaverse*, *2*(1), 8–15.

Felin, T., & Lakhani, K. (2018). What problems will you solve with blockchain? *MIT Sloan Management Review*, *60*(1), 32–38.

Festinger, L. (1962). Cognitive dissonance. *Scientific American*, *207*(4), 93–106. doi:10.1038cientificamerican1062-93 PMID.13892642

Finn, R. L., Wright, D., & Friedewald, M. (2013). Seven types of privacy. *European data protection: Coming of age*, 3-32.

Fjeld, J., Achten, N., Hilligoss, H., Nagy, A., & Srikumar, M. (2020). *Principled artificial intelligence: Mapping consensus in ethical and rights-based approaches to principles for AI.* Berkman Klein Center Research Publication.

Fleming, R. B., & Morgan, R. C. (2012). Standards for Financial Decision-Making: Legal, Ethical, and Practical Issues. *Utah L. Rev.*, 1275.

Flowers, J. C. (2019, March). Strong and Weak AI: Deweyan Considerations. In AAAI spring symposium: Towards conscious AI systems (Vol. 2287, No. 7). AAAI.

Formosa, P., Wilson, M., & Richards, D. (2021). A principlist framework for cybersecurity ethics. *Computers & Security, 109*, 102382. doi:10.1016/j.cose.2021.102382

Fountaine, T., McCarthy, B., & Saleh, T. (2019). Building the AI-powered organization. *Harvard Business Review, 97*(4), 62–73.

FrankenfieldJ. (2018). *Chatbot.* Available at: https://www.investopedia.com/terms/c/chatbot.asp

Frank, M. R., Sun, L., Cebrian, M., Youn, H., & Rahwan, I. (2018). Small cities face greater impact from automation. *Journal of the Royal Society, Interface, 15*(139), 20170946. doi:10.1098/rsif.2017.0946 PMID:29436514

Frey, C. B., & Osborne, M. A. (2017). The future of employment: How susceptible are jobs to computerisation? *Technological Forecasting and Social Change, 114*, 254–280. doi:10.1016/j.techfore.2016.08.019

Friedman, B., Harbers, M., Hendry, D. G., van den Hoven, J., Jonker, C., & Logler, N. (2021). Eight grand challenges for value sensitive design from the 2016 Lorentz workshop. *Ethics and Information Technology, 23*(1), 5–16. doi:10.100710676-021-09586-y

Fritz-Morgenthal, S., Hein, B., & Papenbrock, J. (2022). Financial risk management and explainable, trustworthy, responsible AI. *Frontiers in Artificial Intelligence, 5*, 779799. doi:10.3389/frai.2022.779799 PMID:35295866

Fu, R., Huang, Y., & Singh, P. V. (2020). AI and Algorithmic Bias: Source, Detection, Mitigation and Implications. SSRN *Electronic Journal.* doi:10.2139/ssrn.3681517

Fukuyama, M. (2018). Society 5.0: Aiming for a New Human-Centered Society. *Japan Spotlight.* https://www.jef.or.jp/journal/

G20. (2023, March 6). *G20- Background Brief.* Retrieved September 25, 2023, from https://www.g20.org/content/dam/gtwenty/about_g20/overview/G20_Background_Brief_06-03-2023.pdf

Gandhi, S., & Gervet, E. (2016). Now that your products can talk, what will they tell you? *MIT Sloan Management Review, 57*(3), 49–50.

Gan, I., & Moussawi, S. (2022). A Value Sensitive Design Perspective on AI Biases. *Proceedings of the 55th Hawaii International Conference on System Sciences.* 10.24251/HICSS.2022.676

Gao, P., Meng, F., Mata, M. N., Martins, J. M., Iqbal, S., Correia, A. B., & Farrukh, M. (2021). Trends and future research in electronic marketing: A bibliometric analysis of twenty years. *Journal of Theoretical and Applied Electronic Commerce Research, 16*(5), 1667–1679. doi:10.3390/jtaer16050094

Gao, S., & Zang, Z. (2016). An empirical examination of users' adoption of mobile advertising in China. *Information Development, 32*(2), 203–215. doi:10.1177/0266666914550113

García-Herrero, A., & Tan, J. (2020). Deglobalisation in the context of United States-China decoupling. *Policy Contribution, 21*, 1–16.

Garg, P., Raj, R., Kumar, V., Singh, S., Pahuja, S., & Sehrawat, N. (2023). Elucidating the role of consumer decision making style on consumers' purchase intention: The mediating role of emotional advertising using PLS-SEM. *Journal of Economy and Technology, 1*, 108 118. doi:10.1016/j.ject.2023.10.001

Garg, S., & Sushil. (2022). Impact of de-globalization on development: Comparative analysis of an emerging market (India) and a developed country (USA). *Journal of Policy Modeling, 44*(6), 1179–1197. doi:10.1016/j.jpolmod.2022.10.004

Gauri, D. K., Jindal, R. P., Ratchford, B., Fox, E., Bhatnagar, A., Pandey, A., Navallo, J. R., Fogarty, J., Carr, S., & Howerton, E. (2021). Evolution of retail formats: Past, present, and future. *Journal of Retailing*, *97*(1), 42–61. Advance online publication. doi:10.1016/j.jretai.2020.11.002

Gautam, V., & Sharma, V. (2017). The Mediating Role of Customer Relationship on the Social Media Marketing and Purchase Intention Relationship with Special Reference to Luxury Fashion Brands. *Journal of Promotion Management*, *23*(6), 872–888. Advance online publication. doi:10.1080/10496491.2017.1323262

Gebru, T., Morgenstern, J., Vecchione, B., Vaughan, J. W., Wallach, H., Iii, H. D., & Crawford, K. (2021). Datasheets for datasets. *Communications of the ACM*, *64*(12), 86–92. doi:10.1145/3458723

Gerbert, P., & Spira, M. (2019). Learning to love the AI bubble. *MIT Sloan Management Review*, *60*(4), 8–10.

Ghaffari, F., Gilani, K., Bertin, E., & Crespi, N. (2021). Identity and access management using distributed ledger technology: A survey. *International Journal of Network Management*, *32*(2), e2180. doi:10.1002/nem.2180

Gilani, K., Bertin, E., Hatin, J., & Crespi, N. (2020). A survey on blockchain-based identity management and decentralized privacy for personal data. In *2020 2nd Conference on Blockchain Research & Applications for Innovative Networks and Services (BRAINS)* (pp. 97-101). IEEE. 10.1109/BRAINS49436.2020.9223312

Gillis, A. S. (2023, July 31). *Algorithm*. https://www.techtarget.com/whatis/definition/algorithm

Giuffrida, I. (2019). Liability for AI decision-making: Some legal and ethical considerations. *Fordham Law Review*, *88*, 439.

Gnizy, I., & Shoham, A. (2017). Reverse Internationalization: A Review and Suggestions for Future Research. *Advances in Global Marketing*, 59–75. doi:10.1007/978-3-319-61385-7_3

Gongane, V. U., Munot, M. V., & Anuse, A. D. (2022). Detection and moderation of detrimental content on social media platforms: Current status and future directions. *Social Network Analysis and Mining*, *12*(1), 129. doi:10.100713278-022-00951-3 PMID:36090695

Goodman, B., & Flaxman, S. (2017). European Union regulations on algorithmic decision-making and a "right to explanation". *AI Magazine*, *38*(3), 50–57. doi:10.1609/aimag.v38i3.2741

Govindarajan, V., & Ramaswamy, S. (2021). Disruptive Business Models in the Digital Era. *Harvard Business Review*, *99*(1), 73–81.

Govt of Canada. (2021, March 25). *Comprehensive Economic and Trade Agreement (CETA) 2nd Meeting of the CETA Joint Committee*. Retrieved September 18, 2023, from https://www.international.gc.ca/trade-commerce/trade-agreements-accords-commerciaux/agr-acc/ceta-aecg/2021-03-25-joint_report-rapport_conjoint.aspx?lang=eng

Grady, P. (2023). *ChatGPT Amendment Shows the EU is Regulating by Outrage*. Available online at: https://datainnovation.org/2023/02/chatgpt-amendment-shows-the-eu-is-regulating-by-outrage/

Grant, M. J., & Booth, A. (2009). A typology of reviews: An analysis of 14 review types and associated methodologies. *Health Information and Libraries Journal*, *26*(2), 91–108. doi:10.1111/j.1471-1842.2009.00848.x PMID:19490148

Grunstein, J. (2022). Globalization, Real and Imagined. *Orbis*, *66*(4), 502–508. doi:10.1016/j.orbis.2022.08.005

Gudmundsdottir, G. (2010). From digital divide to digital equity: Learners' ICT competence in four primary schools in Cape Town, South Africa. *International Journal of Education and Development Using ICT*, *6*(2), 84–105.

Gunning, D., & Aha, D. (2019). DARPA's explainable artificial intelligence (XAI) program. *AI Magazine*, *40*(2), 44–58. doi:10.1609/aimag.v40i2.2850

Gupta, A., & Sengupta, S. (2020). DeFi and the Future of Finance: An Exploratory Study. SSRN Electronic Journal. 10.2139/ssrn.3675187

Gupta, S., & Sharma, V. (2023). Transparency and Trust: Opportunities in Decentralized Marketplaces. *Journal of Digital Economy*, *7*(2), 189–208.

Gurbaxani, V., & Dunkle, D. (2019). Gearing up for successful digital transformation. *MIS Quarterly Executive*, *18*(3), 209–220. doi:10.17705/2msqe.00017

Gursoy, D., Chi, O. H., Lu, L., & Nunkoo, R. (2019). Consumers acceptance of artificially intelligent (AI) device use in service delivery. *International Journal of Information Management*, *49*, 157–169. doi:10.1016/j.ijinfomgt.2019.03.008

Guttentag, D. (2015). Airbnb: Disruptive innovation and the rise of an informal tourism accommodation sector. *Current Issues in Tourism*, *18*(12), 1192–1217. doi:10.1080/13683500.2013.827159

Haddouti, S. E., & Kettani, M. D. E. C. E. (2019). *Towards an interoperable identity management framework: a comparative study.* https://doi.org//arXiv.1902.11184 doi:10.48550

Hagiu, A., & Wright, J. (2020). When data creates a competitive advantage. *Harvard Business Review*, *98*(1), 94–101.

Hakami, E., & Hernandez Leo, D. (2020). How are learning analytics considering the societal values of fairness, accountability, transparency and human well-being? A literature review. *LASI-SPAIN 2020: Learning Analytics Summer Institute Spain 2020: Learning Analytics. Time for Adoption? 2020 Jun 15-16; Valladolid, Spain. Aachen: CEUR; 2020,* 121-41.

Hamouda, M. (2018, April 09). Understanding social media advertising effect on consumers' responses: An empirical investigation of tourism advertising on Facebook. *Journal of Enterprise Information Management*, *31*(3), 426–445. Advance online publication. doi:10.1108/JEIM-07-2017-0101

Hancock, A. (2023, September 13). *EU to launch anti-subsidy probe into Chinese electric vehicles.* Retrieved September 30, 2023, from https://www.ft.com/content/55ec498d-0959-41ef-8ab9-af06cc45f8e7

Hanelt, A., Piccinini, E., Gregory, R. W., Hildebrandt, B., & Kolbe, L. M. (2015). Digital Transformation of Primarily Physical Industries - Exploring the Impact of Digital Trends on Business Models of Automobile Manufacturers. *Wirtschaftsinformatik Proceedings, 88.* https://aisel.aisnet.org/wi2015/88

Han, S., Kelly, E., Nikou, S., & Svee, E. O. (2021). Aligning artificial intelligence with human values: Reflections from a phenomenological perspective. *AI & Society*, 1–13.

Hansen, A. M., Kraemmergaard, P., & Mathiassen, L. (2011). Rapid adaptation in digital transformation: A participatory process for engaging IS and business leaders. *MIS Quarterly Executive*, *10*(4), 175–185.

Hanson, R. (2016). The Age of. In *Work, Love, and Life when Robots Rule the Earth (Illustrated edition).* OUP Oxford.

Hao, K. (2020, April 2). *This is how AI bias really happens-and why it's so hard to fix.* Retrieved from https://www.technologyreview.com/s/612876/ this-is-how-ai-bias-really-happensand-why-itsso-hard-to-fix/?utm_source=newsletters&utm_medium=email&utm_campaign=the_algorithm.unpaid. engagement

Hao, Q., & Qin, L. (2020). The design of intelligent transportation video processing system in big data environment. *IEEE Access : Practical Innovations, Open Solutions*, *8*, 13769–13780. doi:10.1109/ACCESS.2020.2964314

Harshini, C. S. (2015). *Influence of social media ads on consumer's purchase intention.* Academic Press.

Hasan Laskar, M. (2023). Examining the emergence of digital society and the digital divide in India: A comparative evaluation between urban and rural areas. *Frontiers in Sociology*, *8*, 1145221. Advance online publication. doi:10.3389/fsoc.2023.1145221

Helberger, N., Araujo, T., & de Vreese, C. H. (2020). Who is the fairest of them all? Public attitudes and expectations regarding automated decision-making. *Computer Law & Security Report*, *39*, 105456. doi:10.1016/j.clsr.2020.105456

Herian, R. (2020). Blockchain, GDPR, and fantasies of data sovereignty. *Law, Innovation and Technology*, *12*(1), 156–174. doi:10.1080/17579961.2020.1727094

Hershbein, B., & Kahn, L. B. (2018). Do recessions accelerate routine-biased technological change? Evidence from vacancy postings. *The American Economic Review*, *108*(7), 1737–1772. doi:10.1257/aer.20161570

History of Artificial Intelligence. (n.d.). Artificial Intelligence. https://www.coe.int/en/web/artificial-intelligence/history-of-ai

Hitt, M. A., Holmes, R. M. Jr, & Arregle, J.-L. (2021). The (COVID-19) pandemic and the new world (dis)order. *Journal of World Business*, *56*(4), 101210. doi:10.1016/j.jwb.2021.101210

Hoban, P. R., & Bucklin, R. E. (2015). Effects of Internet Display Advertising in the Purchase Funnel: Model-Based Insights from a Randomized Field Experiment. *JMR, Journal of Marketing Research*, *52*(3), 375–393. doi:10.1509/jmr.13.0277

Hodges, A. (2009). *Alan Turing and the Turing test*. Springer Netherlands. doi:10.1007/978-1-4020-6710-5_2

Hoffman, D. L., & Fodor, M. (2010). Can you measure the ROI of your social media marketing? *MIT Sloan Management Review*, *52*(1), 41.

Hollebeek, L. D., Glynn, M. S., & Brodie, R. J. (2014). Consumer brand engagement in social media: Conceptualization, scale development and validation. *Journal of Interactive Marketing*, *28*(2), 149–165. doi:10.1016/j.intmar.2013.12.002

Hor'akov'a, Z. (2018). *The Channel of Influence? YouTube Advertising and the Hipster Phenomenon*. Charles University.

Huang, C., Zhang, Z., Mao, B., & Yao, X. (2022). An overview of artificial intelligence ethics. *IEEE Transactions on Artificial Intelligence*.

Huang, L., & Davis, F. D. (2022). User Perceptions of Security in Decentralized Technologies: An Empirical Investigation. *Information Systems Research*, *33*(1), 120–139.

Huang, M. H., & Rust, R. T. (2018). Artificial intelligence in service. *Journal of Service Research*, *21*(2), 155–172. doi:10.1177/1094670517752459

Huawei Investments & Holding Co Ltd. (2022). *2022 Annual Report*. Huawei. https://www.huawei.com/en/annual-report/2022

Hulstaert, L. (2019, March 14). *Machine learning interpretability techniques*. Retrieved from https:// towardsdatascience.com/machine-learninginterpretability-techniques-662c723454f3

Huntingford, C., Jeffers, E. S., Bonsall, M. B., Christensen, H. M., Lees, T., & Yang, H. (2019). Machine learning and artificial intelligence to aid climate change research and preparedness. *Environmental Research Letters*, *14*(12), 124007. doi:10.1088/1748-9326/ab4e55

Hussain, A., Ting, D. H., & Mazhar, M. (2022). Driving Consumer Value Co-creation and Purchase Intention by Social Media Advertising Value. *Frontiers in Psychology*, *13*, 800206. doi:10.3389/fpsyg.2022.800206 PMID:35282229

Hussien, Q. M., & Habeeba, F. A. (2021). Survey on data security techniques in internet of things. *Al-Kunooze Scientific Journal, 2*(2). Retrieved from https://www.iasj.net/iasj/article/221409

Hyundai Motor India. (2023). www.clicktobuy.hyundai.co.in

IAB & PwC Outlook. (2021). *Report Urges Digital Ecosystem to Reset Consumer Value Exchange*. Author.

Ian Bremmer. (2020). Coronavirus and the World Order to Come. *Horizons: Journal of International Relations and Sustainable Development, 16*, 14-23.

Iansiti, M., & Lakhani, K. R. (2020). Competing in the age of AI. *Harvard Business Review, 98*(1), 60–67.

IBM. (2023). *What is data mining?* https://www.ibm.com/topics/data-mining

IBM. (n.d.). *What is Machine Learning?* Retrieved October 8, 2023, from https://www.ibm.com/topics/machine-learning

IMF. (2023). *World Economic Outlook*. IMF.

Imran, A. (2022). Why addressing digital inequality should be a priority. *The Electronic Journal on Information Systems in Developing Countries, 89*(3), e12255. doi:10.1002/isd2.12255

insideBIGDATA. (2022, February 107 *Reasons For Bias In AI and What To Do About It*. https://insidebigdata.com/2022/02/09/7-reasons-for-bias-in-ai-and-what-to-do-about-it/

Isip, M., & Lacap, J. P. (2021). *Social Media Use and Purchase Intention: The Mediating Roles of Perceived Risk and Trust*. Academic Press.

Jackson, L., Kuhlman, C., Jackson, F., & Fox, P. K. (2019). Including vulnerable populations in the assessment of data from vulnerable populations. *Frontiers in Big Data, 2*, 19. doi:10.3389/fdata.2019.00019 PMID:33693342

Janeček, V. (2018). Ownership of personal data in the Internet of Things. *Computer Law & Security Report, 34*(5), 1039–1052. doi:10.1016/j.clsr.2018.04.007

Janga, B., Asamani, G. P., Sun, Z., & Cristea, N. (2023). A Review of Practical AI for Remote Sensing in Earth Sciences. *Remote Sensing (Basel), 15*(16), 4112. doi:10.3390/rs15164112

Janiesch, C., Zschech, P., & Heinrich, K. (2021). Machine learning and deep learning. *Electronic Markets, 31*(3), 685–695. doi:10.100712525-021-00475-2

Jensen, B. (2020, September 1). *AI4ALL: Diversifying the Future of Artificial Intelligence*. https://hai.stanford.edu/news/ai4all-diversifying-future-artificial-intelligence

Jobin, A., Ienca, M., & Vayena, E. (2019). The global landscape of AI ethics guidelines. *Nature Machine Intelligence, 1*(9), 389–399. doi:10.103842256-019-0088-2

Johns Hopkins. (2023, March 10). *Johns Hopkins University & Medicine Coronavirus Resource Centre*. Retrieved September 30, 2023, from https://coronavirus.jhu.edu/map.html

Johnson, Z., & Leonard, T. (2020). *Performance branding: Borrow from the past to win the future*. Henry Stewart Publications.

Johnson, A., & Smith, B. (2020). Security Vulnerabilities in Decentralized Ecosystems: A Literature Synthesis. *Journal of Cybersecurity, 9*(4), 421–438.

Jones, S., & Pippin, J. S. (2019). Learning about the land: What can tax appraisal data tell us about heirs' properties? *SRS-244*, 3-8.

Kamalaldin, A., Linde, L., Sjödin, D., & Parida, V. (2020). Transforming provider-customer relationships in digital servitization: A relational view on digitalization. *Industrial Marketing Management, 89*, 306–325. doi:10.1016/j.indmarman.2020.02.004

Kamila, M. K., & Jasrotia, S. S. (2023). Ethical issues in the development of artificial intelligence: recognizing the risks. *International Journal of Ethics and Systems*.

Kampani, N., & Jhamb, D. (2020). Analyzing the role of e-crm in managing customer relations: A critical review of the literature. *Journal of Critical Review*, *7*(4), 221–226.

Kane, G. C. (2015). Enterprise social media: Current capabilities and future possibilities. *MIS Quarterly Executive*, *14*(1), 1–16.

Kaplan, A., & Haenlein, M. (2019). Siri, Siri, in my hand: Who's the fairest in the land? On the interpretations, illustrations, and implications of artificial intelligence. *Business Horizons*, *62*(1), 15–25. doi:10.1016/j.bushor.2018.08.004

Karunarathne, E. A. C. P., & Thilini, W. A. (2022). Advertising Value Constructs' Implication on Purchase Intention: Social Media Advertising. *Management Dynamics in the Knowledge Economy, 10*(3), 287-303. DOI doi:10.2478/mdke-2022-0019

Kasilingam, D. L. (2020). Understanding the attitude and intention to use smartphone chatbots for shopping. *Technology in Society*, *62*, 101280. doi:10.1016/j.techsoc.2020.101280

Kathiravan, C. (2017). Effectiveness of advertisements in social media. *Asian Academic Research Journal of Multidisciplinary.*, *4*, 179–190.

Kaur, B., Kaur, J., Pandey, S. K., & Joshi, S. (2020). E-service Quality: Development and Validation of the Scale. *Global Business Review*.

Kaurin, D. (2019). *Data protection and digital agency for refugees*. Academic Press.

Kaushik, M., Pathak, S., Bhatt, P. K., Singh, S. K., Tripathi, A., & Pandey, P. K. (2022). Artificial Intelligence (AI). *Intelligent System Algorithms and Applications in Science and Technology*, 119-133.

Kaya, T., Aktas, E., Topçu, İ., & Ulengin, B. (2010). Modeling toothpaste brand choice: An empirical comparison of artificial neural networks and multinomial probit model. *International Journal of Computational Intelligence Systems*, *3*(5), 674–687. doi:10.2991/ijcis.2010.3.5.15

Keller, K. L. (2009). Building strong brands in a modern marketing communication environment. *Journal of Marketing Communications*, *15*(2/3), 139–155. doi:10.1080/13527260902757530

Kelley, S. (2023, March 6). Removing Demographic Data Can Make AI Discrimination Worse. *Harvard Business Review*. https://hbr.org/2023/03/removing-demographic-data-can-make-ai-discrimination-worse

Kernighan, B. W. (2021). *Understanding the digital world: What you need to know about computers, the internet, privacy, and security*. Princeton University Press.

Khan, R., Taqi, M., & Saba, A. (2021). The role of digitization in automotive industry: The Indian perspective. *International Journal of Business Ecosystem & Strategy, 3*(4), 20-29. doi:10.36096/ijbes.v3i4.277

Khanna, P., Kumar, S., & Gauba, R., & Aditya. (2022, October). Non-Fungible Tokens' Marketplace: A Secured Blockchain-Based Decentralized Framework for Online Auction. In *International Conference on Computing, Communications, and Cyber-Security* (pp. 841-856). Singapore: Springer Nature Singapore. 10.1007/978-981-99-1479-1_62

Kim, D., Suh, Y. H., & Lee, J. (2018). The impact of digital signage on customer envy, store patronage, and purchase decision. *Information & Management*, *55*(6), 735–747.

Kim, H.-M., Li, P., & Lee, Y. R. (2020). Observations of deglobalization against globalization and impacts on global business. International Trade. *Politics and Development*, *4*(2), 83–103. doi:10.1108/ITPD-05-2020-0067

Kim, W. C., & Mauborgne, R. (2014). *Blue Ocean Strategy: How to Create Uncontested Market Space and Make Competition Irrelevant*. Harvard Business Review Press.

King, J., & Gonzales, A. L. (2023). The influence of digital divide frames on legislative passage and partisan sponsorship: A content analysis of digital equity legislation in the U.S. from 1990 to 2020. *Telecommunications Policy*, *47*(7), 102573. doi:10.1016/j.telpol.2023.102573

Kiritchenko, S., Nejadgholi, I., & Fraser, K. C. (2021). Confronting abusive language online: A survey from the ethical and human rights perspective. *Journal of Artificial Intelligence Research*, *71*, 431–478. doi:10.1613/jair.1.12590

Kitzmann, H., Yatsenko, V., & Launer, M. (2021). *Artificial intelligence and wisdom*. Academic Press.

Knop, K., & Mielczarek, K. (2018). Using 5W-1H and 4M Methods to Analyze and Solve the Problem with the Visual Inspection Process Case Study. In *MATEC Web of Conferences* (Vol. 183, p. 03006). EDP Sciences.

Köchling, A., & Wehner, M. C. (2020, November 1). *Discriminated by an algorithm: a systematic review of discrimination and fairness by algorithmic decision-making in the context of HR recruitment and HR development*. Business Research. doi:10.1007/s40685-020-00134-w

Kolotouchkina, O., Barroso, C., & Sanchez, J. L. (2022). Smart cities, the digital divide, and people with disabilities. *Cities (London, England)*, *123*, 103613. doi:10.1016/j.cities.2022.103613

Kordzadeh, N., & Ghasemaghaei, M. (2022). Algorithmic bias: Review, synthesis, and future research directions. *European Journal of Information Systems*, *31*(3), 388–409. doi:10.1080/0960085X.2021.1927212

Korovkin, V., Park, A., & Kaganer, E. A. (2022). Towards conceptualization and quantification of the digital divide. *Information Communication and Society*, *26*(1), 2268–2303. Advance online publication. doi:10.1080/1369118X.2022.2085612

Kozlowski, D., Larivière, V., Sugimoto, C. R., & Monroe-White, T. (2022a). Intersectional inequalities in science. *Proceedings of the National Academy of Sciences of the United States of America*, *119*(2), e2113067119. doi:10.1073/pnas.2113067119 PMID:34983876

Kozlowski, D., Murray, D. S., Bell, A., Hulsey, W., Larivière, V., Monroe-White, T., & Sugimoto, C. R. (2022b). Avoiding bias when inferring race using name-based approaches. *PLoS One*, *17*(3), e0264270. doi:10.1371/journal.pone.0264270 PMID:35231059

Kraus, S., Breier, M., Lim, W. M., Dabić, M., Kumar, S., Kanbach, D., Mukherjee, D., Corvello, V., Piñeiro-Chousa, J., Liguori, E., Palacios-Marqués, D., Schiavone, F., Ferraris, A., Fernandes, C., & Ferreira, J. J. (2022). Literature reviews as independent studies: Guidelines for academic practice. *Review of Managerial Science*, *16*(8), 2577–2595G. doi:10.100711846-022-00588-8

Kraus, S., Durst, S., Ferreira, J., Veiga, P., Kailer, N., & Weinmann, A. (2022). Digital transformation in business and management research: An overview of the current status quo. *International Journal of Information Management*, *63*, 102466. doi:10.1016/j.ijinfomgt.2021.102466

Kreimer, I. (2018). *How to Get Started with AI-Powered Content Marketing*. Available at: https://www.singlegrain.com/artificial-intelligence/how-to-get-started-with-ai-powered-content-marketing/

Kreps, G. L., & Neuhauser, L. (2010). New directions in eHealth communication: Opportunities and challenges. *Patient Education and Counseling*, *78*(3), 329–336. doi:10.1016/j.pec.2010.01.013 PMID:20202779

Kreps, G. L., & Neuhauser, L. (2013). Artificial intelligence and immediacy: Designing health communication to personally engage consumers and providers. *Patient Education and Counseling*, *92*(2), 205–210. doi:10.1016/j.pec.2013.04.014 PMID:23683341

Krishna, T. H. (2020, December 2). *Entry barriers for women are amplified by AI in recruitment algorithms, study finds*. Newsroom. https://www.unimelb.edu.au/newsroom/news/2020/december/entry-barriers-for-women-are-amplified-by-ai-in-recruitment-algorithms,-study-finds#:~:text=Research%20showed%20even%20basic%20algorithms,resum%C-3%A9s%2C%20according%20to%20new%20research

Kuhnert, F., Stürmer, C., & Koster, A. (2017-2018). Five trends transforming the Automotive Industry. *PricewaterhouseCoopers GmbH Wirtschaftsprüfungsgesellschaft*. www.pwc.com/auto

Kumar, S., Chavan, M., & Pandey, N. (2023). Journal of International Management: A 25-year review using bibliometric analysis. *Journal of International Management*, *29*(1), 100988. doi:10.1016/j.intman.2022.100988

Kumar, V., Rajan, B., Venkatesan, R., & Lecinski, J. (2019). Understanding the role of artificial intelligence in personalized engagement marketing. *California Management Review*, *61*(4), 135–155. doi:10.1177/0008125619859317

Lábaj, M., & Majzlíková, E. (2023). How nearshoring reshapes global deindustrialization. *Economics Letters*, *230*, 111239. doi:10.1016/j.econlet.2023.111239

Lacity, M. C., & Reynolds, P. (2014). Cloud services practices for small and medium-sized enterprises. *MIS Quarterly Executive*, *13*(1), 31–44.

Lamé, G. (2019). Systematic Literature Reviews: An Introduction. *Proceedings of the Design Society: International Conference on Engineering Design.*, *1*(1), 1633–1642. doi:10.1017/dsi.2019.169

Lamrhari, S., El Ghazi, H., Oubrich, M., & El Faker, A. (2022). A social CRM analytic framework for improving customer retention, acquisition, and conversion. *Technological Forecasting and Social Change*, *174*, 121275. doi:10.1016/j.techfore.2021.121275

LaPointC.YaleS. O. M. (2022). Property tax sales, private capital, and gentrification in the US. doi:10.2139/ssrn.4219360

Lead, A. (2021, February 28). *Citroën India launches 'La Maison Citroen' phygital showrooms in Bangalore*. Automotive Lead. https://automotiveleadnews.com/2021/02/28/citroen-india-launches-la-maison-citroen-phygital-showrooms-in-bangalore/

Lee, H. H., Fiore, A. M., & Kim, J. (2006). The role of the technology acceptance model in explaining effects of image interactivity technology on consumer responses. *International Journal of Retail & Distribution Management*, *34*(8), 621–644. doi:10.1108/09590550610675949

Lee, J. E., & Watkins, B. (2016). YouTube vloggers' influence on consumer luxury brand perceptions and intentions. *Journal of Business Research*, *69*(12), 5753–5760. doi:10.1016/j.jbusres.2016.04.171

Lee, J., & Hong, I. (2016). Predicting positive user responses to social media advertising: The roles of emotional appeal, informativeness, and creativity. *International Journal of Information Management*, *36*(3), 360–373. doi:10.1016/j.ijinfomgt.2016.01.001

Leetaru, K. (2019, January 21). *Why Is AI And Machine Learning So Biased? The Answer Is Simple Economics*. Retrieved from https://www.forbes. com/sites/kalevleetaru/2019/01/20/why-is-ai-andmachine-learning-so-biased-the-answer-is-simpleeconomics/#51eb3979588c

Leidner, J. L., & Plachouras, V. (2017, April). Ethical by design: Ethics best practices for natural language processing. In *Proceedings of the First ACL Workshop on Ethics in Natural Language Processing* (pp. 30-40). 10.18653/v1/W17-1604

Leong, C. M., Loi, A., & Woon, S. (2022). The influence of social media eWOM information on purchase intention. *Journal of Marketing Analytics.*, *10*(2), 1–13. doi:10.105741270-021-00132-9

Leslie, D. (2019). *Understanding artificial intelligence ethics and safety.* arXiv preprint arXiv:1906.05684.

Leviathan, Y., & Matias, Y. (2018). *Google Duplex: An AI system for accomplishing real-world tasks over the phone.* Academic Press.

Liberties, E. U. (2021, May 18). *Algorithmic Bias: Why and How Do Computers Make Unfair Decisions?* Retrieved October 8, 2023, from https://www.liberties.eu/en/stories/algorithmic-bias-17052021/43528

Light, T. H. (2022). Frankenstein's monster: Constructing a legal regime to regulate race and place. *Southern Cultures, 28*(3), 74–89. doi:10.1353cu.2022.0027

Li, H., & Lo, H. Y. (2015). Do You Recognize Its Brand? The Effectiveness of Online In-Stream Video Advertisements. *Journal of Advertising, 44*(3), 208–218. doi:10.1080/00913367.2014.956376

Lim, W. M., Gupta, S., Aggarwal, A., Paul, J., & Sadhna, P. (2021). How do digital natives perceive and react toward online advertising? Implications for SMEs. *Journal of Strategic Marketing*, 1–35. doi:10.1080/0965254X.2021.1941204

Lim, X. J., Radzol, A. M., Cheah, J., & Wong, M. W. (2017). The impact of social media influencers on purchase intention and the mediation effect of customer attitude. *Asian Journal of Business Research, 7*(2), 19–36. doi:10.14707/ajbr.170035

LinkedIn. (2023). https://www.linkedin.com/company/%C5%A1koda-sga-cars-dealer/?originalSubdomain=in

Lipton, Z. C. (2018). The mythos of model interpretability: In machine learning, the concept of interpretability is both important and slippery. *ACM Queue; Tomorrow's Computing Today, 16*(3), 31–57. doi:10.1145/3236386.3241340

Liu, Y., Zhao, Z., Guo, G., Wang, X., Tan, Z., & Wang, S. (2017, August). An identity management system based on blockchain. In *2017 15th Annual Conference on Privacy, Security and Trust (PST)*. IEEE. 10.1109/PST.2017.00016

Livemint.com. (2019). *Emerging tech helps reshape how vehicles are showcased.* Retrieved from https://www.livemint.com/technology/tech-news/emerging-tech-helps-reshape-how-vehicles-are-showcased-11573143279548.html

Li, Y., & Jin, F. (2020). Exploring Disruptive Innovation: A Comprehensive Review. *International Journal of Innovation Management, 24*(6), 2050052.

Lundberg, S. M., & Lee, S. I. (2017). A unified approach to interpreting model predictions. *Advances in Neural Information Processing Systems*, 30.

Lv, Z., & Xie, S. (2022). Artificial intelligence in the digital twins: State of the art, challenges, and future research topics. *Digital Twin, 1*, 12. doi:10.12688/digitaltwin.17524.2

Lynch, C. (2001). The battle to define the future of the book in the digital world. *First Monday, 6*(6). Advance online publication. doi:10.5210/fm.v6i6.864

Lyu, W., & Liu, J. (2021). Artificial Intelligence and emerging digital technologies in the energy sector. *Applied Energy, 303*, 117615. doi:10.1016/j.apenergy.2021.117615

Ma, Y., Wang, Z., Yang, H., & Yang, L. (2020). Artificial intelligence applications in the development of autonomous vehicles: A survey. *IEEE/CAA Journal of Automatica Sinica, 7*(2), 315-329.

Mahadevan, K., & Joshi, S. (2022). Omnichannel retailing: A bibliometric and network visualization analysis. *Benchmarking, 29*(4), 1113–1136. doi:10.1108/BIJ-12-2020-0622

Makridakis, S. (2017). The forthcoming Artificial Intelligence (AI) revolution: Its impact on *society and firms. *Futures, 90*, 46–60. doi:10.1016/j.futures.2017.03.006

Ma, L., & Sun, B. (2020). Machine learning and AI in marketing–Connecting computing power to human insights. *International Journal of Research in Marketing*, *37*(3), 481–504. doi:10.1016/j.ijresmar.2020.04.005

Manceski, G., & Petrevska Nechkoska, R. (2023). Conceptualisation of Decentralized Blockchain-Based, Open-Source ERP Marketplaces: Disruptive Decentralized Technologies for Co-Creation. In *Facilitation in Complexity: From Creation to Co-creation, from Dreaming to Co-dreaming, from Evolution to Co-evolution* (pp. 175–202). Springer International Publishing. doi:10.1007/978-3-031-11065-8_7

Manohar, S., Sharma, V., & Mittal, A. (2023). Reinforcing Requirements and Stimulating the Purchase Intentions: Growing Location Based Mobile Targeting Techniques. In Enhancing Customer Engagement Through Location-Based Marketing (pp. 56-65). IGI Global.

Manyika, J. (2022, November 17). What do we do about the biases in AI? *Harvard Business Review*. https://hbr.org/2019/10/what-do-we-do-about-the-biases-in-ai

Markides, C. (2019). In Search of Ambidextrous Business Models. *MIT Sloan Management Review*, *61*(4), 22–29.

Markides, C. (2020). The Dark Side of Disruptive Innovation. *MIT Sloan Management Review*, *61*(2), 22–30.

Marquardt, K. (2017). Smart services–characteristics, challenges, opportunities and business models. In *Proceedings of the International Conference on Business Excellence* (Vol. 11, No. 1, pp. 789-801). 10.1515/picbe-2017-0084

Martinho, A., Herber, N., Kroesen, M., & Chorus, C. (2021). Ethical issues in focus by the autonomous vehicles industry. *Transport Reviews*, *41*(5), 556–577. doi:10.1080/01441647.2020.1862355

Matt, C., Hess, T., Benlian, A., & Wiesbock, F. (2016). Options for formulating a digital transformation strategy. *MIS Quarterly Executive*, *15*(2), 123–139.

Maurer, B., Nelms, T. C., & Swartz, L. (2013). 'When Perhaps the Real Problem is Money Itself!': The Practical Materiality of Bitcoin. *Social Semiotics*, *23*(2), 261–277. doi:10.1080/10350330.2013.777594

Mavrogiorgou, A., Kiourtis, A., Makridis, G., Kotios, D., Koukos, V., Kyriazis, D., Soldatos, J. K., Fatouros, G., Drakoulis, D., Maló, P., Serrano, M., Isaja, M., Lazcano, R., Vera, J. M., Fournier, F., Limonad, L., Perakis, K., Miltiadou, D., Kranas, P., & Troiano, E. (2023, July). FAME: Federated Decentralized Trusted Data Marketplace for Embedded Finance. In *2023 International Conference on Smart Applications, Communications and Networking (SmartNets)*. IEEE. 10.1109/SmartNets58706.2023.10215814

Max, R., Kriebitz, A., & Von Websky, C. (2021). Ethical considerations about the implications of artificial intelligence in finance. *Handbook on ethics in finance*, 577-592.

Mayrhofer, M., Matthes, J., Einwiller, S., & Naderer, B. (2020). User-generated content presenting brands on social media increases young adults' purchase intention. *International Journal of Advertising*, *39*(1), 166–186. doi:10.1080/02650487.2019.1596447

Mazouzi, D., & Alit, N. (2023). Factors Influencing Consumers' Attitudes and Intentions Towards Online Shopping - A Survey of a Sample of Consumers in Algeria. *Malaysian Journal of Consumer and Family Economics.*, *31*(1), 788–814. doi:10.60016/majcafe.v31.29

McAfee, A. (2011). What every CEO needs to know about the cloud. *Harvard Business Review*, *89*(11), 124–132.

McCarthy, J., Minsky, M. L., Rochester, N., & Shannon, C. E. (2006). A proposal for the dartmouth summer research project on artificial intelligence, august 31, 1955. *AI Magazine*, *27*(4), 12–12.

McConaghy, M., McMullen, G., Parry, G., McConaghy, T., & Holtzman, D. (2017). Visibility and digital art: Blockchain as an ownership layer on the Internet. *Strategic Change*, *26*(5), 461–470. doi:10.1002/jsc.2146

McGrath, R. G., & Nerkar, A. (2019). Real Options Reasoning and a New Look at the R&D Investment Strategies of Biopharmaceutical Firms Facing Uncertainty. *Organization Science*, *30*(3), 495–515.

Meenakshi, N. (2023). Post-COVID reorientation of the Sharing economy in a hyperconnected world. *Journal of Strategic Marketing*, *31*(2), 446–470. doi:10.1080/0965254X.2021.1928271

Mehrabi, N., Morstatter, F., Saxena, N., Lerman, K., & Galstyan, A. (2021, July 13). A Survey on Bias and Fairness in Machine Learning. *ACM Computing Surveys*, *54*(6), 1–35. doi:10.1145/3457607

Mende, M., Scott, M. L., van Doorn, J., Grewal, D., & Shanks, I. (2019). Service robots rising: How humanoid robots influence service experiences and elicit compensatory consumer responses. *JMR, Journal of Marketing Research*, *56*(4), 535–556. doi:10.1177/0022243718822827

Mhlanga, D. (2020). Industry 4.0 in finance: The impact of artificial intelligence (ai) on digital financial inclusion. *International Journal of Financial Studies*, *8*(3), 45. doi:10.3390/ijfs8030045

Minchev, Z. (2017). Security challenges to digital ecosystems dynamic transformation. *Proc. of BISEC*, 6-10.

Ministry of External Affairs. (2023, September 9). *G20 New Delhi Leaders' Declaration*. Retrieved September 25, 2023, from https://www.mea.gov.in/bilateral-documents.htm?dtl/37084/G20_New_Delhi_Leaders_Declaration

Miranda, E. (2011). Timeboxing planning: Buffered Moscow rules. *Software Engineering Notes*, *36*(6), 1–5. doi:10.1145/2047414.2047428

Mishra, A., Shukla, A., Rana, N. P., & Dwivedi, Y. K. (2021). From "touch" to a "multisensory" experience: The impact of technology interface and product type on consumer responses. *Psychology and Marketing*, *38*(3), 385–396. doi:10.1002/mar.21436

Miśkiewicz, J., & Ausloos, M. (2010). Has the world economy reached its globalization limit? *Physica A*, *389*(4), 797–806. doi:10.1016/j.physa.2009.10.029

Mitchell, M. (2019). Artificial intelligence: A guide for thinking humans. Academic Press.

Mitchell, W. J., Borroni-Bird, C. E., & Burns, L. D. (2010). *Reinventing the Automobile: Personal Urban Mobility for the 21st Century*. The MIT Press. doi:10.7551/mitpress/8490.001.0001

Mittelstadt, B. D., Allo, P., Taddeo, M., Wachter, S., & Floridi, L. (2016). The ethics of algorithms: Mapping the debate. *Big Data & Society*, *3*(2). doi:10.1177/2053951716679679

MoCA. (2022, June 30). *India-Australia Economic Cooperation and Trade Agreement (INDAUS ECTA) between the Government of the Republic of India and the Government of Australia*. Retrieved September 30, 2023, from https://commerce.gov.in/international-trade/trade-agreements/ind-aus-ecta/

Mocker, M., Weill, P., & Woerner, S. L. (2014). Revisiting complexity in the digital age. *MIT Sloan Management Review*, *55*(4), 73–81.

MoFA. (2022, July 22). *Comprehensive Economic Partnership Agreement between Japan and the Republic of India*. Retrieved October 1, 2023, from https://www.mofa.go.jp/region/asia-paci/india/epa201102/index.html

Moher, D., Liberati, A., Tetzlaff, J., & Altman, D. G.Prisma Group. (2010). Preferred reporting items for systematic reviews and meta-analyses: The PRISMA statement. *International Journal of Surgery*, *8*(5), 336–341. doi:10.1016/j.ijsu.2010.02.007 PMID:20171303

Monroe-White, T. (2022). Emancipating data science for Black and Indigenous students via liberatory datasets and curricula. *IASSIST Quarterly*, *46*(4). Advance online publication. doi:10.29173/iq1007

Moore, R., Vitale, D., & Stawinoga, N. (2018). *The Digital Divide and Educational Equity*. ACT Research & Center for Equity in Learning.

Moor, J. (n.d.). The dartmouth college artificial intelligence conference: The next fifty years. *AI Magazine*, *27*(4).

Moreland, M. P. (2012). Mistakes about Intention in the Law of Bioethics. *Law and Contemporary Problems*, *75*, 53.

Morse, L., Teodorescu, M. H. M., Awwad, Y., & Kane, G. C. (2021). Do the ends justify the means? Variation in the distributive and procedural fairness of machine learning algorithms. *Journal of Business Ethics*, 1–13.

Moslehpour, M., Ismail, T., Purba, B., & Lin, P-K. (2020). *The Effects of Social Media Marketing, Trust, and Brand Image on Consumers' Purchase Intention of GO-JEK in Indonesia*. . doi:10.1145/3387263.3387282

Moslehpour, M., Dadvari, A., Nugroho, W., & Do, B.-R. (2021). The dynamic stimulus of social media marketing on purchase intention of Indonesian airline products and services. *Asia Pacific Journal of Marketing and Logistics*, *33*(2), 561–583. doi:10.1108/APJML-07-2019-0442

MotorK. (2023). https://www.kia.com/in/vr/showroom/index.html#/showroom

Motors, M. G. (2019, October 31). *MG Motor Unveils India's First Digital Car-Less Showroom: Mg Digital Studio*. MG Motor. https://www.mgmotor.co.in/media-center/newsroom/mg-motor-indias-first-digital-car-less-showroom

Mougayar, W. (2016). *The Business Blockchain: Promise, Practice, and Application of the Next Internet Technology*. Wiley.

Mühle, A., Grüner, A., Gayvoronskaya, T., & Meinel, C. (2018). A survey on essential components of a self-sovereign identity. *Computer Science Review*, *30*, 80–86. doi:10.1016/j.cosrev.2018.10.002

Mukherjee, B. E., Mazar, O., Aggarwal, R., & Kumar, R. (2016). *Exclusion from Digital Infrastructure and Access*. www.defindia.org/publication-2

Mukherjee, K., & Banerjee, N. (2017). Effect of Social Networking Advertisements on Shaping Consumers' Attitude. *Global Business Review*, *18*(5), 1291–1306. doi:10.1177/0972150917710153

Müller, V. C., & Bostrom, N. (2016). Future progress in artificial intelligence: A survey of expert opinion. *Fundamental issues of artificial intelligence*, 555-572.

Muneeb Ali. (2016). Decentralized Apps: What Are They and Could They Take Over? *Forbes*. Retrieved from https://www.forbes.com/sites/muneebali/2016/06/08/decentralized-apps-what-are-they-and-could-they-take-over/

Munir, A. B., Mohd Yasin, S. H., & Muhammad-Sukki, F. (2015). Big data. big challenges to privacy and data protection. *International Scholarly and Scientific Research & Innovation, 9*(1).

Muntinga, D. G., Moorman, M., & Smit, E. G. (2011). Introducing COBRAs: Exploring motivations for brand-related social media use. *International Journal of Advertising*, *30*(1), 13–46. doi:10.2501/IJA-30-1-013-046

Mutlag, A. A., Abd Ghani, M. K., Arunkumar, N., Mohammed, M. A., & Mohd, O. (2019, January). Enabling technologies for fog computing in healthcare IoT systems. *Future Generation Computer Systems*, *90*, 62–78. doi:10.1016/j.future.2018.07.049

Nadikattu, R. R. (2020). Implementation of new ways of artificial intelligence in sports. *Journal of Xidian University*, *14*(5), 5983–5997.

Nagel, M. (2019). *Exploring digital innovations: Mapping 3D printing within the textile and sportswear industry.* https://www.divaportal.org/smash/get/diva2:1369648/FULLTEXT01.pdf

Nakamoto, S. (2008). *Bitcoin: A Peer-to-Peer Electronic Cash System.* Retrieved from https://bitcoin.org/bitcoin.pdf

Nakamoto, S. (2009). *Bitcoin: A Peer-to-Peer Electronic Cash System.* Retrieved from https://bitcoin.org/bitcoin.pdf

Narayanan, A., Bonneau, J., Felten, E., Miller, A., & Goldfeder, S. (2016). *Bitcoin and Cryptocurrency Technologies: A Comprehensive Introduction.* Princeton University Press.

Nasir, V. A., Keserel, A. C., Surgit, O. E., & Nalbant, M. (2021). Segmenting consumers based on social media advertising perceptions: How does purchase intention differ across segments? *Telematics and Informatics*, *64*, 101687. doi:10.1016/j.tele.2021.101687

Natarajan, T., Balakrishnan, J., Balasubramanian, S., & Manickavasagam, J. (2015). Examining beliefs, values, and attitudes towards social media advertisements: Results from India. *International Journal of Business Information Systems*, *20*(4), 427. doi:10.1504/IJBIS.2015.072738

Nazir, S., & Kaleem, M. (2021). Advances in image acquisition and processing technologies transforming animal ecological studies. *Ecological Informatics*, *61*, 101212. doi:10.1016/j.ecoinf.2021.101212

Needham, J. (2013). *Disruptive possibilities: How big data changes everything.* O'Reilly Media, Inc.

NFHS. (2021). *Compendium of Fact Sheets India and 14 States/UTs (Phase-11).* https://main.mohfw.gov.in/sites/default/files/NFHS-5_Phase-II_0.pdf

Nicholas Taleb, N. (2015). The black swan: The impact of the highly improbable. Victoria, 250, 595-7955.

Nicholas, N. (2008). The black swan: The impact of the highly improbable. *Journal of the Management Training Institut*, *36*(3), 56.

Nili, A., Barros, A., & Tate, M. (2019). The public sector can teach us a lot about digitizing customer service. *MIT Sloan Management Review*, *60*(2), 84–87.

Nitika Sharma. (2023, September 8). *Understanding Algorithmic Bias: Types, Causes and Case Studies.* Analytics Vidhya. Retrieved October 12, 2023, from https://www.analyticsvidhya.com/blog/2023/09/understanding-algorithmic-bias/

Nkonde, M. (2021, September 13). Is AI Bias a Corporate Social Responsibility Issue? *Harvard Business Review.* https://hbr.org/2019/11/is-ai-bias-a-corporate-social-responsibility-issue

Nobanee, H., Al Hamadi, F. Y., Abdulaziz, F. A., Abukarsh, L. S., Alqahtani, A. F., AlSubaey, S. K., & Almansoori, H. A. (2021). A bibliometric analysis of sustainability and risk management. *Sustainability (Basel)*, *13*(6), 3277. doi:10.3390u13063277

Obermeyer, Z., Powers, B., Vogeli, C., & Mullainathan, S. (2019). Dissecting racial bias in an algorithm used to manage the health of populations. *Science*, *366*(6464), 447–453. doi:10.1126cience.aax2342 PMID:31649194

OECD. (2013). *Organization for Economic Co-operation Development.* OECD.

OECD. (2020). *Rural Well-being Geography of Opportunities.* https://doi.org/https://doi.org/10.1787/d25cef80-en

OECD. (2020). *The Digital of Science, Technology and Innovation: Key Developments and Policies.* https://www.oecd.org/science/inno/the digitalisation-of-science-technology-and-innovation-b9e4a2c0-en.htm

Okunola, O. M., Rowley, J., & Johnson, F. (2017). The multi-dimensional digital divide: Perspectives from an e-government portal in Nigeria. *Government Information Quarterly*, *34*(2), 329–339. doi:10.1016/j.giq.2017.02.002

Olga, B., & Vlad, M. (2014). Remarking as a Tool in Online Advertising. *Ovidius University Annals, Series Economic Sciences, 14*(2).

Olphert, C. W., Damodaran, L., & May, A. J. (2005, August). Towards digital inclusion–engaging older people in the 'digital world'. In *Accessible Design in the Digital World Conference 2005* (pp. 1-7). Academic Press.

Omowole, A. (2022, November 8). *Research shows AI is often biased. Here's how to make algorithms work for all of us.* World Economic Forum. https://www.weforum.org/agenda/2021/07/ai-machine-learning-bias-discrimination/

Osburg, T. (2017). *Sustainability in a digital world needs trust.* Springer International Publishing. doi:10.1007/978-3-319-54603-2

Osiyevskyy, O., & Dewald, J. (2015). Transformative Potential of Decentralized Marketplaces: A Paradigm Shift in Exchange Processes. *Journal of Business Research, 68*(7), 1458–1466.

Osterwalder, A., & Pigneur, Y. (2010). *Business model generation: a handbook for visionaries, game changers, and challengers (1).* John Wiley & Sons.

Osterwalder, A., & Pigneur, Y. (2021). *Business Model Generation: A Handbook for Visionaries, Game Changers, and Challengers.* John Wiley & Sons.

Ostrom, E. (2010). Beyond Markets and States: Polycentric Governance of Complex Economic Systems. *The American Economic Review, 100*(3), 641–672. doi:10.1257/aer.100.3.641

Overcoming the Tension Between Data Sovereignty and Accelerated Digital Transformation. (2022). Retrieved from https://tinyurl.com/mr2a4ymt

Oxfam India. (2022). *Digital Divide India Inequality Report 2022.* Author.

Paetz, P. (2014). A Disruptive Business Model. In *Disruption by Design.* Apress. doi:10.1007/978-1-4302-4633-6_9

Pagallo, U. (2017). The legal challenges of big data: Putting secondary rules first in the field of EU data protection. *Eur. Data Prot. L. Rev., 3*(1), 36–46. doi:10.21552/edpl/2017/1/7

Pais, N., & Ganapathy, N. (2021). *The Influence of Instagram on Consumer Purchase Intention.* Academic Press.

Palmié, M., Miehé, L., Oghazi, P., Parida, V., & Wincent, J. (2022). The evolution of the digital service ecosystem and digital business model innovation in retail: The emergence of meta-ecosystems and the value of physical interactions. *Technological Forecasting and Social Change, 177*, 121496. doi:10.1016/j.techfore.2022.121496

Pangarkar, A., Arora, V., & Shukla, Y. (2022). Exploring phygital omnichannel luxury retailing for immersive customer experience: The role of rapport and social engagement. *Journal of Retailing and Consumer Services, 68*, 103001. doi:10.1016/j.jretconser.2022.103001

Pantano, E., Priporas, C. V., & Sorace, S. (2017). Enhancing the showrooming experience: A conceptual framework for augmenting mirrors in clothing e-stores. *Journal of Retailing and Consumer Services, 35*, 1–10.

Park, S. S., Tung, C. D., & Lee, H. (2021). The adoption of AI service robots: A comparison between credence and experience service settings. *Psychology & Marketing, 38*(4), 691-703.

Parks, A. M. (2021). Unfair collection: Reclaiming control of publicly available personal information from data scrapers. *Michigan Law Review, 120*, 913.

Paschen, J., Kietzmann, J., & Kietzmann, T. C. (2019). Artificial intelligence (AI) and its implications for market knowledge in B2B marketing. *Journal of Business and Industrial Marketing, 34*(7), 1410–1419. doi:10.1108/JBIM-10-2018-0295

Passey, D., Shonfeld, M., Appleby, L., Judge, M., Saito, T., & Smits, A. (2018). Digital Agency: Empowering Equity in and through Education. *Technology. Knowledge and Learning*, *23*(3), 425–439. doi:10.100710758-018-9384-x

Payne, A. L., Stone, C., & Bennett, R. (2023). Conceptualising and Building Trust to Enhance the Engagement and Achievement of Under-Served Students. *The Journal of Continuing Higher Education*, *71*(2), 134–151. doi:10.1080/07377363.2021.2005759

Pazzanese, C. (2020). Ethical concerns mount as AI takes bigger decision-making role in more industries. *The Harvard Gazette, 26*.

Pedersen, A. B., Risius, M., & Beck, R. (2019). A ten-step decision path to determine when to use blockchain technologies. *MIS Quarterly Executive*, *18*(2), 1–17. doi:10.17705/2msqe.00010

Pelau, C., Dabija, D. C., & Ene, I. (2021). What makes an AI device human-like? The role of interaction quality, empathy and perceived psychological anthropomorphic characteristics in the acceptance of artificial intelligence in the service industry. *Computers in Human Behavior*, *122*, 106855. doi:10.1016/j.chb.2021.106855

Petropoulos, G. (2018). The impact of artificial intelligence on employment. *Praise for Work in the Digital Age*, *119*, 121.

Phillips-Brown, M. (2023). *Algorithmic neutrality.* arXiv preprint, 2303.05103.

PIB. (2020, July 23). *Restrictions on Public Procurement from certain countries.* Retrieved September 18, 2023, from https://pib.gov.in/PressReleasePage.aspx?PRID=1640778

PIB. (2022, March 27). *Comprehensive Economic Partnership Agreement (CEPA) between India and the United Arab Emirates (UAE) Unveiled.* Retrieved September 30, 2023, from https://www.pib.gov.in/PressReleaseIframePage.aspx?PRID=1810279

PIB. (2023, September 9). *India-Middle East-Europe Economic Corridor promises to be a beacon of cooperation, innovation, and shared progress.* Retrieved October 2, 2023, from https://pib.gov.in/PressReleaseIframePage.aspx?PRID=1955842

Pichai, S. (2020, January 20). *Why Google thinks we need to regulate AI.* Retrieved from https://www.ft.com/content/3467659a-386d-11ea-ac3c-f68c10993b04

Pickering, J., Hickmann, T., Bäckstrand, K., Kalfagianni, A., Bloomfield, M., Mert, A., Ransan-Cooper, H., & Lo, A. Y. (2022). Democratising sustainability transformations: Assessing the transformative potential of democratic practices in environmental governance. *Earth System Governance*, *11*, 100131. doi:10.1016/j.esg.2021.100131

Pierre, Pitt, Plangger, & Shapiro. (2012). *Marketing meets Web 2.0, social media, and creative consumers: Implications for international marketing strategy.* doi:10.1016/j.bushor.2012.01.007

Pöhn, D., Grabatin, M., & Hommel, W. (2021). eID and self-sovereign identity usage: An overview. *Electronics (Basel)*, *10*(22), 2811. doi:10.3390/electronics10222811

Popat, A., & Tarrant, C. (2023). Exploring adolescents' perspectives on social media and mental health and well-being–A qualitative literature review. *Clinical Child Psychology and Psychiatry*, *28*(1), 323–337. doi:10.1177/13591045221092884 PMID:35670473

Porter, M. E., & Heppelmann, J. E. (2014). How smart, connected products are transforming competition. *Harvard Business Review*, *92*(11), 64–88.

Porter, M. E., & Heppelmann, J. E. (2015). How smart, connected products are transforming companies. *Harvard Business Review*, *93*(10), 96–116.

Power, J. D. (2022, November 9). *In Era of Digital Information Search, Vehicle Shoppers in India Still Sensitive to Product Discovery at Showrooms*. JD Power. https://www.jdpower.com/business/press-releases/2022-india-sales-satisfaction-index-study-ssi

Pramod, A., Naicker, H. S., & Tyagi, A. K. (2021). Machine Learning and Deep Learning: Open Issues and Future Research Directions for the Next 10 Years. *Computational Analysis and Deep Learning for Medical Care*, 463–490.

Pramod, A., Naicker, H. S., & Tyagi, A. K. (2021). Machine learning and deep learning: Open issues and future research directions for the next 10 years. *Computational analysis and deep learning for medical care: Principles, methods, and applications*, 463-490.

Prasad, C., Rao, B. S., Pujari, J. J., & Hema, C. (2023, August). Developing a Non-Fungible Token-Based Trade Marketplace Platform Using Web 3.0. In *2023 5th International Conference on Inventive Research in Computing Applications (ICIRCA)* (pp. 312-316). IEEE. 10.1109/ICIRCA57980.2023.10220823

Prathapagiri, V. G. (2019). Digital Divide and Its Dimensions. *Advances in Electronic Government, Digital Divide, and Regional Development*, (April), 79–100. doi:10.4018/978-1-5225-5412-7.ch004

Qamar, Y., & Samad, T. A. (2022). Human resource analytics: A review and bibliometric analysis. *Personnel Review, 51*(1), 251–283. doi:10.1108/PR-04-2020-0247

Qiang, V., Rhim, J., & Moon, A. (2023). No such thing as one-size-fits-all in AI ethics frameworks: A comparative case study. *AI & Society*, 1–20. doi:10.100700146-023-01653-w

Qian, Y. (2022). Navigating the Challenges of Decentralized Technologies: A Comprehensive Review. *Journal of Information Technology, 37*(2), 215–234.

Quach, S., Thaichon, P., Martin, K. D., Weaven, S., & Palmatier, R. W. (2022). Digital technologies: Tensions in privacy and data. *Journal of the Academy of Marketing Science, 50*(6), 1299–1323. doi:10.100711747-022-00845-y PMID:35281634

Quinn, A. (2007). Moral virtues for journalists. *Journal of Mass Media Ethics, 22*(2-3), 168–186. doi:10.1080/08900520701315764

Raina, A., & Palaniswami, M. (2021). The ownership challenge in the Internet of Things world. *Technology in Society, 65*, 101597. doi:10.1016/j.techsoc.2021.101597

Raith, M., & Panni, M. F. (2020). Beyond Disruption: A Systematic Review of Business Model Innovation Research. *Journal of Business Research, 110*, 377–389.

Rajam, V., Reddy, A. B., & Banerjee, S. (2021). Explaining caste-based digital divide in India. *Telematics and Informatics, 65*, 101719. doi:10.1016/j.tele.2021.101719

Rasmussen, L. (2018). Parasocial interaction in the digital age: An examination of relationship building and the effectiveness of YouTube Celebrities. *J. Soc. Media Soc., 7*, 280–294.

Rasool, A., Shah, F. A., & Islam, J. U. (2020). Customer engagement in the digital age: A review and research agenda. *Current Opinion in Psychology, 36*, 96–100. doi:10.1016/j.copsyc.2020.05.003 PMID:32599394

Ratchford, B. T. (2019). The impact of digital innovations on marketing and consumers. *Marketing in a Digital World, 16*, 35–61. doi:10.1108/S1548-643520190000016005

RCEP. (2022, January 1). *RCEP Agreement enters into force*. Retrieved October 2, 2023, from https://rcepsec.org/rcep-agreement-enters-into-force/

Reamer, F. G. (2019). Ethical theories and social work practice. The Routledge handbook of social work ethics and values, 15-21.

Reddy, S. K., & Reinartz, W. (2017). Digital transformation and value creation: Sea change ahead. *GfK Marketing Intelligence Review*, *9*(1), 10–17. doi:10.1515/gfkmir-2017-0002

Regan, P. M. (2002). Privacy as a common good in the digital world. *Information Communication and Society*, *5*(3), 382–405. doi:10.1080/13691180210159328

Resta, P., Laferriere, T., McLaughlin, R., & Kouraogo, A. (2018). Issues and Challenges Related to Digital Equity: An Overview. In Second Handbook of Information Technology in Primary and Secondary Education (pp. 987–1004). Academic Press.

Reuters . (2023 , September 12). GlobalFoundries opens $4 billion Singapore chip fabrication plant. *Economics Times*. Retrieved September 26, 2023, from https://economictimes.indiatimes.com/tech/technology/globalfoundries-opens-4-billion-singapore-chip-fabrication-plant/articleshow/103606474.cms

Reviglio, U. (2022). The untamed and discreet role of data brokers in surveillance capitalism: A transnational and interdisciplinary overview. *Internet Policy Review*, *11*(3), 1–27. doi:10.14763/2022.3.1670

Ribeiro, M. T., Singh, S., & Guestrin, C. (2016, August). " Why should i trust you?" Explaining the predictions of any classifier. In *Proceedings of the 22nd ACM SIGKDD international conference on knowledge discovery and data mining* (pp. 1135-1144). 10.1145/2939672.2939778

Ribot, J. C. (2014). Cause and Response: Democracy, Rights, and Nature. *Policy Matters*, *21*, 94–105.

Ring, M., Ai, D., Maker-Clark, G., & Sarazen, R. (2023). Cooking up Change: DEIB Principles as Key Ingredients in Nutrition and Culinary Medicine Education. *Nutrients*, *15*(19), 4257. doi:10.3390/nu15194257 PMID:37836541

Rinuan, C., & Bohlin, E. (2011). *Understanding the digital divide: A literature survey and ways forward.* Innovative ICT Applications- Emerging Regulatory, Economic and Policy Issues.

Robbins, P., O'Gorman, C., Huff, A., & Moeslein, K. (2021). Multidexterity—A new metaphor for open innovation. *Journal of Open Innovation*, *7*(1), 99. doi:10.3390/joitmc7010099

Rodriguez, P. R. (2017). *Effectiveness of YouTube Advertising: A Study of Audience Analysis.* Rochester Institute of Technology.

Romansky, R. (2017). A survey of digital world opportunities and challenges for user's privacy. *International Journal on Information Technologies and Security*, *9*(4), 97–112.

Romansky, R. P., & Noninska, I. S. (2020). Challenges of the digital age for privacy and personal data protection. *Mathematical Biosciences and Engineering*, *17*(5), 5288–5303. doi:10.3934/mbe.2020286 PMID:33120553

Ross, J. W., Sebastian, I. M., & Beath, C. M. (2017). How to develop a great digital strategy. *MIT Sloan Management Review*, *58*(2), 7–9.

Roubini, N. (2020). The Specter of Deglobalization and the Thucydides Trap. Horizons. *Journal of International Relations and Sustainable Development*, *15*, 130–139.

Roy, S. K., Singh, G., Sadeque, S., Harrigan, P., & Coussement, K. (2023). Customer engagement with digitalized interactive platforms in retailing. *Journal of Business Research*, *164*, 114001. doi:10.1016/j.jbusres.2023.114001

Rudin, C. (2019). Stop explaining black box machine learning models for high stakes decisions and use interpretable models instead. *Nature Machine Intelligence*, *1*(5), 206–215. doi:10.103842256-019-0048-x PMID:35603010

Russell, S. (2022). Artificial intelligence and the problem of control. *Perspectives on Digital Humanism*, 19.

Russell, S., & Norvig, P. (2016). *Artificial Intelligence: A Modern Approach*. Pearson.

Saha, V., Mani, V., & Goyal, P. (2020). Emerging trends in the literature of value co-creation: A bibliometric analysis. *Benchmarking*, *27*(3), 981–1002. doi:10.1108/BIJ-07-2019-0342

Sahni, N. S., Narayanan, S., & Kalyanam, K. (2019). An Experimental Investigation of the Effects of Retargeted Advertising: The Role of Frequency and Timing. *Journal of Marketing Research*. doi:10.1177/0022243718813987

Salesforce. (2023). *OEM EVOLUTION: A new customer journey for new customer expectations*. https://www.salesforce.com/content/dam/web/en_us/www/documents/industries/manufacturing/manufacturing-auto-oem-evolution_v2.pdf

Salim. (2023). The Effect of Informativeness, Photo Colour, Visual Aesthetic, and Social Presence towards Customer Purchase Intention on Glory of fats Instagram. *International Journal of Science and Business*, *18*(1), 96–107. doi:10.5281/zenodo.7569471

Samson, R., Mehta, M., & Chandani, A. (2014). Impact of online digital communication on customer buying decision. *Procedia Economics and Finance*, *11*, 872–880. doi:10.1016/S2212-5671(14)00251-2

Santana, C., & Albareda, L. (2022). Blockchain and the emergence of Decentralized Autonomous Organizations (DAOs): An integrative model and research agenda. *Technological Forecasting and Social Change*, *182*, 121806. Advance online publication. doi:10.1016/j.techfore.2022.121806

SAS Institute Inc. (2023). *American Honda Drives Personalization at Scale with SAS*. https://www.sas.com/en_gb/customers/american-honda.html

Schiavi, G. S., & Behr, A. (2018). Emerging technologies and new business models: A review on disruptive business models. *Innovation & Management Review*, *15*(4), 338–355. doi:10.1108/INMR-03-2018-0013

Schiavi, G. S., Behr, A., & Marcolin, C. B. (2019). Conceptualizing and qualifying disruptive business models. *RAUSP Management Journal*, *54*(3), 269–286. doi:10.1108/RAUSP-09-2018-0075

Schiff, D., Rakova, B., Ayesh, A., Fanti, A., & Lennon, M. (2020). *Principles to practices for responsible AI: closing the gap*. arXiv preprint arXiv:2006.04707.

Schneider, S., & Kokshagina, O. (2021). Digital transformation: What we have learned (thus far) and what is next. *Creativity and Innovation Management*, *30*(2), 384–411. doi:10.1111/caim.12414

Schulhof, V., van Vuuren, D., & Kirchherr, J. (2022). The Belt and Road Initiative (BRI): What Will it Look Like in the Future? *Technological Forecasting and Social Change*, *175*, 121306. doi:10.1016/j.techfore.2021.121306

Schultz, D. E., & Block, M. P. (2015). Beyond brand loyalty: Brand sustainability. *Journal of Marketing Communications*, *21*(5), 340–355. doi:10.1080/13527266.2013.821227

Selbst, A. D., Boyd, D., Friedler, S. A., Venkatasubramanian, S., & Vertesi, J. (2019, January). Fairness and abstraction in sociotechnical systems. In *Proceedings of the conference on fairness, accountability, and transparency* (pp. 59-68). 10.1145/3287560.3287598

Sen, A. (2017, August 9). *When artificial intelligence goes wrong*. Mint. https://www.livemint.com/Technology/VXCMw0Vfilaw0aIInD1v2O/When-artificial-intelligence-goes-wrong.html

Sewpersadh, N. S. (2023). Disruptive business value models in the digital era. *Journal of Innovation and Entrepreneurship*, *12*(1), 1–27. doi:10.118613731-022-00252-1 PMID:36686335

Shalender, K., & Shanker, S. (2023). Building Innovation Culture for the Automobile Industry: Insights from the Indian Passenger Vehicle Market. *Constructive Discontent in Execution: Creative Approaches to Technology and Management*, 131.

Shalender, K., & Yadav, R. K. (2019). Strategic flexibility, manager personality, and firm performance: The case of Indian Automobile Industry. *Global Journal of Flexible Systems Managment*, *20*(1), 77–90. doi:10.100740171-018-0204-x

Sharma, A., & Bansal, A. (2023). Digital Marketing in the Metaverse: Beginning of a New Era in Product Promotion. In Applications of Neuromarketing in the Metaverse (pp. 163-175). IGI Global.

Sharma, S. (2023, July 5). Discounted Russian crude imports saved Indian refiners $7 billion. *Indian Express*. https://indianexpress.com/article/business/commodities/discounted-russian-crude-imports-saved-indian-refiners-7-billion-8751745/

Sharma, R. (2017). The Boom Was a Blip: Getting Used to Slow Growth. *Foreign Affairs*, *96*(3), 104–114.

Sharma, V., & Sood, D. (2022). The role of artificial intelligence in the insurance industry of India. In *Big data analytics in the insurance market* (pp. 287–297). Emerald Publishing Limited. doi:10.1108/978-1-80262-637-720221017

Sherman, J. (2021). Data brokers and sensitive data on US individuals. *Duke University Sanford Cyber Policy Program, 9*.

Sheth, A., Unnikrishnan, S., Bhasin, M., & Raj, A. (2021). *How India Shops Online 2021: A post pandemic view of online shopping*. https://www.bain.com/insights/how-india-shops-online-2021/

Sheth, A., Unnikrishnan, S., Bhasin, M., & Raj, A. (2022). *How India Shops Online 2022: An insight into the e-retail landscape and the emerging trends shaping the market*. https://www.bain.com/insights/how-india-shops-online-2022-report/

Shirer, M. (2023). *IDC Survey Finds Data Sovereignty and Compliance Issues Shaping IT Decisions*. Retrieved from https://www.idc.com/getdoc.jsp?containerId=prUS50134623

Sigelman, M., Bittle, S., Markow, W., & Francis, B. (2019). *The hybrid job economy: How new skills are rewriting the dna of the job market*. Burning Glass Technologies.

Silberg, J., & Manyika, J. (2019, June 6). *Tackling bias in artificial intelligence (and in humans)*. McKinsey & Company. https://www.mckinsey.com/featured-insights/artificial-intelligence/tackling-bias-in-artificial-intelligence-and-in-humans

Silva & Kenney. (1960). Algorithms, Platforms, and Ethnic Bias: An Integrative Essay. *Phylon, 55*(1-2), 9–37. https://www.jstor.org/stable/26545017

Silvia, S. (2019). The Importance of Social Media and Digital Marketing to Attract Millennials' Behavior as a Consumer. *Journal of International Business Research and Marketing*, *4*(2), 7–10. doi:10.18775/jibrm.1849-8558.2015.42.3001

Simmons-Jenkins, G., Miller, S. E., & Murray, E. J. (2022, June). One feeds the other: The cycle of environmental and social injustice impacts on Gullah/Geechee Heritage Sites. *Spark: Elevating Scholarship on Social Issues.*

Simshaw, D. (2022). Access to AI justice: Avoiding an inequitable two-tiered system of legal services. *Yale Journal of Law & Technology*, *24*, 150–226.

Singh, J., Flaherty, K., Sohi, R. S., Deeter-Schmelz, D., Habel, J., Le Meunier-FitzHugh, K., Malshe, A., Mullins, R., & Onyemah, V. (2019). Sales profession and professionals in the age of digitization and artificial intelligence technologies: Concepts, priorities, and questions. *Journal of Personal Selling & Sales Management*, *39*(1), 2–22. doi:10.1080/08853134.2018.1557525

Singh, S. (2010). Digital Divide in India: Measurement, Determinants and Policy for Addressing the Challenges in Bridging the Digital Divide. *International Journal of Innovation in the Digital Economy*, *1*(2), 1–24. doi:10.4018/jide.2010040101

Sinha, A., & Katira, D. (2020). *Digital Identity A Survey of Technologies*. Retrieved from https://digitalid.design/tech-survey-2020.html

Siriwardana, A. (2020). Social Media Marketing: A Literature Review on Consumer Products. *Proceedings of the International Conference on Business & Information (ICBI)*. doi:10.2139/ssrn.3862924

Smith, A. N., Fischer, E., & Yongjian, C. (2012). How does brand-related user-generated content differ across YouTube, Facebook, and Twitter? *Journal of Interactive Marketing*, *26*(2), 102–113. doi:10.1016/j.intmar.2012.01.002

Smith, B. D., Khojandi, A., & Vasudevan, R. K. (2023). Bias in reinforcement Learning: A review in healthcare applications. *ACM Computing Surveys*, *56*(2), 1–17. doi:10.1145/3609502

Smith, H. (2020). Algorithmic bias: Should students pay the price? *AI & Society*, *35*(4), 1077–1078. doi:10.100700146-020-01054-3 PMID:32952313

Smith, K. T. (2019). Mobile advertising to Digital Natives: Preferences on content, style, personalization, and functionality. *Journal of Strategic Marketing*, *27*(1), 67–80. doi:10.1080/0965254X.2017.1384043

Smith, K. T., Jones, A., Johnson, L., & Smith, L. M. (2019). Examination of cybercrime and its effects on corporate stock value. *Journal of Information, Communication and Ethics in Society*, *17*(1), 42–60. doi:10.1108/JICES-02-2018-0010

Sokolova, K., & Kefi, H. (2020). Instagram and YouTube bloggers promote it, why should I buy? How credibility and parasocial interaction influence purchase intentions. *Journal of Retailing and Consumer Services*, *53*, 1–9. doi:10.1016/j.jretconser.2019.01.011

Soni, N., Sharma, E. K., Singh, N., & Kapoor, A. (2020). Artificial intelligence in business: From research and innovation to market deployment. *Procedia Computer Science*, *167*, 2200–2210. doi:10.1016/j.procs.2020.03.272

Soni, V. D. (2020). Emerging roles of artificial intelligence in ecommerce. *International Journal of Trend in Scientific Research and Development*, *4*(5), 223–225.

Sood, K., Dhanaraj, R. K., Balusamy, B., Grima, S., & Uma Maheshwari, R. (Eds.). (2022). *Big Data: A game changer for insurance industry*. Emerald Publishing Limited. doi:10.1108/9781802626056

Spath, D., Gausemeier, J., Dumitrescu, R., Winter, J., Steglich, S., & Drewel, M. (2022). Digitalisation of Society. In A. Maier, J. Oehmen, & P. E. Vermaas (Eds.), *Handbook of Engineering Systems Design*. Springer. doi:10.1007/978-3-030-81159-4_5

Spiekermann, S. (2012). The challenges of privacy by design. *Communications of the ACM*, *55*(7), 38–40. doi:10.1145/2209249.2209263

Spremić, M., & Šimunic, A. (2018, July). Cyber security challenges in the digital economy. In *Proceedings of the World Congress on Engineering* (Vol. 1, pp. 341-346). International Association of Engineers.

Stahl, B. C., & Stahl, B. C. (2021). Ethical issues of AI. *Artificial Intelligence for a better future: An ecosystem perspective on the ethics of AI and emerging digital technologies*, 35-53.

Stein, A., & Ramaseshan, B. (2016). Towards the identification of customer experience touch point elements. *Journal of Retailing and Consumer Services*, *30*, 8–19. doi:10.1016/j.jretconser.2015.12.001

Stelzner, M. (2018). *Predictive Analytics: How Marketers Can Improve Future Activities*. Available at: https://www.socialmediaexaminer.com/predictive-analytics-how-marketers-can-improve-future activities-Chris-Penn/

Stieglitz, S., & Brockmann, T. (2012). Increasing organizational performance by transforming into a mobile enterprise. *MIS Quarterly Executive*, *11*(4), 189–204.

Stone, R. W. (2001). The use and abuse of game theory in international relations: The theory of moves. *The Journal of Conflict Resolution*, *45*(2), 216–244. doi:10.1177/0022002701045002004

Strobel, M., & Shokri, R. (2022). Data privacy and trustworthy machine learning. *IEEE Security and Privacy*, *20*(5), 44–49. doi:10.1109/MSEC.2022.3178187

Strydom, M. J., & Buckley, S. B. (2020). Big Data Intelligence and Perspectives in Darwinian Disruption. In *AI and Big Data's Potential for Disruptive Innovation* (pp. 1–43). IGI Global. doi:10.4018/978-1-5225-9687-5.ch001

Stubbs, W., & Cocklin, C. (2021). Sustainable Business Models: A Systematic Literature Review. *Organization & Environment*, *34*(3), 288–317.

Suarez, F. F., & Kirtley, J. (2012). Dethroning an established platform. *MIT Sloan Management Review*, *53*(4), 35–41.

Sundararajan, A. (2019). *The Sharing Economy: The End of Employment and the Rise of Crowd-Based Capitalism*. The MIT Press.

Sung, Y., Kim, E., & Choi, S. M. (2018). Me and brands: Understanding brand-selfie posters on social media. *International Journal of Advertising*, *37*(1), 14–28. doi:10.1080/02650487.2017.1368859

Sun, S., & Wang, Y. (2010). Examining the role of beliefs and attitudes in online advertising: A comparison between the USA and Romania. *International Marketing Review*, *27*(1), 87–107. doi:10.1108/02651331011020410

Swan, M. (2015). *Blockchain: Blueprint for a New Economy*. O'Reilly Media.

SweeneyL. (2013). Discrimination in online ad delivery. Social Science Research Network. https://doi.org/ doi:10.2139/ssrn.2208240

Sweidan, N. S., & Areiqat, A. (2020). The Digital Divide and its Impact on Quality of Education at Jordanian Private Universities Case Study: Al-Ahliyya Amman University. *International Journal of Higher Education*, *10*(3), 1. doi:10.5430/ijhe.v10n3p1

Szabo, N. (1997). Formalizing and Securing Relationships on Public Networks. *First Monday*, *2*(9). Advance online publication. doi:10.5210/fm.v2i9.548

Tandon, U., Mittal, A., & Manohar, S. (2021). Examining the impact of intangible product features and e-commerce institutional mechanics on consumer trust and repurchase intention. *Electronic Markets*, *31*(4), 945–964. doi:10.100712525-020-00436-1

Tan, L., Guo, J., Mohanarajah, S., & Zhou, K. (2021). Can we detect trends in natural disaster management with artificial intelligence? A review of modeling practices. *Natural Hazards*, *107*(3), 2389–2417. doi:10.100711069-020-04429-3

Tapscott, D., & Tapscott, A. (2016). *Blockchain revolution: how the technology behind bitcoin is changing money, business, and the world*. Penguin.

Tapscott, D., & Tapscott, A. (2016). *Blockchain Revolution: How the Technology Behind Bitcoin is Changing Money, Business, and the World*. Penguin.

Tapscott, D., & Tapscott, A. (2017). How blockchain will change organizations. *MIT Sloan Management Review*, *58*(2), 10–13.

Tarafadar, M., Beath, C., & Ross, J. W. (2019). Using AI to enhance business operations. *MIT Sloan Management Review*, *60*(4), 37–44.

Taricani, E., Saris, N., & Park, P. A. (n.d.). *Beyond technology: The ethics of artificial intelligence*. Academic Press.

Teece, D. J. (1986). Profiting from Technological Innovation: Implications for Integration, Collaboration, Licensing and Public Policy. *Research Policy*, *15*(6), 285–305. doi:10.1016/0048-7333(86)90027-2

Teece, D. J. (2020). Disruptive Innovation: Past, Present, and Future. *Long Range Planning*, *53*(6), 101956.

The World Bank. (2021). *World Development Report 2021*. Author.

The World Bank. (2023). *Technical Standards for Digital Identity*. Retrieved from https://tinyurl.com/4eet9zt4

Thiebes, S., Lins, S., & Sunyaev, A. (2021). Trustworthy artificial intelligence. *Electronic Markets*, *31*(2), 447–464. doi:10.100712525-020-00441-4

Thomson, R., & Bailey, C. (2023). *Identifying heirs' property: Extent and value across the South and Appalachia*. Southern Rural Development Center. https://scholarsjunction.msstate.edu/srdctopics-heirsproperty/2/

Throne, R. (2020). Dispossession of land cultures: Women and property tenure among Lowcountry heirs in the Gullah Geechee Corridor. In *Multidisciplinary issues surrounding African Diasporas* (pp. 152–174). IGI Global. doi:10.4018/978-1-5225-5079-2.ch007

Throne, R. (2021). Land as agency: A critical autoethnography of Scandinavian acquisition of dispossessed land in the Iowa Territory. In *Indigenous research of land, self, and spirit* (pp. 118–131). IGI Global.

ThroneR. (2022). Adverse trends in data ethics: The AI Bill of Rights and human subjects protections. *Information Policy & Ethics eJournal*. doi:10.2139/ssrn.4279922

Thuraisingham, B. (2020). Artificial intelligence and data science governance: Roles and responsibilities at the c-level and the board. In *2020 IEEE 21st international conference on information reuse and integration for data science (IRI)* (pp. 314-318). IEEE.

Tizhoosh, H. R., & Pantanowitz, L. (2018). Artificial intelligence and digital pathology: Challenges and opportunities. *Journal of Pathology Informatics*, *9*(1), 38. doi:10.4103/jpi.jpi_53_18 PMID:30607305

Tjepkema, L. (2018). *Why AI is Vital for Marketing with Lindsay Tjepkema*. Available at: https://www.magnificent.com/magnificent-stuff/why-ai-is-vital-for-marketing-with-lindsay-tjepkema

Tjhin, V.U. & Aini, S. (2019). *Effect of E-WOM and Social Media Usage on Purchase Decision in Clothing Industry*. . doi:10.1145/3332324.3332333

Tranfield, D., Denyer, D., & Smart, P. (2003). Towards a methodology for developing evidence-informed management knowledge by means of systematic review. *British Journal of Management*, *14*(3), 207–222. doi:10.1111/1467-8551.00375

Tripathy, B., & Raha, S. (2019). Digital Divide in India. *International Research Journal of Engineering and Management Studies, 3*(5).

Truby, J., Brown, R., & Dahdal, A. (2020). Banking on AI: Mandating a proactive approach to AI regulation in the financial sector. *Law and Financial Markets Review*, *14*(2), 110–120. doi:10.1080/17521440.2020.1760454

Tucker, I. (2018, October 11). "A white mask worked better": why algorithms are not colour blind. *The Guardian*. https://www.theguardian.com/technology/2017/may/28/joy-buolamwini-when-algorithms-are-racist-facial-recognition-bias

Tucker, C. (2023). The economics of privacy: An agenda. In *The economics of privacy*. University of Chicago Press.

Ülker, P., Ülker, M., & Karamustafa, K. (2023). Bibliometric analysis of bibliometric studies in the field of tourism and hospitality. *Journal of Hospitality and Tourism Insights*, *6*(2), 797–818. doi:10.1108/JHTI-10-2021-0291

Umbrello, S., & Yampolskiy, R. V. (2022). Designing AI for explainability and verifiability: A value sensitive design approach to avoid artificial stupidity in autonomous vehicles. *International Journal of Social Robotics, 14*(2), 313–322. doi:10.100712369-021-00790-w

Ungerer, L., & Slade, S. (2022). Ethical considerations of artificial intelligence in learning analytics in distance education contexts. In *Learning Analytics in Open and Distributed Learning: Potential and Challenges* (pp. 105–120). Springer Nature Singapore. doi:10.1007/978-981-19-0786-9_8

Unni, M. *V., S,* R., Kar, R., Bh, R., V, V., & Johnson, J. M. (2023, March 2). Effect of VR Technological Development in the Age of AI on Business Human Resource Management. *2023 Second International Conference on Electronics and Renewable Systems (ICEARS).* 10.1109/ICEARS56392.2023.10085258

Unni. (2020, April). Does Digital and Social Media Marketing Play a Major Role in Consumer Behaviour? *International Journal of Research in Engineering, Science and Management, 3*(4), 272–278. https://www.ijresm.com/Vol.3_2020/Vol3_Iss4_April20/IJRESM_V3_I4_63.pdf

US ITA. (2022, November 4). *Japan - Country Commercial Guide.* Retrieved October 2, 2023, from https://www.trade.gov/country-commercial-guides/japan-market-overview

User Control in Digital Identity. A Guide to Design Principles. (2023). Retrieved from https://digitalprinciples.org/principle/design-with-the-user/

Van Dijck, G. (2022, June 10). Predicting Recidivism Risk Meets AI Act. *European Journal on Criminal Policy and Research, 28*(3), 407–423. Advance online publication. doi:10.100710610-022-09516-8

Van Doorn, J., Mende, M., Noble, S. M., Hulland, J., Ostrom, A. L., Grewal, D., & Petersen, J. A. (2017). Domo arigato Mr. Roboto: Emergence of automated social presence in organizational frontlines and customers' service experiences. *Journal of Service Research, 20*(1), 43–58. doi:10.1177/1094670516679272

Van Esch, P., & Stewart Black, J. (2021). Artificial intelligence (AI): Revolutionizing digital marketing. *Australasian Marketing Journal, 29*(3), 199–203. doi:10.1177/18393349211037684

Vassilakopoulou, P., & Hustad, E. (2023). Bridging Digital Divides: A Literature Review and Research Agenda for Information Systems Research. *Information Systems Frontiers, 25*(3), 955–969. doi:10.100710796-020-10096-3 PMID:33424421

Venkatraman, V. (2019). How to read and respond to weak digital signals. *MIT Sloan Management Review, 60*(3), 47–52.

Verma, B., & Srivastava, A. (2023). Impact of different dimensions of globalisation on firms' performance: An unbalanced panel-data study of firms operating in India. *World Review of Entrepreneurship, Management and Sustainable Development, 19*(3-5), 360–378. doi:10.1504/WREMSD.2023.130618

Versey, H. S., & Throne, R. (2021). A critical review of Gullah Geechee midlife women and heirs' property challenges along the Gullah Geechee Cultural Heritage Corridor. In *Examining international land use policies, changes, and conflicts* (pp. 46–64). IGI Global. doi:10.4018/978-1-7998-4372-6.ch003

Viertola, W. (2018). *To What Extent Does YouTube Marketing Influence the Consumer Behaviour of a Young Target Group.* Metropolia University of Applied Sciences.

Villaronga, E. F., Kieseberg, P., & Li, T. (2018). Humans forget, machines remember: Artificial intelligence and the right to be forgotten. *Computer Law & Security Report, 34*(2), 304–313. doi:10.1016/j.clsr.2017.08.007

Vincent, C. M. (2020). Ethics of Artificial Intelligence and Robotics. *The Stanford Encyclopedia of Philosophy.* Retrieved January 15, 2021 from https://plato.stanford.edu/archives/win2020/entries/ethics-ai/

Vingilisa, E., Yildirim-Yeniera, Z., Vingilis-Jaremkob, L., Seeleya, J., Wickensc, C. M., Grushkaa, D. H., & Fleiterd, J. (2018). Young male drivers' perceptions of and experiences with YouTube videos of risky driving behaviors. *Accident; Analysis and Prevention*, *120*, 46–54. doi:10.1016/j.aap.2018.07.035 PMID:30086437

Vrontis, D., Makrides, A., Christofi, M., & Thrassou, A. (2021). Social media influencer marketing: A systematic review, integrative framework and future research agenda. *International Journal of Consumer Studies*, *45*(4), 617–644. Advance online publication. doi:10.1111/ijcs.12647

Wang, C., & Hu, Y. (2021). Accessibility in Decentralized Models: A Comparative Analysis. *Journal of Information Systems and Technology Management*, *18*, e202112.

Wang, C., Liu, S., Yang, H., Guo, J., Wu, Y., & Liu, J. (2023). Ethical Considerations of Using ChatGPT in Health Care. *Journal of Medical Internet Research*, *25*, e48009. doi:10.2196/48009 PMID:37566454

Wang, H., & Li, X. (2021). Disruption and corporate foresight: The moderating role of strategic flexibility. *Technological Forecasting and Social Change*, *162*, 120341.

Wang, Y., & Genç, E. (2019). Path to effective mobile advertising in Asian markets. *Asia Pacific Journal of Marketing and Logistics*, *31*(1), 55–80. Advance online publication. doi:10.1108/APJML-06-2017-0112

Watkins, R. D., Denegri-Knott, J., & Molesworth, M. (2016). The relationship between ownership and possession: Observations from the context of digital virtual goods. *Journal of Marketing Management*, *32*(1-2), 44–70. doi:10.1080/0267257X.2015.1089308

Watson, H. J. (2017). Preparing for the cognitive generation of decision support. *MIS Quarterly Executive*, *16*(3), 153–169.

Webber, B. L., & Nilsson, N. J. (Eds.). (2014). *Readings in artificial intelligence*. Morgan Kaufmann.

Weber, R. H. (2015). The digital future–A challenge for privacy? *Computer Law & Security Report*, *31*(2), 234–242. doi:10.1016/j.clsr.2015.01.003

Westerman, G., Bonnet, D., & McAfee, A. (2014). The nine elements of digital transformation. *MIT Sloan Management Review*, *55*(3), 1–6.

White & Case LLP. (2017, January 23). *Algorithms and bias: What lenders need to know*. JD Supra. Retrieved October 8, 2023, from https://www.jdsupra.com/legalnews/algorithms-and-bias-what-lenders-need-67308/

White House. (2023, September 9). *FACT SHEET: World Leaders Launch a Landmark India-Middle East-Europe Economic Corridor*. Retrieved October 2, 2023, from https://www.whitehouse.gov/briefing-room/statements-releases/2023/09/09/fact-sheet-world-leaders-launch-a-landmark-india-middle-east-europe-economic-corridor/

Wiens, J., Price, W. N. II, & Sjoding, M. W. (2020). Diagnosing bias in data-driven algorithms for healthcare. *Nature Medicine*, *26*(1), 25–26. doi:10.103841591-019-0726-6 PMID:31932798

Wijewickrema, M. (2023). A bibliometric study on library and information science and information systems literature during 2010–2019. *Library Hi Tech*, *41*(2), 595–621. doi:10.1108/LHT-06-2021-0198

Williams, D. D. (2022). *Digital equity: Difficulties of implementing the 1:1 computing initiative in low-income areas*. Academic Press.

Williams, L. (2015). *Disrupt: Think the unthinkable to spark transformation in your business*. FT Press.

Winkler, T. J., & Kettunen, P. (2018). Five principles of industrialized transformation for successfully building an operational backbone. *MIS Quarterly Executive*, *17*(2), 121–138.

Winkler, T., & Spiekermann, S. (2019). Human values as the basis for sustainable information system design. *IEEE Technology and Society Magazine*, *38*(3), 34–43. doi:10.1109/MTS.2019.2930268

Winters-Michaud, C., Burnett, W., Callahan, S., Keller, A., Williams, M., & Harakat, S. (2023). Land-use patterns on heirs' property in the American South. *Applied Economic Perspectives and Policy*, •••, 1–15. doi:10.1002/aepp.13354

Wirtz, J., Patterson, P. G., Kunz, W. H., Gruber, T., Lu, V. N., Paluch, S., & Martins, A. (2018). Brave new world: Service robots in the frontline. *Journal of Service Management*, *29*(5), 907–931. doi:10.1108/JOSM-04-2018-0119

Witt, M. A. (2019). De-Globalization: Theories, Predictions, and Opportunities for International Business Research. SSRN *Electronic Journal*. doi:10.2139/ssrn.3315247

Witt, M. A., Lewin, A. Y., Li, P. P., & Gaur, A. (2023). Decoupling in international business: Evidence, drivers, impact, and implications for IB research. *Journal of World Business*, *58*(1), 101399. doi:10.1016/j.jwb.2022.101399

World Bank. (2022). *Making the Most of the African Continental Free Trade Area: Leveraging Trade and Foreign Direct Investment to Boost Growth and Reduce Poverty*. World Bank. https://openknowledge.worldbank.org/entities/publication/09f9bbdd-3bf0-5196-879b-b1a9f328b825

World Bank. (n.d.). *World Bank national accounts data, and OECD National Accounts data files*. Retrieved October 5, 2023, from https://data.worldbank.org/indicator/NY.GDP.MKTP.CD?end=2022&locations=CN-US-JP&start=2002

World Trade Organization. (2023a). *Re-globalization for a secure, inclusive and sustainable future*. World Trade Report 2023. https://www.wto.org/english/res_e/publications_e/wtr23_e.htm

World Trade Organization. (2023b). *World Trade Statistical Review 2023*. World Trade Report 2023 https://www.wto.org/english/res_e/booksp_e/wtsr_2023_e.pdf

World Trade Organization. (2023c). Retrieved September 20, 2023, from https://www.wto.org/english/thewto_e/thewto_e.htm

Wu, C., Li, X., Guo, Y., Wang, J., Ren, Z., Wang, M., & Yang, Z. (2022). Natural language processing for smart construction: Current status and future directions. *Automation in Construction*, *134*, 104059. doi:10.1016/j.autcon.2021.104059

Wu, Y., Farrukh, M., Raza, A., Meng, F., & Alam, I. (2021). Framing the evolution of the corporate social responsibility and environmental management journal. *Corporate Social Responsibility and Environmental Management*, *28*(4), 1397–1411. doi:10.1002/csr.2127

Wylde, V., Rawindaran, N., Lawrence, J., Balasubramanian, R., Prakash, E., Jayal, A., Khan, I., Hewage, C., & Platts, J. (2022). Cybersecurity, Data Privacy and Blockchain: A Review. *SN Computer Science*, *3*(2), 127. doi:10.100742979-022-01020-4 PMID:35036930

Xu, Q. (2014). Should I trust him? the effects of reviewer profile characteristics on eWOM credibility. *Computers in Human Behavior, 33*, 136-144. . doi:10.1016/j.chb.2014.01.027

Xu, J., & Feng, Y. (2022). Reap the harvest on blockchain: A survey of yield farming protocols. *IEEE Transactions on Network and Service Management*, *20*(1), 858–869. doi:10.1109/TNSM.2022.3222815

Yakubu, B. M., Khan, M. I., Javaid, N., & Khan, A. (2021). Blockchain-based secure multi-resource trading model for smart marketplace. *Computing*, *103*(3), 379–400. doi:10.100700607-020-00886-7

Yenkar, V., & Bartere, M. (2014). Review on data mining with big data. *International Journal of Computer Science and Mobile Computing*, *3*(4), 97–102.

Yermack, D. (2015). Is Bitcoin a Real Currency? An Economic Appraisal. In *Handbook of Digital Currency* (pp. 31–43). Elsevier. doi:10.1016/B978-0-12-802117-0.00002-3

Yigitcanlar, T., Desouza, K. C., Butler, L., & Roozkhosh, F. (2020). Contributions and risks of artificial intelligence (AI) in building smarter cities: Insights from a systematic review of the literature. *Energies*, *13*(6), 1473. doi:10.3390/en13061473

Yin, J. Z., & Hamilton, M. H. (2018). The conundrum of US-China trade relations through game theory modelling. *Journal of Applied Business and Economics*, *20*(8).

Yin, J., & Qiu, X. (2021). AI technology and online purchase intention: Structural equation model based on perceived value. *Sustainability (Basel)*, *13*(10), 5671. doi:10.3390u13105671

YouTube. (n.d.). *For Press*. Available online: https://www.youtube.com/intl/en-GB/about/press/

Yuan Sun, J. C., & Metros, S. E. (2011). The Digital Divide and Its Impact on Academic Performance. *US-China Education Review*, *2*, 153–161.

Zahra, S. A. (2021). International entrepreneurship in the post Covid world. *Journal of World Business*, *56*(1), 101143. doi:10.1016/j.jwb.2020.101143

Zambodla, N. (n.d.). *Millennials Are Not a Homogenous Group*. Available online: http://www.bizcommunity.com/Article/196/424/172113.html#more

Zeng, F., Tao, R., Yang, Y., & Xie, T. (2017). How Social Communications Influence Advertising Perception and Response in Online Communities? *Frontiers in Psychology*, *8*, 1349. doi:10.3389/fpsyg.2017.01349 PMID:28855879

Zhang, H., & Li, X. (2020). Decentralized Approaches: Unlocking Efficiency in Business Models. *Journal of Business Efficiency and Innovation*, *1*(1), 56–72.

Zhang, T. C., Omran, B. A., & Cobanoglu, C. (2017). Generation Y's positive and negative eWOM: Use of social media and mobile technology. *International Journal of Contemporary Hospitality Management*, *29*(2), 732–761. doi:10.1108/IJCHM-10-2015-0611

Zliobaite, I. (2015). *A survey on measuring indirect discrimination in machine learning*. arXiv preprint arXiv:1511.00148.

Zowghi, D., & da Rimini, F. (2023). *Diversity and Inclusion in Artificial Intelligence*. arXiv preprint arXiv:2305.12728.

Zuboff, S. (1988). *In the age of the smart machine: The future of work and power*. Basic Books, Inc.

Zulfiqar, A., Ghaffar, M. M., Shahzad, M., Weis, C., Malik, M. I., Shafait, F., & Wehn, N. (2021). AI-ForestWatch: Semantic segmentation based end-to-end framework for forest estimation and change detection using multi-spectral remote sensing imagery. *Journal of Applied Remote Sensing*, *15*(2), 024518–024518. doi:10.1117/1.JRS.15.024518

Zwitter, A. J., Gstrein, O. J., & Yap, E. (2020). Digital identity and the blockchain: Universal identity management and the concept of the "Self-Sovereign" individual. *Frontiers in Blockchain*, *3*, 26. doi:10.3389/fbloc.2020.00026

About the Contributors

Balraj Verma works as an Assistant Professor at Chitkara Business School-Doctoral Research Centre with Chitkara University, Rajpura, Punjab. He had his PhD from Jaypee University of Information Technology (JUIT), Waknaghat and holds a master's degree in Business Administration and has 16 years of academic and corporate experience. He has qualified National eligibility test (NET) for teaching in the management discipline. He had been teaching courses like Marketing Management, Strategic Management, Business Statistics and Research Methodology. He believes in quality research and has published many papers in ABDC listed/ Unpaid Scopus Index journals. He has authored/edited books and many book chapters so far published by reputed major presses. He has been actively involved at different levels in organizing Workshops/ Conferences by the Department/ University.

Babita Singla is a professor at Chitkara Business School, Chitkara University, Punjab, India. She has a Ph.D. in management and is UGC-NET qualified. She has over 13 years of experience in teaching, research, and administration. Her areas of expertise are marketing, e-commerce, omnichannel, and retail. In her career, she has been involved in important academic and research assignments such as being the guest editor of a reputed journal, organizing and conducting international and national-level conferences and faculty development programs, and providing guidance for research projects. She has research publications in reputable international and national journals such as Scopus, SCI, etc., and has presented research papers at various national and international conferences. In the short span of 13 years of her career in academia and administration, she has authored and edited several books on retailing, supply chain management, branding, customer relationship management, and product management, covering the course content of various universities nationwide. She has successfully delivered guest sessions at international and national universities. She has over twenty research publications in international and national journals, over 11 publications/ presentations in international and national conferences, including 8 Keynote lectures/Invited Talks and ten books to her credit. Her current research interests are in business management, omnichannel retail, marketing management, and managerial economics. She loves to generate new ideas and devise feasible solutions to broadly relevant problems. She enjoys embracing the lessons learned from failure, stands up, and continues to grow.

Amit Mittal is a Professor of Management / Dean at Chitkara University, Punjab, India, and Dean, Doctoral Research Centre. He has over 19 years of experience in teaching, research, consulting, and academic administration. His areas of research and consulting expertise are graduate employability, leadership, emerging market studies, business research methods, etc. Eight scholars have been awarded a PhD under his supervision. He has published in reputable journals such as Frontiers in Psychology;

Benchmarking; European Business Review; Business: Theory & Practice; International Journal of Quality Research; South Asian Journal of Business Management Cases; International Journal of Human Capital & Information Technology Professionals; International Journal of e-Services & Mobile Applications; International Journal of Business & Globalisation; International Journal of Socio-technology & Knowledge Development, etc. He is also a reviewer for the International Journal of Consumer Studies; International Journal of Quality Research; Frontiers in Psychology; Higher Education, Skills & Work-based Learning; Business: Theory & Practice, etc.

* * *

Yash Pal Azad is an Assistant Professor of Psychology at Eternal University, Sirmaur, India, with expertise in Cognitive Psychology and Organizational Behavior. His research interests span a broad spectrum, encompassing Interdisciplinary approaches to Psychology, Ethical Concerns in Research, Technological Advances in Educational Psychology, Neuro Psychology, Developmental Psychology, Consumer Behavior, Psychological well-being, and Social Psychology. Dr. Azad has demonstrated his scholarly acumen by authoring two edited volumes of research articles and actively serving as a reviewer for numerous reputable journals. Committed to academia and research, he consistently pushes the boundaries of knowledge in psychology and its interdisciplinary applications.

Divya Goswami is presently a Research Scholar at Chitkara University , Punjab (India) who has accumulated a wealth of teaching experience over the past 10 years. She has served as a dedicated faculty member at S.A.Jain (P.G.) College, Ambala City (Haryana), where she has taught a range of courses in commerce, including Financial Accounting, Advertisement Management, Financial Institutions Markets and others. Her teaching philosophy is grounded in fostering critical thinking, analytical skills, and a holistic understanding of commerce principles among students. In her role as a Researcher Scholar, currently at the Center for Financial Technologies and AI, Ms. Divya Goswami explores innovative applications of AI in financial decision-making processes.

Atul Grover is a dynamic individual whose journey through the worlds of ServiceNow, business process consulting, and education has been marked by a deep commitment to personal growth, knowledge sharing, and a belief in the lasting impact of actions and attitude. With a career spanning nearly a decade, Atul G has accumulated invaluable experience as a Techno-Functional Business Analyst, Business Process Consultant, and implementation expert. His dedication to the field is further evidenced by his role at DXC Technology, a Global Elite Partner, where he serves as a Senior Business Process Consultant. Here, he leverages his expertise to make a significant impact in the IT industry. Atul's passion for sharing knowledge and nurturing talent led him to specialize in ServiceNow Techno Functional Training. Over four years, he has trained numerous individuals, sharing his expertise and helping them excel in the complex world of ServiceNow. This dedication to training extends beyond the classroom, as Atul has also published two courses on ServiceNow Learning and authored a book chapter in IGI Global's publication on "Metaverse in Teaching Pedagogy." His association with Chitkara University as a Research Scholar, where he is pursuing a Ph.D. in Metaverse Adoption in the IT Industry, is a testament to his commitment to both theoretical and practical understanding of cutting-edge topics. As "Learn N Grow Together With Atul G," Atul has reached a global audience through his YouTube channel. However, Atul's impact transcends the digital realm. For over 12 years, he has been a People Developer, Trainer, and Guest Speaker. His remarkable record includes delivering 31 Guest Lectures at Deemed Universities, enriching the lives of over 4,000 students. As a testament to his expertise, he also conducted a session as part of the ServiceNow Boot camp at Northeastern University London. Atul's training efforts have not only been limited to the digital world, as he has delivered over 2,000 hours of classroom training to freshers, shaping the careers of more than 2,000 students. Atul G's life story is a testament to the idea that one's actions and attitude leave a lasting impact on the world, regardless of their designation or position. His journey serves as an inspiration to all who seek to Learn Together and Grow Together.

Priya Gupta is a research scholar in the field of Commerce. She is UGC-NET and JRF qualified. She is pursuing a Ph. D. in the Department of Commerce, as a Junior research fellow at Indira Gandhi University, Meerpur, Rewari, India. Her research interests include gender equality, digitalization, innovations, sustainability, financial empowerment and women empowerment. In addition to her research pursuits, she has presented several research papers at national as well as international conferences. In this context, she is engaged in highlighting the evolving landscape of the digital divide in the era of rapid technological advancements.

Amit Kumar is currently working as an Assistant Professor in the Department of Management at Eternal University, Baru Sahib. He has 7+ years of teaching experience and various publications in national and international Journals. His research interests include customer relationship management and consumer behavior.

Sridhar Manohar is currently working in Doctoral Research Center, Chitkara University, completed his doctorate in the area of Services Marketing from VIT Business School, VIT University. He has a Bachelor of Technology and Dual Masters in Business Administration and Organization Psychology. Dr. Sridhar further certified with FDP at IIM-A. He is expertise in Service Marketing, Innovation and Entrepreneurship, Scale Development Process and Multivariate Analytics and interests in teaching Business Analytics, Innovation and Entrepreneurship, Research Methodology and Marketing Management. He has published around 20 research papers that includes Scopus listed and ABDC ranked International Journals like – Society and Business Review, Benchmarking-An International Journal, Electronics Market, Corporate Reputation Review, International Journal of Services and Operations Management, International journal of Business Excellence and presented papers and ideas in numerous international conferences.

Natchimuthu comes with 16+ years of experience in Teaching and Research. Currently, he is working as an Assistant professor at the Department of Commerce, CHRIST (Deemed to be University). Natchimuthu has a PhD from Bharathiar University, Coimbatore India and has Master of Philosophy and Master of Commerce degrees from Madurai Kamaraj University, India. His research interest lies in the field of Financial Econometrics and particularly in the area of Volatility estimation and Machine Learning. Dr Natchimuthu has published research papers in Scopus-indexed and other international Journals. His recent publications include "Volatility spill over among metals", "Calendar anomalies in Indian stock market" and "Is gold price leveraged in India?"

M. Nandhini is currently working as Professor in Department of Management, Karpagam Academy of Higher Education. She started her career in teaching profession since June 2004. She has obtained Doctorate degree in Karpagam Academy of Higher Education. She has 19 years of experience in teaching. She has done Post Graduate degrees in Commerce, Philosophy and Business Management under Bharathiar University. She has also qualified SET in 2012. She has organized 4 conferences successfully. She has conducted training programme on "Rights of Women" under the sponsorship of NHRC. She has published 54 Articles in various journals out of which 12 in scopus indexed journal. Her current research includes cost reduction and cost control with respect to manufacturing concern. She has published two text books, a. Direct Taxation and Tax Planning b. Financial Management.

P. Palanivelu is presently working as Senior Professor in the Department of Management, Karpagam Academy of Higher Education, Coimbatore. He has started his career in teaching since 1996. He has 27 years of teaching experience in Commerce and Management discipline. He has completed M.Phil. and Ph.D. in Bharathiar University. He also qualified with MCA and PGDCA. His area of specialization is Finance. He has published 110 articles in reputed indexed journals.

Kaushikkumar Patel is a distinguished leader in harnessing data-driven strategies within the financial sector, boasting an extensive career that intersects finance and technology. He is pivotal at TransUnion, where he innovates through Big Data, enhancing decision-making processes and financial strategies. His profound knowledge extends to Data Analytics, Financial Technology (FinTech), and Digital Transformation, contributing significantly to advancing tech integration in finance. Based in the United States, Mr. Patel is renowned for his strategic oversight in developing solutions that navigate the complex challenges of data privacy and risk assessment, ensuring compliance and governance in dynamic financial landscapes. His insights have fortified business intelligence, utilizing machine learning and cloud computing solutions to drive organizational success. An influential thought leader, Mr. Patel has shared his expertise and vision through various high-impact publications, shedding light on the transformative power of data in finance. His commitment to excellence was internationally recognized when he was honored with the ET Leadership Excellence Award for his groundbreaking work in Data-Driven Financial Strategies. Mr. Patel's dedication transcends his immediate professional sphere, having a lasting impact on societal well-being. He actively engages in CSR initiatives, leveraging technology to enhance lives and contribute to sustainable development. His unique blend of technical prowess and strategic acumen establishes him as a visionary in his field, continually pushing the boundaries of what's possible at the intersection of finance and technology.

Vinoth S. is currently working with MEASI Institute of Management, Chennai as Assistant Professor and has over 12 years of experience in teaching. He is specialized in marketing and has around 3 years of industry experience. His research interests includes consumer behaviour, green marketing and organisational behaviour. His area of expertise also includes Statistics and Operations research. He has published around 15 research papers in various international journals and conferences.

Nidhi Srivastava is currently working with MEASI Institute of Management, Chennai as Assistant Professor and has over 11 years of experience in teaching. She has 1 year of industry experience in HR domain and her research interests includes organizational behaviour, leadership, human resource management, general management and corporate social responsibility. She is Certified DiSc Trainer and also PhD reviewer at some renowned universities. She has published various research articles indexed in Scopus, ABDC, UGC Care and some renowned peer reviewed journals.

Robin Throne, PhD, is a human research protections administrator and research methodologist. Her research agenda continues to consider doctoral researcher positionality and agency, and voice and land dispossession from various social justice research approaches. She is the author of Autoethnography and Heuristic Inquiry for Doctoral-Level Researchers: Emerging Research and Opportunities (IGI Global, 2019), and editor of Practice-Based and Practice-Led Research for Dissertation Development (IGI Global, 2021), Indigenous Research of Land, Self, and Spirit (IGI Global, 2021) and Social Justice Research Methods for Doctoral Research (IGI Global, 2022).

Sidhi Menon U. is an ambitious MCom student at the Department of Commerce, Christ University, India. Her academic journey began at St. Claret College, where she completed her undergraduate studies with a strong inclination towards research. Her remarkable dedication led to being a rank holder at Bangalore University. Throughout her educational tenure, Sidhi showcased her academic prowess by contributing to several national and international journals, demonstrating her commitment to scholarly pursuits. Prior to pursuing her postgraduate degree, she gained invaluable industrial expertise through a two-year stint at Amazon India, enriching her with practical knowledge and experience. Sidhi's multifaceted background, blending academic excellence, a passion for research, and a robust professional stint, reflects her determination to excel in both academia and industry.

Saravana Krishnan V. is in the noble profession as an educator for the last 12 years. He is a consultant for last 10 years at Multi-ray Financial Services, a business partner of Religare Securities Ltd with a bouquet of financial products like Equity, Commodity, Currency, Mutual fund, Insurance, Financial Planning and Tax planning. Currently having a franchisee of Angel Broking and Assistant Professor at CHRIST Deemed to be University, Bangalore. He is an alumni of IIM Calcutta were he completed one year executive program in applied finance – EPAF (IIM-C). Completed Data Analytics certification program from IIM Rohtak. He holds Bloomberg Market Concepts certification from Bloomberg. He also completed FDP in Econometrics from IIM-K, Kozihode. He has rattling stake towards security analysis, portfolio management, derivatives and tax planning. He is associated as faculty cum trainer of BSEI – Bombay Stock Exchange Institute for GFMP course in southern region – Coimbatore.

Anjali Verma is a research scholar in the field of Economics. She is UGC-NET and JRF qualified. She is pursuing a Ph. D. in the Department of Economics, as a Junior research fellow at Indira Gandhi University, Meerpur, Rewari, India. Her research interests include agriculture, farmer-producer organizations, innovations, digitalization, and financial performance. She has presented several research papers at national and international conferences. In this context, she is working on a comprehensive paper that investigates the role of digital equity in transforming the rapid progress of technology in the contemporary landscape.

Artur Zawadski is the CEO, Sunrise CSP Pty, Australia and Sunrise International Pvt Ltd.

Index

E

F

G

H

I

L

M

O

P

R

S

T

U

V

W

Y

www.ingramcontent.com/pod-product-compliance
Lightning Source LLC
LaVergne TN
LVHW080633011026
849160LV00014B/1201